MOLECULAR DESIGN OF TAUTOMERIC COMPOUNDS

UNDERSTANDING CHEMICAL REACTIVITY

Molecular Design of Tautomeric Compounds

V. I. MINKIN, L. P. OLEKHNOVICH, and Yu. A. ZHDANOV

Institute of Physical and Organic Chemistry
University of Rostov, Rostov on Don, U.S.S.R.

D. Reidel Publishing Company

A MEMBER OF THE KLUWER ACADEMIC PUBLISHERS GROUP

Dordrecht / Boston / Lancaster / Tokyo

0285942⁷

CHEMISTRY

Library of Congress Cataloging in Publication Data

CIP

Minkin, V. I. (Vladimir Isaakovich)
  Molecular design of tautomeric compounds.

  (Understanding chemical reactivity)
  Translation of: Molekuliârnyĭ dizaĭn tautomernykhsistem.
  Includes bibliographies and index.
  1. Molecular theory. 2. Tautomerism. I. Olekhnovich, Lev Petrovich. II. Zhdanov,
IU. A. (IUriĭ Andreevich). III. Title. IV. Series.
QD461.M58613 1987   541.2'2   87–28641
ISBN 90–277–2478–4

Published by D. Reidel Publishing Company,
P.O. Box 17, 3300 AA Dordrecht, Holland.

Sold and distributed in the U.S.A. and Canada
by Kluwer Academic Publishers,
101 Philip Drive, Norwell, MA 02061, U.S.A.

In all other countries, sold and distributed
by Kluwer Academic Publishers Group,
P.O. Box 322, 3300 AH Dordrecht, Holland.

Original title: Molekyulyarnyĭ Dizaĭn Tautomerych Sistem.

First edition: © 1977 University of Rustov.

Printed in the Netherlands

Table of Contents

Preface

Until the early seventies, tautomeric — i.e. fast and reversible rearrangement reactions accompanied by migrations of carbon-centered groups — were practically unknown. For a long time it was assumed that the family of tautomeric reactions was confined to prototropic transformations only. However, the discovery in the fifties of the reversible metallotropic rearrangements showed the domain of migratory tautomerism to be substantially broader.

The synthesis of the metallotropic compounds was based on the substitution of a proton in prototropic compounds by an organometallic group. This approach rarely proved fruitful when attempting to effect tautomeric rearrangements of organic and organometallic groups formed by the elements to the right of carbon in the Periodic Table. By contrast, a novel approach involving an analysis of the steric requirements inherent in the structure of the transition state of a reactive center and an examination of the stereodynamic possibilities has given rise to a target-oriented molecular design of compounds capable of rapid and reversible intramolecular migration of the type indicated. The implementation of this approach, which is the subject of the present book, has already led to the preparation of new tautomeric compounds in which such heavy organic migrants as acyl, aryl, sulfinyl, phosphoryl, arsinyl, and other groups migrate in molecules at a frequency of $10^6-10^9 \, \text{s}^{-1}$ at ambient temperature, i.e., at the rates comparable with protonic migrations.

Special emphasis is laid in this book on the molecular design of carbonotropic and other elementotropic tautomeric systems based on the mechanism of associative nucleophilic substitution. The structural rules developed are directly applicable only to this one, albeit very important, type of mechanism. However, the main principle of molecular design involving the requirement for a close correspondence of the structures of the starting and transition states of a reaction may also be applied to any other type of mechanism of rearrangement reactions.

The first edition of the book appeared in 1977 in the Publishing House of Rostov University (USSR). Since that time, a large body of additional material, both experimental and theoretical, concerning the synthesis and dynamics of new tautomeric systems has been accumulated, thus necessitating an extensive revision of the text. The present version takes account of the available literature data until the middle of 1986, furnishing, we hope, a review embracing all important results obtained so far in the field under consideration, as well as outlining the prospects for significant new developments therein.

We wish to express special thanks to all our research students, whose names are listed in the corresponding references, for their valuable contribution to the experimental investigation of many previously unknown tautomeric reactions described in the book.

Rostov on Don, 1987

V. I. MINKIN
L. P. OLEKHNOVICH
Yu. A. ZHDANOV

1. The Problem Of Tautomerism

1. Tautomerism as Dynamic Isomerism

In 1877 Butlerov [1] obtained an equilibrium mixture of two isomeric diisobuty-lenes by treating tert. butyl alcohol with sulfuric acid

$$(CH_3)_3C-CH=C(CH_3)_2 \rightleftharpoons [(CH_3)_3C-CH_2-\underset{\underset{OH}{|}}{C}(CH_3)_2] \rightleftharpoons (CH_3)_3C-CH_2-\underset{\underset{CH_3}{|}}{C}=CH_2$$

The shift of the double bond in this reaction was explained as the consequence of successive processes of water addition and elimination, i.e., in present-day terms, as an intermolecular rearrangement. However, Butlerov, in the same work, takes the next step by admitting the possibility of equilibrium between isomeric substances in the absence of any additional reagent: ". . . a molecule . . . may behave as a combination of two or more isomeric forms." And then, in the paper 'On Isodibutylene' written at about the same time, he states: "In some cases one also encounters such bodies the mass of which permanently contains noticeable amounts of isomeric particles continually competing with one another and rearranging from one structure into another one." [2] Here, apparently for the first time, in clear terms was stated the notion of equilibrium dynamic isomerism corresponding to the present-day concept of tautomerism.*

The investigation of tautomerism opened some important chapters of structural and theoretical organic chemistry. Solution of the problem of keto—enol equilibria played the key role in the understanding of the acid—base and structure—reactivity relationships [4—6] as well as of the concept of dual reactivity [5]. Even the concept of the hydrogen bond has emerged from studies of prototropic tautomerism [8] whose mechanism is thought to be associated with the nature of spontaneous mutations and the genetic code [9]. Tautomerism is the subject of

---

* The term tautomerism (from ταὐτό — the same — and μερος — part — was coined by Laar [3] in 1885. Laar interpreted tautomerism not as dynamic isomerism but as a possibility that one substance may combine properties of two or more isomers resulting from the fact that one of the groups, usually hydrogen, occupies an intermediate location between two or more possible positions. The development of the term tautomerism has been traced in [4, 5].

numerous reviews and books (see, for instance [4, 5, 10—15]) in which various facets of the structure and transformation mechanism are dealt with in detail. However, the scope of tautomeric compounds and delimitations of the term tautomerism still remain rather vague.

This is due, in the first place, to the uncertainties in a number of terms and notions in the classical structural theory, which may clearly be seen when attempting to translate them into the language of the physical theory of electron—nuclear interactions. Employing this language to reformulate the definition by Butlerov quoted above, which is adopted in organic chemistry, we are confronted with a typical situation. It turns out that this definition of tautomerism has from the physical point of view a purely semantic character and instead of explaining one term, we now have to explain three, since it is necessary to determine which structures can be regarded as isomers and what is their dynamic, i.e. rapid and reversible, transformation. Even when remaining within the framework of the classical theory, we must decide what types of isomeric relations should be associated with tautomerism. Tautomerism is, apparently, only one of the possible consequences of isomerism, all of which must be clearly differentiated.

The distinguishing feature of the tautomeric transformation leading to the transition of a molecule from one isomeric form to another is the change in the sequence of chemical bonds in the molecule during the course of this transformation, i.e. the inevitable rupture of some and formation of other chemical bonds between the atoms of the same molecule (intramolecular tautomerism) or of a molecular aggregate (intermolecular tautomerism).

On this condition, the scope of tautomeric transformations is differentiated from stereodynamic (conformational, inversional or permutational) isomerisms although such a differentiation is, as will be shown below, not very clear-cut. Moreover, it becomes necessary to define in physical terms what a chemical bond is, what is meant by its rupture or its formation and even what are, in general, isomers and the phenomenon of isomerism.

It turns out, however, that this task has no simple solution. Such widespread and effective terms as bond, bond energy, bond order, etc. have no straightforward and unambiguous physical definitions.

1.1. *Molecular Structure and Isomerism*

Isomerism is a very important concept in the classical structural theory, which has experienced a substantial evolution over the years. All compounds having an identical molecular formula are isomers. They are divided into structural (constitutional) isomers and stereoisomers. The former class comprises the compounds whose molecules are made up of nonidentical structural fragments (groups, bonds), e.g. HCN and HNC, 1-propanol and 2-propanol, etc. Unlike the structural isomers, the molecules of stereoisomers consist of an equal number of identical structural fragments but differ in their spatial arrangement. From the above definition of tautomerism, it follows that tautomeric transformations are interconversions of structural (conditional) isomers. The concept of isomerism is, apparently, a derivation from a more general concept of the molecular structure. It is therefore expedient to consider the question of the interrelation between the classical and quantum mechanical definitions of the latter. In itself, the concept of molecular structure does not follow from the first principles of quantum mechanics, since quantum mechanics does not contain an operator corresponding to molecular structure. The solution of the Schroedinger equation for a set of nuclei and electrons corresponding to the molecular formula leads, in the absence of additional approximations, to stationary states and wave functions in which the nuclei as well as electrons are delocalized, and the requirements of permutational symmetry ensure the indiscernibility of identical nuclei as well of electrons. These wave functions cannot be analyzed in terms of molecular structure and chemical bonds [17].

The concept of molecular structure appears in the quantum theory at its lower levels, those concerned with approximations suited to the conditions of the experimental observation of molecules. The principal approximation is the separation of the wave functions of electrons and nuclei, an admissible operation, given the large difference between their masses (Born–Oppenheimer approximation). This approximation constitutes the basis for the theoretical determination of the molecular structure, which is closely connected with the concept of the potential energy surface. The latter is represented by a function of the total energy of a molecular system (minus kinetical energy of its nuclei) of all independent nuclear coordinates q. If the number of these nuclei is N, then

$$E(q) = E(q_1, q_2, q_3, \ldots q_{3N-6}) \tag{1.1}$$

Only those atomic aggregates described as molecular structures are stable, and they are represented as minima on the potential energy surface (1.1) and represent, therefore, the stationary points at which the first energy derivatives over all coordinates are zero and, in addition to that, all the second derivatives (force constants) are positive [18]. To every isomeric form on the potential energy surface there is a corresponding local minimum. The correspondence of the structure to the local minimum on the potential energy surface is the obligatory condition for its stability and, in principle, for the possibility to observe it experimentally. It would, however, be incorrect to think that the quantum mechanically determined molecular structure is simply a point of the configurational space at the minimum of the potential surface with rigorously fixed equilibrium coordinates q_{1e}, q_{2e}. ... Such a purely geometrical definition of molecular structure and, consequently, of isomerism is unsatisfactory. Owing to continuous internal vibrations (corollary of the uncertainty principle), the instantaneous nuclear configuration of the molecule is permanently distorted with respect to the lowest point of the potential energy surface minimum. The extent of the distortion also depends on external factors; in particular, on temperature. Such distortions can be particularly great in a number of structurally nonrigid molecules and ions with large amplitudes of motion along the nonrigid coordinate. They can also be rather extensive at the start of a chemical reaction along the reaction pathway when the structure is still intact.

Interconversions of isomers including structural isomers, corresponding to tautomeric reactions, are described in terms of the potential energy surface transitions from one minimum of the surface to another. Here, the question arises of how to determine the instant (delimit the appropriate region on the potential energy surface) at which the destruction of the old and the appearance of the new molecular structures occurs; how to make the purely geometrical treatment of the potential surface (1.1) compatible with the conventional ideas on molecular structure based on principles of topology which take account, not of molecular mechanics, but of the bond sequence and stereochemical types of coordination centers. The topological approach appears to be the most fruitful in the analysis of the function (1.1) and its gradient $-\nabla E(q)$ allowing one to divide it into a set of nonintersecting domains, or basins. Within each such basin, all separate geometrical configurations, however different, belong to one and the same structure.

The topological approach to the concept of molecular structure has been fully retained and its effectiveness amply demonstrated in Bader's [20—23] quantum mechanical theory of molecular structure based on the topological properties of electron density distribution $\rho(\mathbf{r})$ in molecular systems. This theory — called quantum topology — is, in principle, not limited by the Born—Oppenheimer approximation; it presents an invariant picture of the distribution of the chemical bonds linking various atoms in a molecule and unambiguously determines the type of molecular structure. Topological characteristics of electron density distribution are determined by the properties of its gradient field $\nabla\rho(\mathbf{r}, \mathbf{q})$ where $\mathbf{r}$ is the coordinate of the point in the real space at which ρ has been calculated for the given nuclear configuration $\mathbf{q}$. This field is represented by trajectories $\nabla\rho(\mathbf{r})$, as shown in Fig. 1.1. The gradient paths formed by the trajectories are of two types. The first type comprises the paths passing through the local maxima of the function $\rho(\mathbf{r}, \mathbf{q})$ which lie at the positions of atomic nuclei. The region of the space delimited by the gradient paths going from infinity to a given nucleus is just the basin of the corresponding atom in the molecule. Each pair of adjacent atomic basins is separated from another pair by the nodal surface at whose line of intersection with the molecule plane lies a critical point of the saddle type.

The second type of gradient path, also shown in Fig. 1.1, determines distribution of bonds in the molecular system. It comprises two gradient paths connecting the critical point lying between each pair of the adjacent atomic basins with the

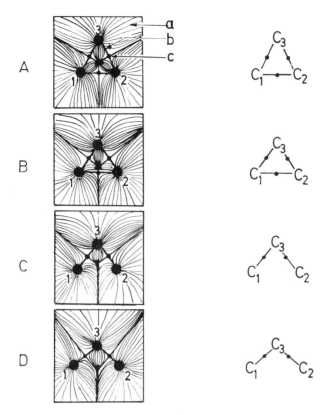

Fig. 1.1. Electron density distributions, gradient paths and molecular graphs of compounds **I** according to *ab initio* (STO—3G) calculations [24]. A molecular fragment is shown corresponding to the cyclopropane ring in **Ib**. (a) The gradient paths delimiting atomic basins; (b) the bonding gradient paths; (c) saddle points on the bonding paths (electron density minima) determining the presence of a bond.

The right portion of the figure shows fragments of the molecular graphs in the region of the atoms C_1, C_2, C_3. A — R_1, R_2 = CN($l_{C_1C_2}$ = 1.543 Å; B — R_1, R_2 = CH$_3$($l_{C_1C_2}$ = 1.770 Å; C — R_1, R_2 = CH$_3$($l_{C_1C_2}$ = 1.826 Å; D — R_1, R_2 = H($l_{C_1C_2}$ = 2.235 Å).

corresponding nuclei. Addition of these two paths produces a line in the space along which the electron density retains its maximal value. This line is called the bonding path. The total network of bonding paths for a given nuclear configuration $Q(q)$ determines the molecular graph, which serves as an analog of structural formulae.

As is evident from calculations, small and, in some cases, fairly large structural variatons do not change the type of the molecular graph and consequently do not distort molecular structure. We can illustrate this using the data of Fig. 1.1 which show electron density distribution and bonding paths for a number of derivatives of the tautomeric system [10]annulene — dinorcaradiene **Ia** ⇌ **Ib**

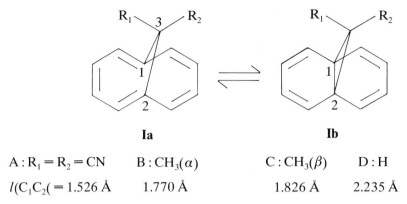

Ia   Ib

A : $R_1 = R_2 = CN$ B : $CH_3(\alpha)$ C : $CH_3(\beta)$ D : H

$l(C_1C_2(= 1.526$ Å   1.770 Å   1.826 Å   2.235 Å

For the dicyano derivative ($R_1 = R_2 = CN$) the distance between the atoms C_1 and C_2 is found by X-ray diffraction methods to be only 1.526 Å. The presence of the simple bond C_1—C_2 and the assignment of the structure to the dinorcaradiene type are certain. For the dimethyl derivative (**I**, $R_1 = R_2 = CH_3$) in the crystal unit cell, two different types of the molecule are present in which the distance C_1—C_2 is 1.770 and 1.826 Å, respectively. Does the bond between atoms still exist in this case, or is it simply a strained molecular skeleton in which the atomic configuration forces the carbon atoms to draw together to a distance less than the sum of their van der Waals radii?

The distribution of electron densities in these molecules, shown in Fig. 1.1, can give an answer to this question Whereas in the molecule with the C—C_2 bond length of 1.770 Å the corresponding bonding path with the saddle point in its center does still exist, in the other molecular form with $l_{C_1-C_2} = 1.826$ Å this path is ruptured and the corresponding saddle point vanishes. (Upon the stretching of the C_1—C_2 bond, the structure passes through the so-called catastrophe configuration [21].) The two types of the molecule are represented by different molecular graphs; hence they are to be considered as isomers **Ib** and **Ia**.

Analysis of topological characteristics of electron density distribution in molecules **I** illustrates the basic possibility of drawing a parallel between classical structural concepts categorized in terms of interatomic bonding and the quantum mechanical electron nuclear concepts. On the potential energy surface (1.1), nuclear subspaces $Q(q)$ can be singled out within which isomeric molecular structures are retained regardless of the degree of deformation.

The concept of molecular structure is thus defined as the equivalent class of molecular graphs in which the same number of bonding paths links the same nuclei in each molecular graph belonging to a given molecular structure [23].

1.2. *Bond Rupture Formation Criterion*

Proceeding from the above topological definition of molecular structure the interconversions of structural (constitutional) isomers, to which tautomeric rearrangements belong, can be described as structural transformations resulting in the change of the type (i.e. the character) of the molecular graph's connectedness. In the language of the classical structural theory this apparently means rupture of the old and formation of new bonds during the course of the transformation. The concept of the bonding path introduced in the topological theory dispels much of the uncertainty in the determination of the chemical bond between the pairs of atomic centers.

From the structural peculiarities of compounds **I** considered above it follows that such geometrical characteristics as internuclear distance cannot be an accurate criterion of the presence of a chemical bond between the chosen pair of atoms in a molecule. Indeed the length of the standard ordinary C—C bond in ethane is 1.532 Å and even its slight stretching (by 0.02—0.04 Å) results, according to the form of the vibrational potential, in its rupture. The longer C—C bonds are indeed not observed in unstrained alkanes. However, for the norcaradienes **Ib** a considerable electron density at the C—C bond is retained up to the distance of 1.783 Å (*Ib*, R_1 = H, R_2 = CH_3) [25]. At the same time in the recently synthesized [1.1.1]-propellane **II** the length of the central C—C bond (1.545 Å), determined by the gas phase electron diffraction method, is close to that for the ethane, although the electron density in the region of this bond is very small and the nature of the bonding in the molecule **II** is rather unique, [27, 28].

II   **III**   **IIIa**

Given that **II** and other [*n*.1.1.]-propellanes (see [27]) do not belong to the biradicals, it is difficult to imagine — within the framework of the classical structural theory — the actual absence of the central C—C bond in these compounds.

It is even more difficult to describe, without abandoning this framework, the existence of the rather strong central C—C bond in [2.2.2.2]-paddlane **III**, as predicted by the MNDO and MINDO/3 calculations [29]. (Bond order is 0.99, length 1.56 Å.) Another geometrical form of [2.2.2.2.]-paddlane, the bond-stretch isomer **IIIa** is energetically unfavourable and in no way corresponds to the minimum on the potential energy surface. The main peculiarity of the C—C bonds in molecules **II** and **III** considered above consists in the fact that they do not belong to normal two-center, two-electron ($2c-2e$) bonds such as the C—C bond in the ethane molecule. If one electron is removed from the bond its length in the radical cation $C_2H_6^+$ increases, according to nonempirical (orbital basis 6—31 G^*) calculations [30], to 1.983 Å. However, even with this stretching, the bond energy is still 38 kcal/mol. These examples indicate that the presence or absence of the valence line in the classical structural formula does not always denote the presence or absence of a chemical bond between corresponding atomic centers. Here we are confronted with that case of rearrangements of compounds having multicenter bonds which is the most complicated to define in the general terms of tautomerism. With these compounds, uncertainly often arises as to whether the isomerization process should be assigned to the stereo- or to the constitutional type.

Thus, the dynamic process of the diene fragment rotation with respect to the iron tricarbonyl group in the π-complex **IV**, detected by the NMR method, should be characterized as structural diastereomerization, if one considers that the iron atom is coordinated to each separate π-bond.

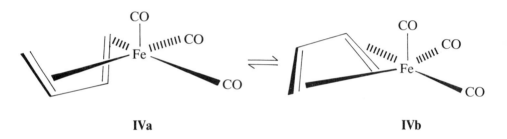

| IVa | IVb |

If, however, the bonding of the diene fragment with the iron tricarbonyl group is considered more correctly, namely as a multicenter bonding comprising all atomic centers of the diene, then the rotation **IVa** ⇌ **IVb** should be assigned to the stereorearrangement class. Among organic compounds there are quite a few similar examples belonging to the area of nonclassical structures [27, 31].

Rigorous assignment of the isomerization process requires a topological analysis of the electron density variation in the interatomic regions when the system is proceeding along the reaction path on the potential energy surface. Clearly, this approach, while retaining its general importance, is not applicable when examining each system individually. This is also true of a number of other formalized schemes for the representation of isomeric transformations [32—34] which are rather remote from the language of structural theory to which the chemists are accustomed. The simplest and most generally acceptable criterion for the assignment of isomerization to the constitutional type (to which tautomerization belongs) is the presence in these processes of bond rupture—bond formation steps in the classical sense of this term.

1.3. *Tautomerism and Rearrangements*

Let us now consider the second part of the definition of tautomerism which states that it is a dynamic, i.e. a rapid, reversible isomerism. The following question is important: in what way does tautomerism differ from the general concept of rearrangements? It is well known that every single-step chemical reaction is reversible, although the equilibrium may be considerably shifted to either side. Hence, the requirement of reversibility is necessary to distinguish tautomerism from rearrangement as well as requirement of rapid reversibility (dynamic features), i.e. a large enough speed to establish equilibrium should be related to certain practically developed criteria for assigning reactions to reversible and rapid classes.

From the physical point of view, this means that first of all standard conditions must be determined (temperature, solvent, pressure) and then the terms reversibility and rapid reversibility must be made dependent upon the sensitivity and the characteristic frequency of the experimental method used for studying the state of the tautomeric system.

Among the methods extensively employed for studying the position of the shifted equilibria, electron spectroscopy (particularly of emission type) and brommetric titration are, apparently, the most sensitive. The improved modification of the latter method helped, for instance, to determine the constant of keto–enol prototropic equilibrium of acetone in water at 25°C $K_T = [E]/[K] = 0.9 \times 10^{-6}$ [35]

$$CH_3-\underset{\underset{O}{\|}}{C}-CH_3 \rightleftharpoons CH_2=\underset{\underset{OH}{|}}{C}-CH_3$$

The free energy of equilibrium

$$\Delta G^0 = -RT \ln K_T \qquad (1.2)$$

amounts at this temperature to approximately 8 kcal/mol.

Often the difference between the free energies of the tautomers exceeds this maximum value and only one of the potentially possible forms is observed experimentally. In such cases the magnitude of ΔG^0 can be sometimes lowered to the required limit by changing the solvent, and/or by varying temperature and concentration.

Another factor characterizing a rearrangement as a tautomeric one is related to the conversion rate, i.e. to the time required for the establishment of equilibrium which is determined by the magnitudes of the free activation energy $\Delta G^{\neq}$ (Fig. 1.2).

$$k = 1/\tau = \kappa k_B T/h \exp(-\Delta G^{\neq}/RT) \qquad (1.3)$$

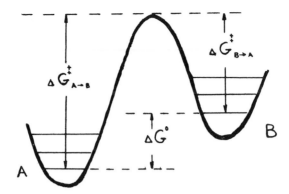

Fig. 1.2. The free energy profile for the equilibrium A ⇌ B.

where k is the reaction rate constant, τ is the effective lifetime of the isomer, κ is the transmission factor (usually taken to be 1), k_B is the Boltzmann constant, and h is the Planck constant.

The magnitude of the free activation energy also depends on temperature

$$\Delta G^{\neq} = \Delta H^{\neq} - T\Delta S^{\neq} \tag{1.4}$$

where $\Delta H^{\neq}$ is the enthalpy and $\Delta S^{\neq}$ is the entropy of activation. These quantities are determined by the depth of the potential energy surface minima corresponding to the isomeric forms and by the curvature of this surface in the region of the minima (see Chapter 3).

Even if at appropriate values of ΔG^0 the conversion $A \rightleftharpoons B$ proceeds at a low rate due to the high magnitude of the free activation energy, the interconversion of isomers is usually considered as being a rearrangement.

Consider now, by way of example, two valence tautomerizations based on one common mechanism; the Cope rearrangement.

1. *Isomerization of 1,5-hexadiene-1,1D^2* [36]

The free activation energy at 25°C is estimated to be 38.2 kcal/mol corresponding to the very slow conversion of one tautomeric form to the other, $\approx 10^{-13}$ sec^{-1} according to (1.3). Taking the half-life period $\tau_{1/2}$ to be the time required for the establishment of equilibrium, we find that it amounts to approximately 10^{12} sec $(0.693\,\tau)$.

2. *Electrocyclic benzene oxide—oxepine rearrangement* [37]

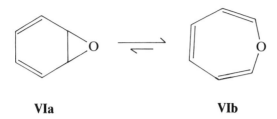

| **VIa** | **VIb** |

The activation energies of the direct and reverse reactions are 9.1 and 7.2 kcal/mol, respectively, and half-conversion periods (at 0°) are $\tau_{1/2\,(\text{VIa})} = 0.5 \times 10^{-7}$ sec^{-1} and $\tau_{1/2\,(\text{VIb})} = 0.3 \times 10^{-6}$ sec^{-1}.

Chemical intuition suggests that the former reaction should be considered as a rearrangement whereas the latter relates to a tautomeric conversion. At the same time, simple extrapolation shows that at 300°C the former reaction proceeds at a fairly rapid rate of $\sim 3.7 \times 10^{-1}$ sec^{-1} and, following the same common sense, it can be assigned to the tautomeric conversions.

Another example is the enolization of 1,3-dicarbonyl compounds. It is known than in the gas phase this proceeds extremely slowly [35], in aprotic neutral solvents its rate is moderate, but in the presence of acids and bases the establishment of equilibrium is practically instant.

It is obviously difficult, perhaps even impossible, to draw a distinct line between the concepts of tautomerism and rearrangement. The difference between them is not qualitative but quantitative.

1.4. *Thermodynamic and Activational Scale of Tautomeric Processes*

All the same, for practical purposes of classification and terminology, it is useful to have an approximate scale for thermodynamic and activational parameters of tautomeric equilibria. In the foregoing, it was emphasized that this scale can easily be shifted when experimental conditions are altered and that the scale also depends on the method by which a potentially tautomeric system is studied.

Over a good many years, however, the organic chemists have developed, in accordance with the natural norms of the biosphere, a simple qualitative way for assessing reaction rates, stability of organic molecules and standard conditions, under which these characteristics are compared. The barrier separating the stable (i.e. slowly reacting, isomerizing) compounds from the unstable ones is habitually established with reference of the stability *at room temperature*, and therefore by the possibility to preparatively isolate a compound under those conditions. In terms of activation energies, it corresponds to the requirement that $\Delta G^{\neq} >$ 20—25 kcal/mol and $K_{25°} < 10^{-4}\,s^{-1}$.

For tautomeric equilibria one can approximately set the upper limit on the magnitude of the free activation energy so that the time for the equilibrium to become established (in a monomolecular reaction) should be commensurable with the usual time of the experiment (not more than 7 hours at room temperature of 300 K):

$$\tau_{1/2} = 25\,000\,s, \qquad k = 0.693/\tau_{1/2} = 3 \times 10^{-5}\,s^{-1}, \qquad \Delta G^{\neq} \approx 25\,kcal/mol.$$

Assuming that even the most sensitive methods cannot detect unstable tautomeric forms if their relative concentration is less than 10^{-7} (practically, this limit is 10^{-4}–10^{-5}), one can also determine the place of tautomeric reactions on the thermodynamic scale [13, 14].

$$\begin{aligned} \Delta G^0_{25} &< 6\,\text{kcal/mol} \\ \Delta G^{\neq}_{25} &< 25\,\text{kcal/mol} \end{aligned} \tag{1.5}$$

Naturally, one should not forget about the relative character of the limits (1.5). Thus, proceeding from the same definitions, changing only the temperature to which they are assigned, one obtains the upper limits of $\Delta G^{\neq}_{-70} < 16$ kcal/mol and $\Delta G^{\neq}_{200} < 32$ kcal/mol.

2. Degenerate Tautomerism

In the acetylacetone *cis*-enol, rapid and reversible migrations take place in solution and in the gas phase

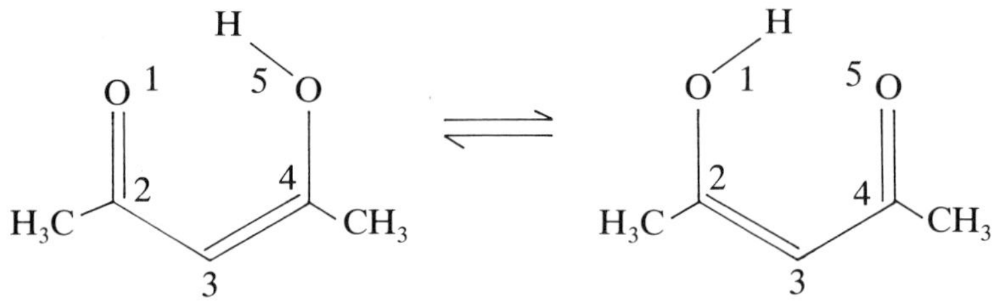

VIIa   **VIIb**

resulting in the molecule transforming, as it were, 'into itself'. Of course, this transformation is accompanied by rupture of some and formation of other (albeit chemically identical) bonds. Therefore, the transformations **VIIa** ⇌ **VIIb** can be described as degenerate isomerism and, considering the fact that it satisfies the criteria (1.5), as degenerate tautomerism.

Such tautomerism is degenerate only in the light of those methods by which individual atomic centers cannot be distinguished and which study properties either of the molecular system as a whole (chemical, electrical, magnetic methods, electron spectroscopy etc.) or of its separate fragments (vibrational spectroscopy). Naturally, these methods cannot detect processes of degenerate tautomerism.

If, however, the method of investigation allows one to individualize different atomic centeres, that is, makes it possible to observe each of them separately, this problem does not arise and degenerate tautomerism can be understood as a particular case of the general problem of tautomerism.

On the potential energy surface, the symmetrical double-well curve shown in fig. 1.5b corresponds to the degenerate tautomerism. A powerful experimental tool for studying degenerate tautomerism is the NMR method which allows one to observe the signals of individual nuclei in the molecular system. Some other methods of this type are given in the Table 1.I. The quantum mechanical calculation of tautomeric reactions can, apparently, be also applied to the theoretical studies of the degenerate tautomeric systems. An impressive amount of data on degenerate tautomeric reactions has been amassed with the aid of these methods.

Table 1.I. Characteristic times of various physical methods [50, 51].

Method	$\tau'(s)$
X-ray diffraction	10^{-18}
Neutron diffraction	10^{-18}
Electron diffraction	10^{-18}
Photoelectron spectroscopy	10^{-18}
UV- and visible spectroscopy	$10^{-14}-10^{-15}$
IR- and Raman spectroscopy	$10^{-11}-10^{-13}$
Mössbauer spectroscopy (Fe)	10^{-7}
ESR spectroscopy	$10^{4}-10^{8}$
Ultrasonic absorption	$10^{-4}-10^{-8}$
NQR spectroscopy	$10^{-1}-10^{-8}$
NMR spectroscopy	$10^{-1}-10^{-6}$
Temperature jump	$10^{-1}-10^{6}$
Usual kinetical methods	Greater than 10^{1}

These elementary explanations are essential for two reasons. First, we should like to emphasize that degenerate tautomerizations possess no basic peculiarities with regard to their mechanism and energetics as compared to the general case of nondegenerate tautomerism. At the same time, owing to the symmetry of the potential surface of degenerate tautomeric interconversions, their energetics is of special interest since the activation energies obtained correspond to the intrinsic activation energy barrier uncontaminated by additional thermodynamical contributions.

Secondly, introduction of a special term for the degenerate tautomeric system is necessary in the light of the definition of tautomerism as a dynamic isomerism. The very term of isomerism which originated in the framework of the classical structural theory cannot contain the notion of degenerate isomerism since it is coined from the Greek ισομερος, that is, containing equal parts. Hence, degenerate tautomers such as **VIIIa**, **VIIb** or the degenerate valence tautomers of bullvalene existing in dynamic equilibrium which were predicted by Doering [38]

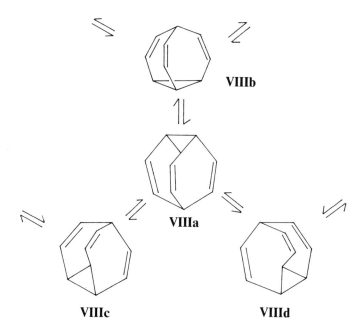

cannot be regarded as isomers in the accepted sense of this term, according to which they are no more than identical compounds. However, at the level of differentiation between individual nuclei, the structures **VIIa**, **VIIb** as well as all 1 209 600 (10!/3) structures of bullvalene* differing in the character of their interatomic bonds are nonidentical to one another, and the equilibrium is a degenerate tautomerism.

At present degenerate isomerizations are defined [41] as chemical reactions in

---

* A suggestion as to the dynamic structure of the then still hypothetical bullvalene was first made in 1961 by Doering at the conference in Leningrad devoted to the centenary of the theory of chemical structure: in bullvalene ". . . atoms are constantly moving over the surface of the molecule, no pair of atoms stays bound to each other for a long time, but the system as a whole maintains its structure. If, however, this hypothetical molecule undergoes, as is expected, Cope rearrangement, it will be difficult to describe its structure in classical terms. But I think that the great Butlerov with his bold and inquisitive mind would be satisfied by this development of his theory of chemical structure" [38]. Two years later an examination of the ¹H NMR spectrum of bullvalene confirmed Doering's prediction: at 100° the bullvalene spectrum consists of a single peak common to all protons [39, 40].

which the initial compound and the product are different only in the permutation of identical atoms. Therefore, degenerate isomerizations are also called permutational isomerism [42], as well as isodynamic transformations [43], automerizations [44] and topomerizations [45]. The last term, based on the stereochemical definition of homo-, enantio- and diastereotopic groups developed by Mislow and Raban [46] as well as by Eliel [47], is most the complete.

Similarly to isomeric transformations, topomerizations are divided into stereo- and constitutional topomerizations. Degenerate tautomerizations belong to the latter. If rearrangements between topomers result in exchange of positions of homotopic groups (which are interchangeable in the operation of the rotation about axis C_n), they are regarded as belonging to the homotopomerizations [45]. An example of this is the degenerate valence tautomerism of cyclobutadiene:

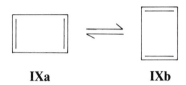

IXa   IXb

Interconversions of topomers characterized by exchange of positions of enantiotopic (interchangeable only in operations of the rotation about the S_n axes) and diastereotopic (noninterchangeable in any operation of symmetry) groups and combined under one common class of heterotopomerizations. Degenerate tautomerizations of the *cis*-enol form of acetylacetone **VIIa** ⇌ **VIIb** and bullvalene can serve as examples of constitutional heterotopomerizations.

3. Methods for Investigating Tautomeric Systems

Since the phenomenon of tautomerism is associated with rather rapid interconversions of isomers and topomers, conventional methods for studying the structure and properties of stable compounds of rigid structure give little information on the characterization of tautomeric systems. A more or less complete description of these systems requires solutions to the following basic problems:

1. Determination of the structure of interconverting tautomers corresponding to distinct minima on the potential energy surface separated by energy barriers for which condition (1.5) is satisfied.
2. Determination of the energetics of these interconversions, i.e. determination of the barrier height and the relative stability of the interconverting isomers (Fig. 1.2).
3. Ascertainment of the nature of the energy barrier, i.e. elucidation of the electronic and steric factors governing its height.

The last problem can be solved on the basis of a purely theoretical analysis. In order to determine the structure of the nonrigid molecules and to evaluate energy barriers there exists an abundant variety of experimental methods the choice of which is, in each case, dictated mainly by the height of corresponding barriers. The magnitude of the energy barrier is — as follows from the relationship (1.3) — directly related to the effective lifetimes τ of interconverting isomers or topomers. Depending on the magnitude of τ, structural nonrigidity can be revealed by some methods and not revealed at all by others. Indeed, some isomeric forms prove to be experimentally indistinguishable, this being the case when their lifetimes τ_0 are shorter than the characteristic time τ' of the experimental method [48].

On condition that

$$\tau' < \tau_0 \tag{1.6}$$

the measurements reveal in principle all the individual structural forms which correspond to the minima of the potential surface because during the measurement time to transitions between them can occur. On the contrary, if

$$\tau' > \tau_0 \tag{1.7}$$

the system can, during this time, transform from one form to another and the experiment is fixing, not separate isomeric or topomeric structures, but a certain averaged picture.

The characteristic time of a given experimental method is determined by the lifetime of the excited state $\Delta t \equiv \tau'$ arising upon absorption of a corresponding quantum of energy. In spectral measurements, information on the value of Δt is contained in the parameter of the natural (that is, free from any instrumental effects) linewidth, which, owing to the uncertainty principle

$$\Delta E \cdot \Delta t \simeq h/2\pi \tag{1.8}$$

cannot be infinitesimal and, considering that $E = h\nu$, depends on t as follows

$$\Delta \nu \simeq 1/2\pi \cdot \Delta t. \tag{1.9}$$

Most accurately, the full form of the spectral line is described by the Lorentz function

$$\mathscr{L}(\omega_{jk} - \omega_0) = \frac{1}{\pi} \left[\frac{1/\tau}{(1/\tau)^2 + (\omega_{jk} - \omega_0)^2} \right] \tag{1.10}$$

where $\omega = 2\pi\nu$, ω_{jk} is the resonance frequency corresponding to transitions between the k-th and j-th energy levels and τ is the relaxation time, i.e., the time required for the restoration of the equilibrium population in the energy levels of the system.

The maximal amplitude of the function (1.10) shown in Fig. 1.3 is ω_{jk} Of

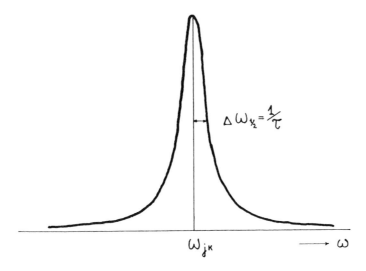

Fig. 1.3. The normalized Lorentz curve of the absorption (1.10). $\Delta\omega_{1/2}$ is the half width at the half height. The curve maximum is at $\mathscr{L}(0) = \tau/\pi$.

importance here is the value of the half-width $\Delta\omega_{1/2}$ of the curve (1.10) at its half-height

$$\tfrac{1}{2}\mathscr{L}(0) = \frac{\tau}{2\pi} = \frac{1}{\pi}\left[\frac{1/\tau}{(1/\tau)^2 + (\Delta\omega)^2}\right] \tag{1.11}$$

whence

$$\Delta\omega_{1/2} = 1/\tau \tag{1.12}$$

or

$$\tau = 1/2\pi\,\Delta\nu_{1/2} \tag{1.13}$$

In rigid structures, where physical processes of excited state decay are the only relaxation mechanism, $\tau = \Delta t = \tau'$. When, however, chemical exchange, i.e. the reaction of interconversion of isomeric forms takes place, the relaxation mechanism becomes more complex, and the quantity τ can now be defined as

$$1/\tau = 1/\tau' + 1/\tau_0 \tag{1.14}$$

At slow rates of chemical exchange when $\tau_0 \gg \tau'$, the isomer interconversion processes do not affect the spectral line form, but as these values draw closer to each other, a broadening of the spectral line is observed. Depending on the magnitude of τ', fast interconversions of isomeric forms can produce considerable broadening of lines in one spectral range without affecting them at all in another. Thus, for the exchange processes to be reflected in the form of the IR or the Raman spectral line, they must proceed at the rates exceeding the rates of diffusion controlled reactions, i.e., interconversions of isomeric forms have to occur more frequently than 10^{10} s^{-1}. Indeed, the minimal natural width of the IR spectrum lines amounts approximately to 0.1 cm^{-1} whence, according to (1.9), the lifetime of the excited state $\Delta t = 5 \times 10^{-11}$ s or $k = 1/\Delta t \simeq 10^{12}$ s^{-1}.

Recently, the possibility of registering such fast processes by the method of IR spectroscopy was demonstrated experimentally for the first time [49]. In the IR spectrum of 1,2-dichloroethane, the presence of two rotational isomers, *gauche* **Xa** and *trans* **Xb**, is seen in the splitting of each of 18 vibrational lines belonging to every conformer

Xa   **Xb**

In the NMR method, the minimal natural linewidth is approximately 0.1 s^{-1} (Hz). Consequently, broadening of the NMR spectral lines allows one to observe, in accordance with (1.9), exchange processes having lifetime less than 2 s or with the rates higher than 0.5 s^{-1}. For the coalescence of single signal peaks belonging to two interconverting isomers or topomers and separated, for example, by 200 Hz (the usual range of chemical shifts in the ^{13}C NMR spectra), the chemical exchange rate has to equal 10^3 s^{-1}.

In Table 1.I characteristic times τ' are given for various physical and physico-chemical methods employed for investigating the dynamics of structurally nonrigid compounds.

In such methods as X-ray, electron, and neutron diffraction the characteristic times τ' are, approximately, 10^{-18} s, which is far less than the time required for any molecular rearrangement and even less than the periods of normal vibrations of individual molecular bonds. Therefore, using these methods, one is able to register only instantaneous configurations of the nonrigid molecule, obtaining only afterwards some averaged picture of the structure. This permits, at the same time, the determination of the populations of separate forms coexisting in the solid or gaseous phases and, having separated the high amplitude motions from other nuclear vibrations, the elucidation of the shape of the corresponding potential energy function. This approach has proved very effective for the quantitative description of structural nonrigidity in systems with extremely low activation barriers for molecular rearrangements.

Fig. 1.4 illustrates the applicability of various methods to studying tautomerization reactions and shows the place of these on energy and kinetic scales.

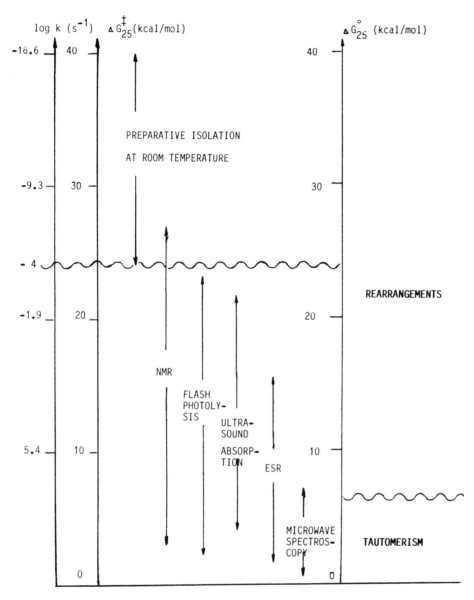

Fig. 1.4. The activation and thermodynamic scales of tautomeric processes and the methods for their investigation.

4. Butlerov's and Laar's Tautomeric systems

Whereas Butlerov interpreted tautomerism as an equilibrium isomerism, Laar believed that no real equilibrium existed between separate forms and the ability of a tautomeric substance to react in two or more directions (in those days only chemical methods were known for investigating a tautomeric system) was the result of continuous intramolecular oscillation of individual atoms (see [5, 52]).

From the physical point of view, the difference between Butlerov's and Laar's tautomeric systems can, in the main, be reduced to the distinction between the forms of adiabatic potential, namely, double-well potential in the former case and single-well in the latter. Fig. 1.5 shows this difference.

Compounds which may conventionally be called Laar's tautomers, can be exemplified by substances with symmetrical hydrogen bonds such as maleate-anion **XI**, monoprotonated 1,8-bis-dimethylaminonaphthalene **XII** (proton sponge).

The single-well potential (Fig. 1.5a) for proton migration in these compounds was assessed by the X-ray structural and electron diffraction data, as well as by quantum mechanical calculations (see the reviews [53, 54]).

XI   **XII**

The double-well potential (Fig. 1.5b) where each separate minimum corresponds to one of the interconverting isomeric forms, is characteristic of the typical tautomeric system $A \rightleftharpoons B$. The level of fluxionality of this system is governed by the height of the energy barrier ΔE.

At small values of ΔE, the distinction between tautomeric systems and the systems of the types **XI** and **XII** becomes rather subtle. Even at a low thermal excitation, all the minima of the potential energy surface divided by small energy barriers are populated. Although the Born—Oppenheimer approximation remains valid when the surface of the lower electron-excited state is sufficiently remote from the saddle point in the $A \rightleftharpoons B$ transition, i.e.

$$\psi(r, q) = \psi_e(r, q_0) \, \psi_n(q) \tag{1.11}$$

where $\psi_e(r, q_0)$ is the electron wave function in the region of the equilibrium nuclear configuration q_0 and $\psi_n(q)$ is the nuclear wave function, the effective nuclear configuration and the characteristics depending on it are not exactly determined by the minima points but have to be described by the square of the nuclear wave function $\psi_n^2(q)$ similarly to the way in which the electron density distribution is described. If the energy of zero vibrations of isomeric forms is smaller than the interconversion barrier of these forms, then, in principle, there

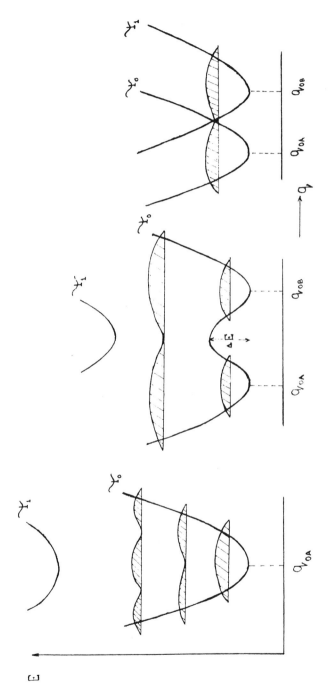

Fig. 1.5.   PES sections along the coordinate of the degenerate isomer interconversion A ⇌ B. Ψ_0, Ψ_1 are the ground and lowest excited electronic states. Shaded portions are the regions of the function $\Psi_n^2(q)$ for respective vibrational states. q_{OA}, q_{OB} are the equilibrium coordinates (a) symmetrical structure A ≡ B, large energy gap between the surfaces Ψ_0 and Ψ_1; (b) structurally nonrigid molecular system, small energy gap between Ψ_0 and Ψ_1; (c) structurally and electronically nonrigid molecular system, intersection of the surfaces Ψ_0 and Ψ_1.

remains the possibility of separately observing and even isolating these forms at low temperatures. At higher temperatures, when the system is excited to higher vibrational levels, its nuclear configuration is completely delocalized between two isomeric structures.

If the investigation of a tautomeric system is performed by a method whose characteristic time is longer than the lifetimes of the individual tautomers (1.7), then two tautomeric forms cannot be distinguished by this method. In this case no difference can be observed between Butlerov's and Laar's tautomeric systems. In order to reveal this difference, one should employ a method satisfying condition (1.9). For instance, using NMR spectroscopy one is unable to make a distinction between the single-well and double-well forms of the adiabatic potential for *cis*-enols **VII**

VIIa, C_s **XIII**, C_{2v} **VIIb**, C_s

(TS, Transition State)

Indeed, the lifetime of a proton in the potential minima of the chelate intramolecular hydrogen bond is less than 10^{-6} s [55], and from (1.9) it is evident that in order to distinguish the signals of two tautomers in the NMR spectrum, the difference between chemical shifts of indicator nuclei should be of the order of 10^6 s^{-1}. This, however, is far beyond the scale of chemical shifts not only of ^{1}H-NMR but even ^{17}O-NMR. Consequently, on the NMR time scale the molecule **VII** possesses C_{2v} symmetry. Data obtained with the aid of IR-spectroscopy and X-ray diffraction are inconclusive; however, studies of isotopomers of **VII** performed using microwave spectroscopy produced convincing evidence in favor of the C_s structure **VIIa, b** with nonequivalent C—O and C—C bonds [56], i.e., in favor of degenerate tautomerism **VIIa** ⇌ **VIIb**. This conclusion is in agreement with the results of the most accurate *ab initio* calculations [57] (see also [18]) in which the extended orbital basis set was employed and the electron correlation was taken into account. The symmetrical C_{2v} form corresponds to the transition state structure for interconversion of degenerate tautomers. The calculated activation energy is only 5 kcal/mol.

A similar conclusion as to the existence of an extremely fast degenerate prototropic tautomerism was made by using the methods with very high time resolution (microwave spectroscopy, laser-excited fluorescence etc.) for a number of other conjugated structures with strong and very short hydrogen bonds [53, 58—62].

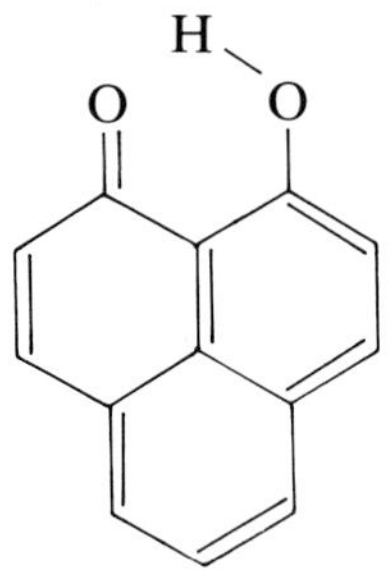

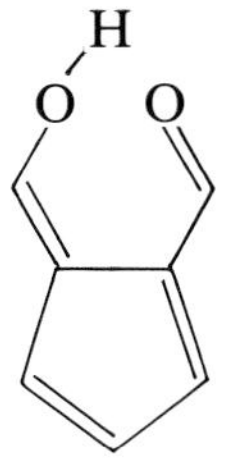

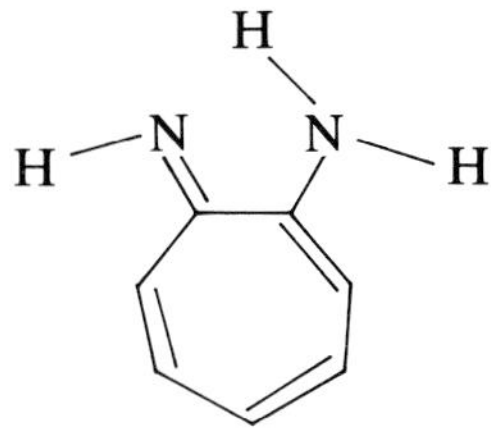

XIV   **XV**   **XVI**

In the compounds of this type, the form of the adiabatic potential corresponds to that shown in fig. 1.5b and both minima belonging to degenerate tautomers have the same depth. In these systems, the atomic shifts (including the heavy atoms of C, O and N) occuring in the course of a transition from one potential surface minimum to another are not infrequently very small (0.1—0.2 Å) which is comparable to the de Broglie wavelengths of the atoms [63, 64]. This leads to a fairly high contribution of the tunneling process (underbarrier transition) to the observed reaction rate. The tunneling frequency for the **VIIa** ⇌ **VIIb** reaction calculated from microwave spectroscopy data [56] is 6.3×10^{11} s^{-1} whereas the rate constant for interconversion of the C_s tautomers of 3-hydroxy-2-propen-2-al evaluated from equation (1.3) for the classical overbarrier path with an activation energy of 5 kcal/mol amounts at ambient temperature to only 1.5×10^9 s^{-1}, i.e. it is lower by two orders of magnitude.

For another degenerate valence tautomerization mentioned earlier,

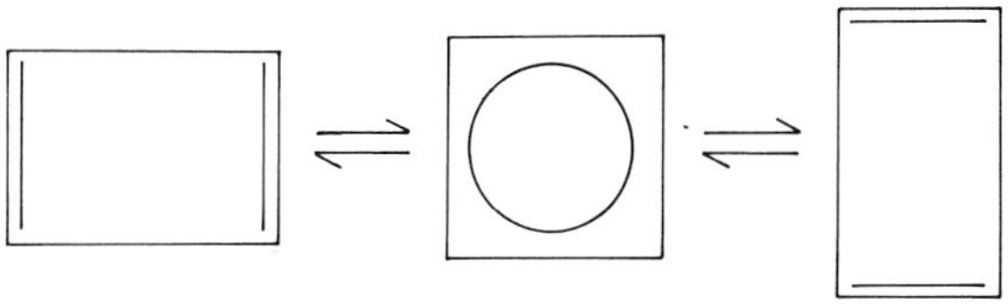

IXa, D_{2h} **XVII**, D_{4h} **IXb**, D_{2h}
                     (TS)

tunneling is the principal reaction path at temperatures less than 0°C [65] and even at 77°C. The rate of tunneling **IXa** ⇌ **IXb** is, according to MINDO/3 calculations [66], higher by three orders of magnitude than the rate of the classical overbarrier reaction.

The instability of the symmetrical structures **XIII** and **XVII** with respect to **VII** and **IX**, respectively, is, from the physical point of view, nothing other than a manifestation of the second order Jahn—Teller effect [67, 68] associated with the drawing together of the surfaces of the ground (ψ_0) and the excited (ψ_1) electron states in the region of the symmetrical form.

When a vibration of appropriate symmetry is present, more stable structural forms of a lower symmetry appear, between which rapid interconversions take place (Jahn—Teller isomers [69]).

Fig. 1.5c shows a basically different — from the physical point of view — type of dynamic isomerism which may also be characterized as electronic nonrigidity [70]. Its distinctive feature is the intersection (or very close proximity) of the potential surface of the electron excited state ψ_1 and the ground state surface ψ_0 in the latter's minimum at q_{0A}. In the intersection region of the potential energy surfaces ψ_0 and ψ_1, the electronic states become degenerate, giving rise to the first order Jahn—Teller effect and, consequently, to the dynamic instability of the corresponding molecular system [68].

This means that in the region of the approaching potential energy surfaces, the rates of the motion of nuclei become comparable to those of the electrons. The Born—Oppenheimer approximation loses its validity, and in order correctly to describe the properties of the systems of the type shown in fig. 1.5c, the so-called vibronic approximation should be used

$$\psi(r, q) = c_0 \Psi_{0e}(r, q) \Psi_{0n}(q) + c_1 \Psi_{1e}(r, q) \Psi_{1n}(q) \qquad (1.12)$$

From Eq. (1.12) and Fig. 1.5c, it follows that the relative weight of electron wave functions ψ_{0e} and ψ_{1e} depends on the vibrational level and the corresponding nuclear wave functions. Hence, by varying the temperature and thus by varying the population of the vibrational levels, one can vary both the nuclear configuration and the electronic properties of a molecular system. The dynamic equilibrium in binuclear complexes of the metallic ions with variable valency **XVIII**, in which the adiabatic mechanism of electron transfer through π-conjugated bridges is manifested [71], may serve as an example of electronic nonrigidity.

$$(NH_3)_5Co^{2+} \underline{\quad\quad} N \bigcirc \!\!-\!\!X\!\!-\!\! \bigcirc N \underline{\quad\quad} {}^{3+}Ru(NH_3)_5 \quad \rightleftharpoons$$

XVIIIa

$$(NH_3)_5Co^{3+} \underline{\quad\quad} N \bigcirc \!\!-\!\!X\!\!-\!\! \bigcirc N \underline{\quad\quad} {}^{2+}Ru(NH_3)_5$$

XVIIIb $X = CH_2, S, CH = CH, C \equiv C$

In these reactions, along with the geometric rearrangement which can be regarded as a valence tautomerism, a drastic change in the electron configuration occurs which is clearly reflected in the electronic absorption and luminiscence spectra. Depending on the nature of the bridge group in the complexes **XVIII**, the rate of the interconversion **XVIIIa** ⇌ **XVIIIb** changes over a very wide range, corresponding to the transition from Butlerov's to Laar's tautomerism.

5. Tautomerism and the Mechanisms of Organic Reactions

A target oriented approach to the tautomeric structures inevitably requires an investigation of the reaction mechanism which the tautomeric transformation is based on.

A reaction mechanism is the sum of the elementary stages, each of which makes its contribution to the transformation of reagents into products. A full knowledge of this mechanism presupposes the determination of the sequence of elementary stages in this transformation (stoichiometric mechanism) as well as the gaining of information on the nature of all elementary stages. In the elementary act of a chemical reaction, as many molecules (ions or radicals) take part as are envisaged by the equation of the corresponding elementary stage, and the mechanism of an elementary act, which is called the intimate or intrinsic mechanism, may be defined as a path of the chemical reaction over the potential energy surface from the valley of reagents to the valley of the products, passing the saddle point of the transition state. The concept of the reaction mechanism is thereby clarified as a logical development and an extension of the concept of structure to nonequilibrium distances and nonequilibrium nuclear configurations of the system of molecules (one molecule), ions, radicals etc. Thus, mechanism is the concept of structure extended into time as a fourth dimension.

It is clear that the elucidation of the intrinsic mechanism of the elementary acts is the key to an understanding of the entire mechanism. If the kinetics and energetics of elementary stages of any chemical transformation are known, we can calculate the rate, the heat and other parameters of the reaction as a whole.

The great importance of the chemical transformations which are commonly termed tautomeric lies in the fact that they proceed, as a rule, as elementary stages. This rule has, apparently, no exceptions in those cases in which we deal with intramolecular tautomeric transformations, i.e. with the chemical reactions proceeding as a rearrangement of the nuclear skeleton of a single molecule. These transformations retain all the general features of a chemical reaction and may serve as the most convenient and adequate model when studying the similar mechanisms of more complex intermolecular reactions.

Historically, the study of the mechanisms governing the elementary acts of chemical transformations could only be tackled much later than the period during which vast and varied data on the results of chemical reactions were accumulated and the development of the phenomenological theory of reaction mechanisms, particularly with regard to organic reaction [52, 72]. Despite the unsatisfactory physical foundation of this theory, it proved possible to carry out within its framework a detailed classification of reactions of organic compounds according to their main types. Moreover, owing to extensive studies of the influence of reagent structure, solvent, temperature and catalysis on the reaction kinetics, and to the study of the stereochemical results of reactions, a deep understanding was often achieved of the structural peculiarities of transition states as the key structures determining the energetics and the stereochemical result of a given reaction.

Of special importance are the transition states of the slowest, i.e. the limiting stage, of a reaction because the relative energy of this stage determines the rate of the whole process. The same is also true of the stage at which the structure of the final product is formed. Therefore, when studying the mechanism of a reaction, the principle efforts, both experimental and theoretical, are directed towards identifying these stages and elucidating their intimate mechanisms.

Thus, in the case of the pinacolic rearrangement of glycols, the crucial step is the rearrangement of the carbenium ion **XIX**, formed under the action of a strong acid.

$$R^1{-}\underset{\underset{\textstyle OH}{|}}{\overset{\overset{\textstyle R_1}{|}}{C}}{-}\underset{\underset{\textstyle OH}{|}}{\overset{\overset{\textstyle R_2}{|}}{C}}{-}R_2 \quad \xrightarrow{H^+} \quad R_1{-}\underset{\underset{\textstyle +}{}}{\overset{\overset{\textstyle R_1}{|}}{C}}{-}\underset{\underset{\textstyle OH}{|}}{\overset{\overset{\textstyle R_2}{|}}{C}}{-}R_2 \quad \longrightarrow$$

XIX

$$R_1{-}\underset{\underset{\textstyle R_2}{|}}{\overset{\overset{\textstyle R_1}{|}}{C}}{-}\underset{\underset{\textstyle OH}{|}}{\overset{\overset{\textstyle +}{}}{C}}{-}R_2 \quad \xrightarrow{-H^+} \quad R_1{-}\underset{}{\overset{\overset{\textstyle R_1}{|}}{C}}{-}\underset{\underset{\textstyle O}{\|}}{\overset{}{C}}{-}R_2$$

XX

Reactions of the **XIX** → **XX** type can be carried out as a reversible dynamic, i.e. tautomeric, transformation in which the limiting elementary stage is singled out. Isomerizations of carbenium cations generated in superacidic media — for instance, the degenerated tautomeric rearrangement of the pentamethylethenium cation — can serve as an example of such transformations [31, 73].

$$CH_3{-}\underset{\underset{\textstyle CH_3}{|}}{\overset{\overset{\textstyle CH_3}{|}}{C}}{-}\underset{\underset{\textstyle +}{}}{\overset{\overset{\textstyle CH_3}{|}}{C}}{-}CH_3 \quad \rightleftharpoons \quad CH_3{-}\underset{\underset{\textstyle +}{}}{\overset{\overset{\textstyle CH_3}{|}}{C}}{-}\underset{\underset{\textstyle CH_3}{|}}{\overset{\overset{\textstyle CH_3}{|}}{C}}{-}CH_3$$

XXIa   **XXIb**

Evidently, the tautomeric transformations should currently be one of the most important research targets for the branch of theoretical chemistry concerned with mechanisms of organic reactions. They are also the most appropriate model for the study of elementary stages of these reactions.

The problem of tautomerism has yet another aspect resulting from the historical process of the development of theoretical organic chemistry: the vast material accumulated in the study of reaction mechanisms should be of use in the design new tautomeric systems. The deliberate construction of such systems has been greatly facilitated by advances in applied quantum chemistry.

The most significant feature of tautomeric reactions is the ease with which they proceed due to low activation barriers between the reagents and products. One may postulate an obvious analogy between tautomeric reactions and the most perfectly organized chemical transformations, i.e., enzymatic reactions. This analogy cannot be purely formal; it undoubtedly springs from the similarity of mechanisms governing the elementary stages of transformations. A look at this problem both from above and from below will certainly be productive.

6. Classification of Tautomeric Reactions

There are several classification schemes for tautomeric processes which are based either on structural features, the topological properties of the bond network, or on the characteristics of the reaction mechanism.

Application by Knorr and Wislicenus of the theory of electrolytes to the analysis of the nature of reactions of the tautomeric forms has led to tautomeric transformations being regarded as, at least, the two-step reactions whose first step is ionization of the rupturing bond [4]. Depending on whether a migrating group belongs to the cationoid or the anionoid type, tautomeric processes have been divided into cationotropic and anionotropic ones [4, 72]. To these types of tautomerism there have been added the valence and the ring-chain types of tautomerism which do not fit into framework of the cationotropic and anionotropic reactions, and each of the above mentioned principal types is further dissected into a series of subdivisions according to the structural peculiarities of the molecular fragment to which the migrating group is attached.

W. Hückel noted [74] that it was expedient to classify tautomeric processes on the basis of the intramolecular transformation mechanism. The main types of tautomerism are, in this case, the reversible intramolecular reactions, which proceed in accordance with the mechanisms of the substitution and addition reactions. Ingold [72], too, drew an analogy between a number of tautomeric transformations and the substitution, elimination and addition reactions.

In the classification of the tautomeric processes developed by the English school, ideas emerge of the desirability of taking account of the structure of the carrier system of the migrating group; to single out, for example, the triad prototropic systems

$$\text{H}\diagdown_{\displaystyle \text{X}{-}\text{Y}{=}\text{Z}} \;\rightleftharpoons\; \text{X}{=}\text{Y}{-}\text{Z}\diagup^{\displaystyle \text{H}} \tag{1.13}$$

Depending on the type of X, Y and Z, such widespread tautomeric types as the keto—enol, amido—imidol, thiol—thione, amidine, and others, should be assigned to the triad systems [4, 16, 52, 72, 74].

Symbolic equations of the type (1.13) representing a generalized chemical transformation covering various groups of compounds formed according to a common structural principle are one of the first (if not the first, see [75]) examples of a formal equation describing and predicting a series of chemical reactions. Equations of this type [76—78], which can be developed on the basis of the broader formalism of graph theory, are gaining acceptance as a useful tool for systematization, classification, and also prediction of hitherto unknown chemical reactions. In particular, attention should be drawn to the development of a rather effective artificial-intelligence-based computer program [80] for the elaboration of new chemical reactions, including isomerizations.

An obvious shortcoming of this approach to the classification of reactions of various types is the fact that the formal equations similar to (1.13) contain only indications as to the structure of the initial and the final products, without specifying the mechanism of their interconversion which, of course, may vary widely for different compounds, and the unification of which, based on the common reaction scheme would, in this case, be rather formal. This approach, with its basis in formal logic, is nevertheless justified if only for the reason that it can serve as a starting point for a systematic search for new tautomeric systems based on combinatorial theory.

Zefirov and Trach [12] recently described a detailed scheme for the classification of tautomeric transformations based on the formalization of symbolic equations. The essence of their approach consists in the generalization of similar equations in accordance with the topology of electron transfer, which determines the course of the reaction. Within each formally established class of tautomeric reactions further systematization is performed by taking account of the transformation types. Six such types of tautomerism are singled out: (1) sigmatropic (2) electrocyclic, (3) electron, (4) addition, (5) cycloaddition, and (6) cyclodismutational. The full classification hierarchy of tautomeric reactions looks, in descending order, as follows: topology of electron transfer — transformation type — topological type — reaction type. In order to completely assign a tautomeric transformation, the same authors [12] further divide the scheme into isodesmic and allodesmic subgroups.

The proposed detailed systematization of the reaction types which may form the basis for tautomeric transformations allows one to derive a number of symbolic equations generalizing all known as well as all conceivable tautomeric reactions. There is certainly no guarantee that a reaction fitting some definite type will be tautomeric, i.e. fast and reversible. This additional condition has to be satisfied independently. Like any formalized scheme, the classification [12] is not free from certain shortcomings such as the necessity to consider electron tautomerism (resonance) as being among the number of the main tautomeric processes. A formal logical approach fixes only the character of redistribution of ordinary and multiple bonds in a system having a migrant group attached to it, without taking notice of the main characteristic of a tautomeric reaction, i.e. the rupture and formation of bonds with the migrating group. Therefore, a simple classification is needed which should clearly show the most important peculiarity of tautomeric (that is, rapid and reversible) reactions. This peculiarity undoubtedly lies in the mechanism of the tautomeric transformation and, as was noted above, consists in the fact that only those rapid and reversible reactions may be classified as tautomeric which are accomplished as a result of the rupture and formation of chemical bonds.

It seems therefore expedient to classify tautomeric transformations in accordance by following the types of the rupturing and forming chemical bonds [72, 74]. Therefore a classification not of the phenomenon but rather of the terminology is achieved; however, apart from it being necessary in itself, this approach is also justified because of the impossibility of separating the tautomeric reactions from the related class of transformations.

In the majority of the tautomeric systems, the processes of bond rupture—formation can be described as the migration of a given atomic group between two or more atomic centers in the molecule. Certainly, the most important migratory group is the proton, and the main type of tautomerism is prototropic tautomerism, also called prototropism (from the Greek $\tau\rho\varepsilon\pi\varepsilon\iota\nu$ — to change).

6.1. *Prototropy*

The proton, being the particle with no electron surrounding it, possesses the minimal atomic mass and makes only limited demands on the nature of the alternative center of localization and on the trajectory of the reaction path. In case steric hindrances inhibit intramolecular transition, the proton may prefer the intermolecular migration mechanism.

In either case, the proton transfer from one center to another is ensured by the unique mechanism of hydrogen bonding:

$$A-H\cdots B \underset{k_2}{\overset{k_1}{\rightleftharpoons}} A\cdots H-B \qquad (1.14)$$

Passing over the detailed illustrations of extremely varied prototropic systems (see the Reviews [4, 10, 81—83]), let us direct our attention to the most important feature of prototropic reactions, *viz.*, their extremely high rates.

The proton transfer reactions belong to the so-called superfast reactions. The proton transfer rate constants k_1, k_2 in Eq. (1.14) for the oriented system of reagents attain rates of 10^{13}—10^{14} s^{-1} [84, 85] which is substantially higher than the diffusion rates. They are comparable to the vibration frequency of the bonds A—H (A = C, O, N . . .). The rate constants for the superfast proton transfer were most accurately determined for intermolecular reactions [86], but a number of observations (see, for instance, Chapter 5 on the rate of tautomerization **XIII**—**XVI**) suggest that intramolecular proton transfers also occur at frequencies which are almost as high.

Prototropic equilibrium between the benzenoid and the quinonoid forms of methyl salicylate in the electron excited state is established even faster. Analysis of data on the fluorescence of frozen solutions yields the value of $K_1 \simeq 10^{14}$ s^{-1} at room temperature [85, 87].

The extremely high mobility of the proton moving over the bridge of hydrogen bonds could not be disregarded by Nature, which applied this unique mechanism to the implementation of such an extraordinary achievement of evolution as the genetic code which, in essence, is a proton code. Prototropic tautomeric transformations in the DNA base pairs are, most probably, responsible for spontaneous and induced mutations [9].

With extremely low activation barriers corresponding to very high proton migration frequencies, the tunnel effects of under barrier transfer of the light proton begin to play an important role [9, 84, 85, 88]. At the same time, the frequency of transformations in the (1.44)-type systems depends a great deal on the acidity of the A—H bonds and on the basicity of the electron-donating center B [84, 86]. At low acidity, e.g. with the CH acids, the proton transfer reaction proceeds at a relatively low rate. Thus, the lifetimes of the enol and ketone forms of acetylacetone, as evaluated by the method of dual nuclear resonance, amount, in the prototropic transformation to 14.2 s and 3.8 s, respectively [83] (*cf.* reaction **VIIa** → **VIIb**). Interconversion between some prototropic ketone and enol tautomers of 1,3-dicarbonyl compounds proceed at such low rates that they can be separated by preparative methods at ambient temperature [4, 5].

In view of the limits described in (1.5), such transformations can formally be regarded as nontautomeric. Thus, the range of frequencies of prototropic migrations in prototropic systems amounts to 17—18 orders of magnitude (in s^{-1} units).

6.2. *Metallotropy*

Until recently, the concept of tautomerism was, on the whole, identified with the notion of prototropy. Development of the chemistry of organometallic compounds and, particularly, an understanding of the specific problems of their structure arising from lability of bonds formed by metals have resulted in the discovery of a number of fast and reversible (on the scale of the tautomeric processes) rearrangements of organometallic and, generally, organoelement compounds. Probably the first example of metallotropic tautomerism is the multicenter reaction of circumambulation of organometallic groups in the cyclopentadiene ring, discovered by Piper and Wilkinson [89]:

$$MR_n = Fe(\eta^5 - C_5H_5)(CO)_2; Cr(\eta_5 - C_5H_5)(NO)_2; Hg(\eta^1 - C_5H_5); Cu(PEt_3).$$

The authors of [89] did not at the time have the capability of measuring the NMR spectra of other than the hydrogen nuclei, nor of performing their measurements at other than room temperature. Neither did they have available the data of the X-ray diffraction study of the structure, but all available spectral and chemical evidence argued for the classical monohapto (η^1) structure of compounds **XXII**. Intuition suggested the correct solution to Wilkinson and Piper: There occurred fast, reversible shifts of the organometallic group along the perimeter of the cyclopentadiene ring at a rate exceeding the expected chemical shift (Δv), i.e. 200—300 s^{-1}.

Straightforward proof of fast and reversible 1,2-migrations of organometallic groups in **XXII**-type compounds was supplied ten years later, thanks to the analysis of the low-temperature ^{1}H-NMR spectrum of compound **XXII** (MR$_n$ =

$Fe(\eta^5 - C_5H_5)(CO)_2)$ and the X-ray diffraction study of its molecular structure in the crystalline state [90]. Soon after this, degenerate tautomerizations of the (1.15) type of compounds **XXII** with many diverse migrants MR_n (M = Si, Ge, Sn, B, P, Mo, Ru . . .) were studied, see the Reviews [91, 92]. The term 'fluxional molecules' was suggested [93] for compounds in which, as is the case with **XXII**, fast and reversible intramolecular rearrangements occur.

Nesmeyanov and Kravtsov [94] were the first to find examples of tautomeric systems in which fast and reversible migrations of organometallic groups occur between two heteroatoms. They introduced the term metallotropy to denote the dynamic equilibria arising from the processes of rupture—formation of bonds with the metal center of the organometallic group. They found the first instances of nondegenerate metallotropy and pointed out that, similarly to prototropy, the metallotropic transformations can have an intermolecular character. Thus, studies of electronic absorption spectra of solutions of arylmercury derivatives of nitro-sophenoles revealed an equilibrium between the nitrosophenol and quinoneoxime forms:

$$\text{(1.16)}$$

Metallotropic rearrangements of the (1.15) and (1.16) type are designated [95, 96] as σ, σ- (or sigmatropic) transformations, this term pointing to the type of the rupturing and forming bond. Also possible are the σ, π and π, π-transformations.

6.2.1. σ, σ-TRANSFERS

σ, σ-Metallotropy closely resembles prototropy, and the general approach to obtaining the expected σ, σ-metallotropic systems consists, in essence, in the replacement of hydrogen in prototropic compounds by organometallic moieties [7, 97]. Of special interest is the mechanism of metallotropic migrations which, at least in the case of intramolecular migrations, possesses features relating it to proto-tropy.

This mechanism has been examined in great detail in (1.15) type reactions using an analysis of the full line form in the dynamic NMR spectra. The following mechanistic possibilities have been reliably ruled out: intermolecular exchange, 1,3-shift of migrating group, rearrangement through intermediates or transition states of the η^5-structural type. The reaction proceeds according to the scheme of the 1,2-shift (1,5-sigmatropic displacement thermally allowed by the orbital symmetry rules) [91, 92]. The free activation energy of the metalloorganic group displacement is 10—15 kcal/mol lower than that for the (also possible) prototropic isomerization.

The preferential occurence of the 1,2-shift is explained by the ease with which the additional coordination of metal to the adjacent center takes place. The role of this coordination, which is an analog of the hydrogen bond bridge, becomes particularly evident in the case of migration between the heteroatomic centers which provide, as the donor orbitals, not the π-orbitals, as is the case with the cyclopentadiene system (1.15), but the localized orbitals of the electron lone-pairs.

Phenylmercury derivatives of *o*-hydroxyaldimines may serve as an example [98]. In solutions of these compounds, an equilibrium of two forms is observed

(1.17)

XXIIIa   **XXIIIb**

Convincing proof of equilibrium (1.17) is the negative thermochromism of compounds **XXIII** found in frozen solutions, which is caused by the shift of the equilibrium towards the more deeply colored quinonoid tautomer **XXIIIb**.

The reason for the rather low energy barrier for the **XXIIIa** ↔ **XXIIIb** inter-conversion is the intramolecular coordination of mercury to the center of the forming σ-bond, which is confirmed by X-ray study of the molecular structure of the phenylmercury derivative of methyl salicylaldimine [99] which exists in the crystal as the **XXIIIb** form (Fig. 1.6).

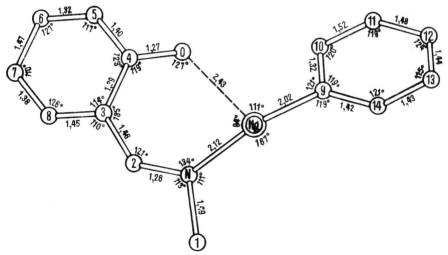

Fig. 1.6. Molecular structure of the phenylmercury derivative of *N*-methyl salicylamine **XXIIIb** (R = CH₃) according to an X-ray study [99].

Tautomeric system (1.17) is an example of unsymmetrical double-well potential in metallotropic migration. There are instances, particularly typical of organo silicon compounds, to which there corresponds a symmetrical double-well potential of the type shown in Fig. 1.5b. Degenerate silylotropy of acetylacetone derivatives belongs to this type [100]

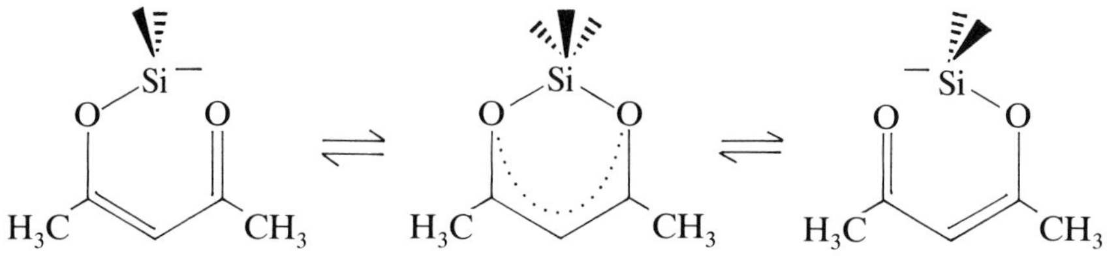

$$\tag{1.18}$$

XXIVa   **XXIVc**   **XXIVb**

The barrier separating the degenerate unsymmetrical forms **XXIVa** and **XXIVb** amounts, according to the NMR data, to 13.8 kcal/mol. This is due to the fairly moderate propensity of silicon to pentacoordination and it, consequently, a result of the transition structure **XXIVc**. But in other organometallic derivatives of acetylacetone such as antimony [101, 102], boron [103], lead and tin [104] derivatives as well as in transition metal-chelate acetylacetonates [105, 106], where strong intramolecular coordination exists, the bonds are totally averaged, and symmetrical structures **XXV** and **XXVI** are stable.

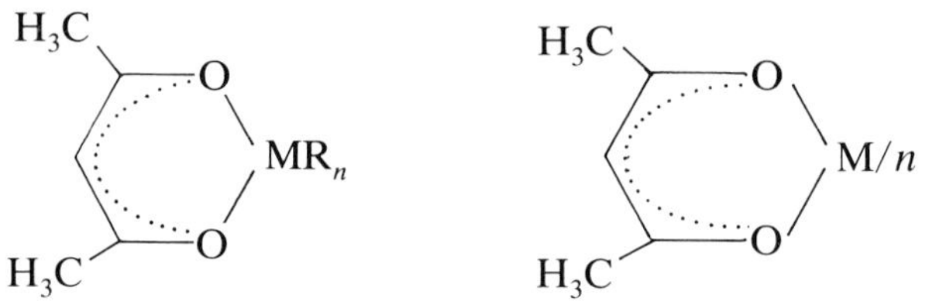

XXV   **XXVI**

$MR_n = SbPh_2Cl_2; BR_2; SnPh_3; \frac{1}{2} PbR_2. M = Zn, Co, Ni, Cu, Be, Cr, Al, Rh, \ldots$

Fig. 1.5 a shows the form of the adiabatic potential corresponding to this case.

Thus, the character of the σ, σ-metallotropic and prototropic systems is fairly similar.

6.2.2. σ, π-TRANSFERS

One distinctive feature of metals and many heavy elements is their ability to form π-bonds as well as the ordinary σ-bonds. Therefore, the possibility arises of the existence of a type of reversible transformation in which the migration of the organometallic group proceeds in one direction with the rupture of the σ- and formation of the π-bond, whereas for the migration in the opposite direction this process is reversed.

In the Reviews [95, 96, 107] various theoretically possible topological types of these transformations have been examined. The metallotropic σ, π-transformations do actually occur; for instance, in the case of the degenerate tautomerization of the dicobalt cluster **XXVII** [108]

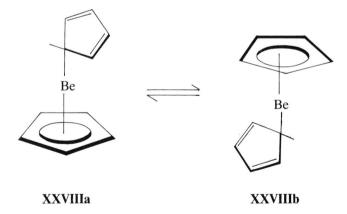

XXVIIa   **XXVIIb**

η^5-Cp—π-cyclopentadienyl

An interesting example of exceptionally fast σ, π-metallotropy is the fluxional molecule of beryllocene. The structure of this molecule was the subject of numerous theoretical and experimental studies (using gas electron diffraction and X-ray diffraction, see the Reviews [27, 109]). According to these studies, the molecule possesses η^1, η^5-structure **XXVIII** in which exchange of cyclopentadiene rings proceeds, judging from the results of studies of dielectric losses in the microwave region [110], at the rate of approximately 10^9 s^{-1}.

XXVIIIa   **XXVIIIb**

Another example of σ, π-metallotropy is given by the dynamic processes related to the migration of molybdenum and tungsten η^5-cyclopentadiencarbonyl moieties within the azaallyl system [96]

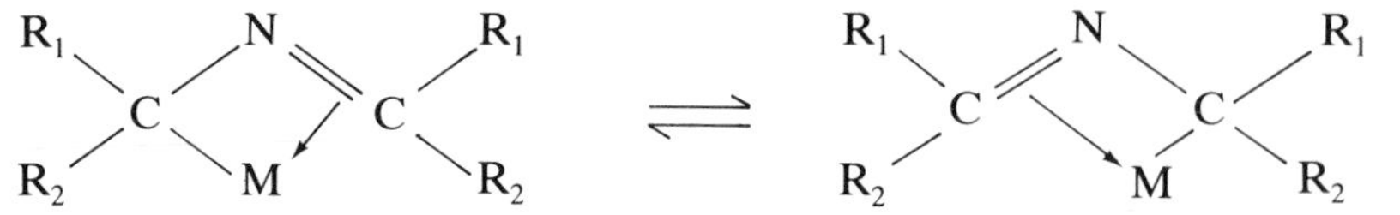

XXIXa   **XXIXb**

$$M = (\eta^5\text{-}C_5H_5Mo(CO)_2, \quad (\eta^5\text{-}C_5H_5W(CO)_2, \quad R_1R_2 = —C_6H_4CX_3(X = H, F)$$

It has been claimed [95] that σ, π-metallotropy has no prototropic analogs. This view is scarcely correct, since a proton as well as other migrants, the derivatives of main group elements, are also able, albeit to a much lesser degree than the transition metals, to form π-complexes. Computational and experimental data indicate that the complexes **XXXc** not only are necessary intermediates in Wagner—Meerwein type rearrangements (1.19), but, in some cases, are more stable than the corresponding classical cations **XXXa, b** (*sec.*-butyl cation, protonated cyclopropane and others) [111—114].

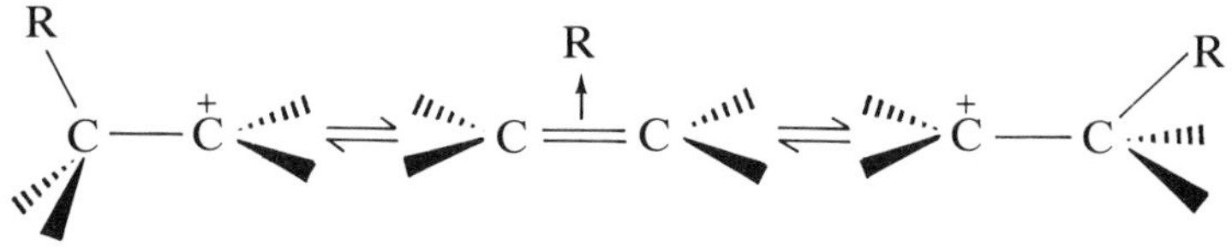

(1.19)

XXXa   **XXXc**   **XXXb**

$$R = H, SiH_3, CH_3, Cl, F$$

6.2.3. π, π-Transfers

Apparently, these transformations are indeed specific for metallotropic systems since they require migration of a metal-containing group between two different π-complexes.

Probably the best studied is the π, π-metallotropy of π-complexes of cyclooctatetraene [95, 115, 116]. A simpler instance of metallotropic tautomerism of this type is represented by the degenerate tautomerism of allene iron tetracarbonyl complex [117].

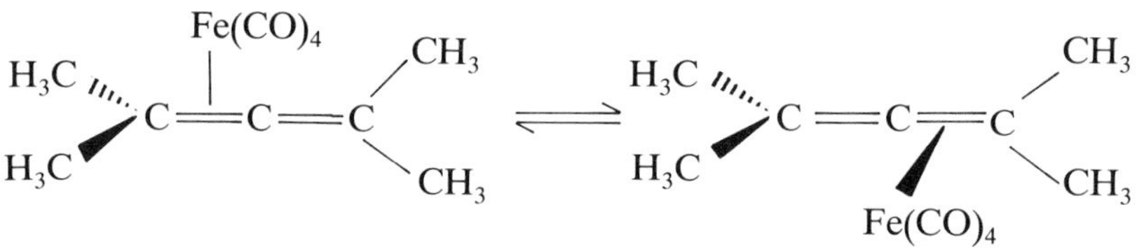

XXXIa   **XXXIb**

Allene derivatives have two orthogonal π-systems, only one of which serves as a donor in the formation of a π-complex with the iron tetracarbonyl moiety. This is clearly evidenced by the ^{1}H-NMR spectrum of complexes **XXXI**: at lower temperatures the methyl groups are anisochronic (Three signals are observed with 2:1:1 intensities) Upon raising the temperature, the NMR spectra show fast intramolecular exchange **XXXIa** ⇌ **XXXIb**, which is revealed in the averaged common signal of the protons of all methyl groups. The most important mechanism of the π, π-metallotropy (similarly to the σ, π-case) is the mechanism of the 1,2-shift of the organometallic group. If, however, its coordination is due to multicenter π-orbitals, as is the case with the π-cyclooctatetraene complexes, then the mechanisms of 1,3- and 1,4-shifts are possible. A more complicated mechanism, not fitting the scheme of the *i, j*-shifts, governs the interesting tautomeric transformation of the bis-cyclooctatetraene titanium complex.

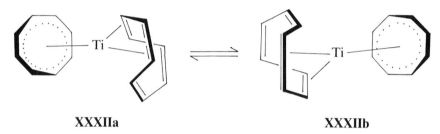

| **XXXIIa** | **XXXIIb** |

X-ray structural data point to a nonequivalence of the cyclooctatetraene rings, only one of which is planar [115]. The dynamic NMR spectra [118] reveal fast degenerate interconversions **XXXIIa** ⇌ **XXXIIb**. Similarly to the σ, π-metallotropy, the π, π-metallotropy belongs to the type of pericyclic reactions in which there occurs a change in the character of the migrating center coordination with respect to the polyene system. R. Hoffmann suggested [119, 120] the term haptotropic rearrangements for these reactions. Rearrangements of the π-complexes in which the organometallic group moves over the plane of the conjugated system belong to the reactions of this type. Computational [120] and experimental (see the Review [121]) data indicate presence of intermediate forms, as in the case of tautomeric rearrangements of fluorenyl π-complexes.

Cr(CO)₃   ⁻Cr(CO)₃   Cr(CO)₃

XXXIIIa   **XXXIIIc**   **XXXIIIb**

The activation energies of the metallotropic rearrangement of the K-derivative of **XXXIII** in tetrahydrofuran amount to 16 kcal/mol at the ($\eta^6-\eta^5$) stage and 13.8 kcal/mol at the ($\eta^5-\eta^6$) stage, where the η^5-structure **XXXIIIc** is an observable intermediate [122].

6.3. *Anionotropy*

The singling out of this type of tautomerism contradicts the original logic of classifying the tautomeric reactions in accordance with the type of rupturing and forming bonds, as well as with the type of the migrating group. Indeed, from this point of view, tautomeric transformations associated with the migrations of anion and anionoid groups fall quite clearly into the group of elementotropic reactions. Separate examination of this type is, however, warranted by the specificity of its mechanism.

The possibility of anionotropy follows from the differentiation of tautomeric systems into the cationotropic and anionotropic ones [4, 72]; however, numerous examples of anionotropy cited in a number of papers [4, 72, 74] belong, in fact, to irreversible (in the sense required by the concept of tautomerism, see Section 1.3) rearrangements. Examples of intermolecular triad anionotropic tautomerism are also known, such as following equilibria:

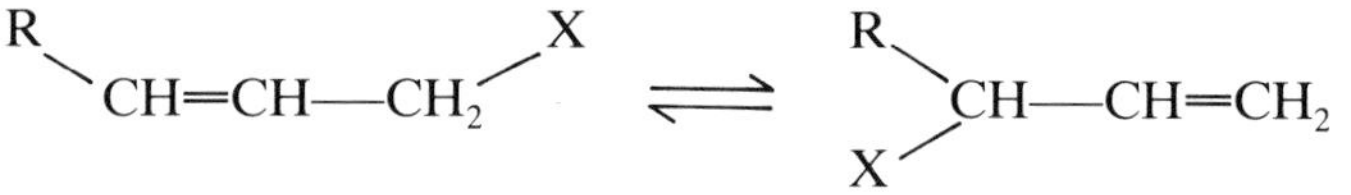

$$\underset{R}{\overset{}{\diagdown}}CH{=}CH{-}CH_2\overset{X}{\diagup}\quad\rightleftharpoons\quad\underset{X}{\overset{R}{\diagdown}}CH{-}CH{=}CH_2 \qquad (1.20)$$

X = Hal [72], SPh [123]

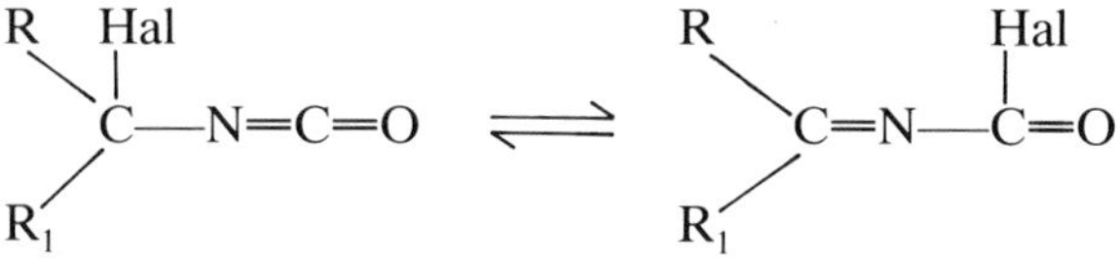

$$\underset{R_1}{\overset{R\quad Hal}{\diagdown}}C{-}N{=}C{=}O\quad\rightleftharpoons\quad\underset{R_1}{\overset{R}{\diagdown}}C{=}N{-}\overset{Hal}{C}{=}O \qquad (1.21)$$

Hal = Cl, Br [124, 125]

Examples of intramolecular anionotropy with an anion released through dissociation which, apparently, migrates inside the tight ion pair from one center of the initial compound to the other have been detected only recently. The overall scheme of the reaction is represented by the following equations:

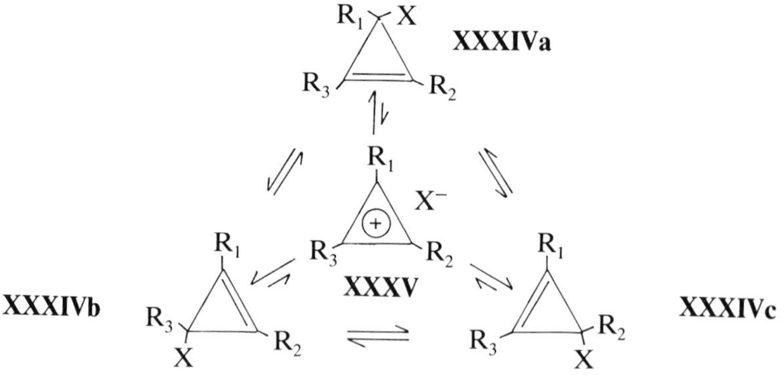

$$(1.22)$$

$R_1 = Cl, R_2, R_3 = H, X = Cl$ [126]
$R_1 = R_2 = R_3 = Me, X = N_3$ [127]
$R_1 = R = R_3 = t\text{-Bu}, X = N_3$ [128].

Both the ionization equilibrium with the formation of the intermediate ion pair of the cyclopropenium cation **XXXV** and the processes of the nondegenerate [126] as well as degenerate [126—128] exchange are evidenced by the NMR spectra.

For 7-substituted cycloheptatrienes **XXXVI**, detailed studies of the dynamic NMR spectra using the saturation transfer experiments permitted a very distinct separation of the anionotropic rearrangements proceeding through the sigmatropic shifts from those proceeding through the ionization—recombination reaction [129]

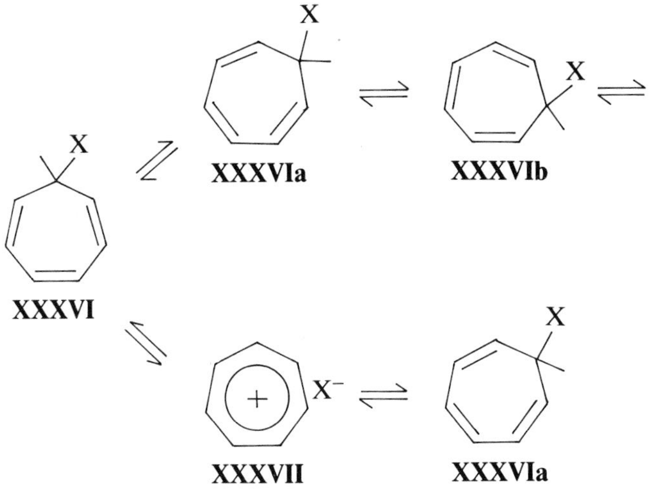

(1.23)

The mechanism of the reaction is very sensitive to the polarity of the solvent and to the type of the anion. The NCS shift in tropyl isothiocyanate (X = NCS) is a purely ionic random process [130] but a preferred 1,3-shift mechanism was found in tropyl azide (X = N_3) [131].

In the case of intramolecular prototropic and metallotropic transformations, the direction of a tautomeric reaction is controlled by the character of coordination (hydrogen or donor—acceptor bonds) with the center accepting the migrant in motion. In ion pairs **XXXV** and **XXXVII**, the choice of an atomic center to which the migration is directed is, in the first place, determined by the character of the electrostatic potential of the cation which in anionotropic systems must have several minima divided by low energy barriers. Thus, in (1.22) and (1.23) type systems the reaction path of the anion group shift is determined by electrostatics.

6.4. *Valence Tautomerism*

In the types of tautomerism cosidered above, the processes of the reversible rupture and formation of bonds could roughly be reduced to the process of some atomic group migrating between two or more centers. However, there exists a considerable number of various isomerization reactions (both degenerate and nondegenerate) which are characterized by the reversible rupture of some and formation of other bonds in the molecule occuring without the transfer of any group or changes in the connectivity of the atomic skeleton. Many of these reactions fit the thermodynamic and activation energy scale of tautomeric processes and should be regarded as belonging to these. These reactions constitute the range of valence tautomeric systems.

Notions of valence tautomerism have been rather confused for some time. The term itself was introduced rather a long time ago (see [4]) to denote those tautomeric transformations which are characterized only by an interchange of bonds in the molecule. From this formal structural point of view, both the reasonance of the Kekulé structures of benzene and the dynamic process of degenerate isomerization of cyclobutadiene **XIa** ⇌ **XIb** belong to the class of valence tautomerism.

The essential difference between these processes consists in the form of the adiabatic potential of the system (Fig. 1.5). The necessary condition for a system to be assigned to the valence tautomeric category is the presence of two or more minima on the potential energy surface of the system as a function of all its nuclear coordinates and, hence, the presence of an energy barrier between the structures which are in the minima.

The nature of valence tautomeric transformation is determined by a molecular vibration accompanied by synchronous variations in internuclear distances and bond angles and, as a corollary thereof, by instant (on the time scale of nuclear motions) displacement of electrons.

In the case of the valence tautomerism, it is particularly clear that the concept of tautomerism needs elucidating from two independent positions of classical structural theory. Such elucidation requires the selection of a specific structural characteristic (rupture formation of bonds without transfer of groups) and then the specification of the adiabatic potential form necessary for singling the resonance (Laar's tautomerism) out of the phenomenon under consideration.

Some examples of valence tautomerism classified by the types of bond-making and bond-breaking are given below.

6.4.1. π-VALENCE TAUTOMERISM

In tautomeric rearrangements of this type, the σ-bond framework is fully retained, but π-bond shifts occur within it. In the terms of nuclear motions, this can be envisaged as a synchronous pulsation of the atomic core of a molecule resulting, upon transition from one minimum of adiabatic potential to another, in the shortening of the long bonds (σ-bond → σ, π-bond) and lengthening of the short bonds (σ, π-bond → σ-bond).

Bond-shift isomerisations of cyclobutadiene and cyclooctatetraene may serve as examples of the π-valence tautomerism. In the case of the latter compound, the

NMR spectra reveal two dynamic processes, i.e. ring inversion (RI) and shift of π-bonds (BS) [115, 132, 133]. The former process is conformational isomerization, whereas the latter, being even faster, is an example of π-valence tautomerism. The free activation energy of the bond-shift reaction **XXXVIIIa** $\rightleftharpoons$ **XXXVIIIc** $G_{25}^{\neq} = 13.9\,\text{kcal/mol}$

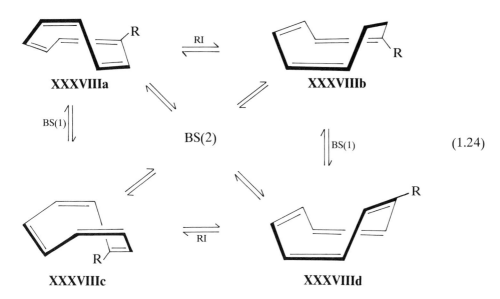

$$\text{(1.24)}$$

The shape of the potential energy surface of reaction (1.24) is fairly complex. It has been a commonly accepted view that the D_{8h} structure with average C—C bonds is the transition state for the bond-shift reaction whereas the flat D_{4h} structure with alternating C—C bonds serves as the transition state for ring inversion transformations [69, 133]. Recent calculations [134], however, have proved that the D_{4h} structure does not possess the properties of the transition state of a reaction, see [18].

6.4.2. σ, π-VALENCE TAUTOMERISM

The most frequently occuring type of the valence tautomerism is characterized not only by redistribution of the π-bonds but also by changes in the framework of σ-bonds during a reaction.

The σ, π-valence tautomeric interconversions follow a reaction scheme inherent in intramolecular pericyclic reactions, chiefly the sigmatropic and electrocyclic ones. A striking example of the σ, π-valence tautomerism proceeding, similarly to a Cope's rearrangement, as a series of iterated 3,3-sigmatropic shifts, is the degenerate isomerization of bullvalene.

Valence tautomerizations of barbaralenes [135] and semibulvalenes **XXXIX**, similar in their mechanisms, are characterized by particularly high rates.

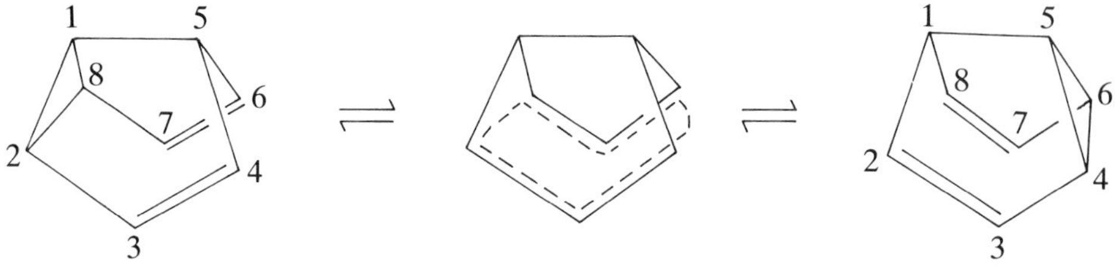

XXXIXa   **XXXIXc**   **XXXIXb**

The free energy of the interconversion **XXXIXa** ⇌ **XXXIXb** at −150° is a mere 5.5 kcal/mol [136]. Theoretical calculations [137—139] predict symmetrical homo-aromatic or biradical structure **XXXIXc** for the transition state of the rearrange-ment. A substitution of basal hydrogens by electron-withdrawing groups (CN and others) will lead to stabilization of the symmetrical structure **XXXIXc** with respect to **XXXIXa, b**, presumably to give a single minimum on the potential surface (cf. Section 4).

Among heterocyclic compounds, there are numerous cases of the σ, π-valence tautomerism, such, for instance, as 1,6,6a, S^{IV}-trithiapentalenes, the so-called no-bond resonance compounds:

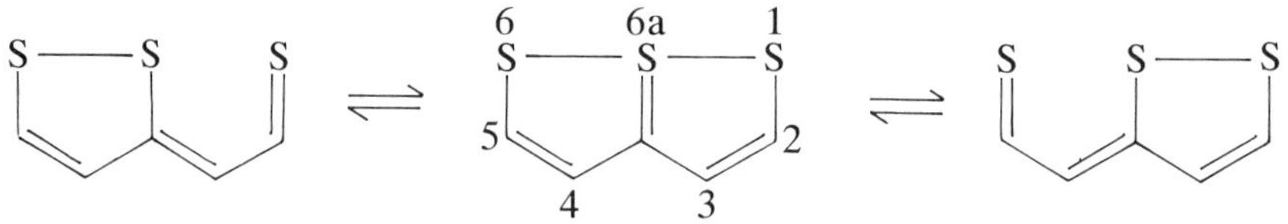

XLa   **XLc**   **XLb**

In most compounds of this class the lengths of the S—S bonds are not equal, differing at most by 0.3 Å in crystals [140], but neither any unsymmetrical nature nor any kinetic evidence of rapid **XLa** ⇌ **XLb** transformations in solution have been found in spite of careful NMR studies [141]. Only recently such transforma-tions were reported [142] to have been observed in the transition metal carbonyl complexes of compounds **XL**.

In some symmetrically substituted trithiapentalenes such as the 2,5-dimethyl derivative, the lengths of the S_1—S_{6a} and S_6—S_{6a} bonds are equal and the sym-metrical structure **XLc** is stable both in solution and in the crystalline state. One interesting example, the compound **XLI**, was reported [143] in which all neighbor-ing bonds are equal in length.

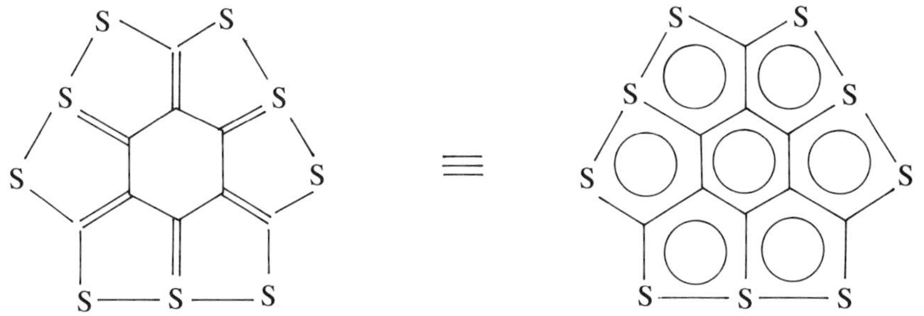

XLI

6.4.3. σ-VALENCE TAUTOMERISM

This type of tautomeric equilibria comprises transformations which do not affect the system of π-bonds. They are characterized by the fact that the rupture or formation of the σ-bond between two centers leads to transition from one stable nuclear configuration to the other which is separated from the former by an energy barrier whose value is within the scale of tautomeric processes.

One carefully studied example of σ-valence tautomerism is the equilibrium between phosphorane **XLIIa** and phosphonium **XLIIb** isomeric forms of the benzyl adduct [144, 145]

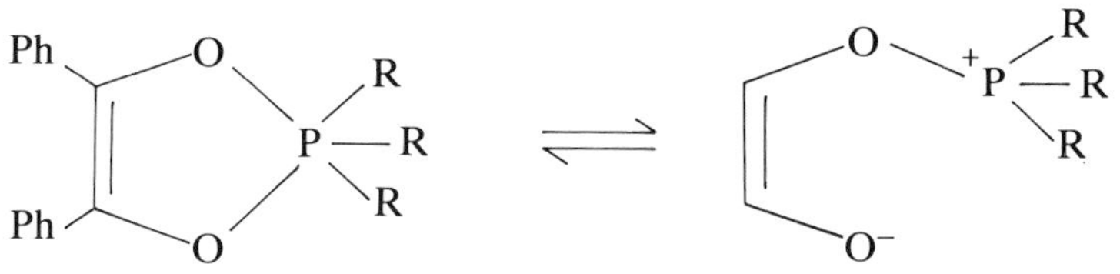

XLIIa   **XLIIb**

Compound **XLII** was isolated in two crystalline modifications corresponding to structures **XLIIa** and **XLIIb**. In solution, both forms come rapidly to equilibrium. The position of the equilibrium is readily identifiable by use of the ^{31}P-NMR spectra, owing to a considerable difference between the values of chemical shifts of the pentacoordinated and tetracoordinated charged phosphorus.

Stohrer and Hoffmann [36, 137] drew attention to an essential condition for the stretching of the σ-bond to result in the formation of the potential curve with the double minimum, i.e. of the bond-stretch isomerism: such stretching has to stabilize the antisymmetrical antibonding σ^*-orbital of the bond subjected to stretching and to destabilize the symmetrical bonding σ-orbital so that they intersect at a certain length of the stretched bond. Intersection of orbitals of different symmetries produces a barrier of bond-stretching with the maximum at the crossing point of the σ and σ^* orbitals leading thus to a symmetry-forbidden reaction, i.e. the reaction of interconversion of structures separated by this barrier.

In the reaction **XLIIa** $\rightleftharpoons$ **XLIIb**, stretching of the P—O bond in phosphorane results in its heterolytic rupture and in the stabilization of the bipolar form. If the bond is formed by identical atoms, its rupture occurs homolytically. For strained saturated compounds this may lead to stabilization of the valence isomer of biradical type. Calculations [36, 146, 147] have shown that the biradical **XLIIIb** is a stable isomer [2.2.2] of propellane **XLIIIa**.

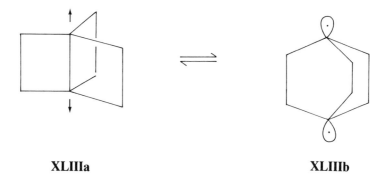

XLIIIa   **XLIIIb**

If the radical center in a bond-stretch isomer happens to be in the conjugated position, both σ-valence isomers may have approximately equal energies, and the energy barrier dividing them may be small (22—23 kcal/mol [148]) such as in anthracene derivatives.

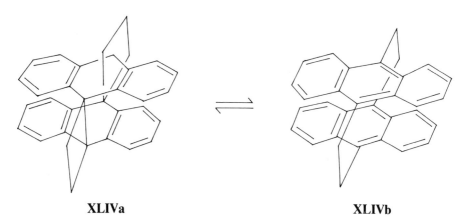

XLIVa   **XLIVb**

Interestingly, the idea to search for the equilibria including biradical (diyl) formation of the σ-bonds originated from a purely synthetic basis. It belongs to Wittig (see the Review [149]) who also gave practical examples of its implementation.

Finally, we may note as an essential feature of the valence tautomerism the fact that the reaction pathway of a tautomeric transformation is predetermined either by the direction of the single σ-bond (σ- and σ, π-valence tautomerism) or by the directions and the mutual orientation of the whole system of the σ-bonds (π-valence tautomerism).

7. Conclusion

The phenomenon of tautomerism, the dynamic isomerism of non-degenerate and (from the chemical point of view) degenerate types, is not isolated by any clear-cut physical boundaries from the broader domain of single-step chemical reactions. Tautomerism is a structural concept inseparably linked with the notion of chemical bonding in the frameworks of the classical theory of chemical structure and the theory of valency. Assignment of a chemical transformation to the tautomeric type requires that the reaction, regardless of its structural features, should be characterized by the thermodynamic and activation parameters corresponding to the conventional scale of tautomeric processes. This scale is not rigidly delimited, it rather depends on environmental conditions; primarily on temperature and solvent.

Intramolecular tautomeric reactions proceed as elementary stages and are, therefore, of particular interest for modelling and studying mechanisms of the reactions which form the basis of the tautomeric process. The relative ease (low activation barriers) with which tautomeric transformations proceed is accounted for by the fact that the molecular structure of a tautomer is already prepared for the transformation into another tautomeric form which is then attained through adjustment to the stereoelectronic demands of the reaction pathway. In prototropic tautomers, this pathway is defined by the formation of the hydrogen bond between the centers of the breaking and forming bonds. Metallotropic and, in general, elementotropic migrations of cationoid groups result from the ability of many elements to increase their coordination number owing to the donor-acceptor interactions with the η or π-donor groups of the distant molecular fragments. Orientation of the interacting centers predetermines the pathway for the elementoorganic group transfer. In the case of the anionotropic transformations, the direction of anion migration is primarily related to the form of the electrostatic potential of the cation and is dictated by the Coulomb interactions. Finally, the character and the direction of reversible intramolecular rearrangements assigned to the valence tautomerism are determined by the σ-bond frameworks of tautomers.

REFERENCES

1. Butlerov A., *Ann.*, **189**, 44 (1877).
2. Butlerov A. M., *Zh. Russ. Phys-Khim. Obschz.* **9**, 69 (1877).
3. Laar K., *Ber.* **18**, 648 (1885).

4. Baker J. W., *Tautomerism*. G. Routledge. London. 1934.
5. Kabachnik M. I., *Zh. Vsesoyuzn. Khim. D. I. Mendeleev Obschz.*, **7**, 263 (1962).
6. Leffler J. E., Grunwald E. *Rates and Equilibria of Organic Reactions*. J. Wiley. New York. 1963.
7. Nesmeyanov A. N., *J. Organometal. Chem.*, **100**, 161 (1975).
8. Huggins M. L., *Angew. Chem.*, **83**, 155 (1971).
9. Löwdin P.-O., *Adv. Quant. Chem.*, **2**, 213 (1965); Pullman B., Pullman A., *Adv. Heterocycl. Chem.*, **13**, 77 (1971); Cooper W. G., *Intern. J. Quant. Chem.*, **14**, 71 (1978).
10. Elguero J., Marzin C., Katritzky A. R., Linda P., *The Tautomerism of Heterocycles*. Adv. Heterocycl. Chem. Supplement 1. Academic Press. New York. 1976.
11. Walter R. E., *Uspekhi Khimii*, **51**, 1374 (1982).
12. Zefirov N. S., Trach S. S., *Zh. Org. Khim.*, **12**, 697 (1976).
13. Minkin V. I., Olekhnovich L. P., Zhdanov Yu. A., *Acc. Chem. Res.*, **14**, 210 (1981).
14. Minkin V. I., *Soviet Sci. Rev., B. Chem.*, **7**, 51 (1985).
15. Mamaev V. P., Lapachev V. V., *Soviet Sci. Rev., B. Chem.*, **7**, 1 (1985).
16. Slanina Z., *Theoretical Aspects of the Phenomenon of Isomerism in Chemistry* (in Russian). Mir Publ. Moscow. 1984. *Adv. Quant. Chem.*, **13**, 89 (1981).
17. Wooley R. G., *J. Amer. Chem. Soc.*, **100**, 1073 (1978); *Israel J. Chem.*, **19**, 30 (1980).
18. Minkin V. I., Simkin B. Ya., Minyaev R. M. *Quantum Chemistry of Organic Compounds. Mechanisms of Reactions* (in Russian). Khimiya. Moscow. 1986.
19. Mezey P. G., *Theor. Chem. Acta*, **54**, 95 (1980); *J. Mol. Struct. THEOCHEM*, **103**, 81 (1983).
20. Bader R. F. W., Anderson S. G., Duke A. J., *J. Amer. Chem. Soc.*, **101**, 1389 (1979).
21. Bader R. F. W., Nguyen-Dang T. T., *Adv. Quant. Chem.*, **14**, 63 (1981).
22. Tal Y., Bader R. F. W., Nguyen-Dang T. T., Ojha M., Anderson S. G., *J. Chem. Phys.*, **74**, 5162 (1981).
23. Bader R. F. W., Tang T.-H., Tal Y., Biegler-König F. W., *J. Amer. Chem. Soc.*, **104**, 940, 946 (1982).
24. Gatti C., Barzaghi M., Simonetta M., *J. Amer. Chem. Soc.*, **107**, 878 (1985).
25. Bianchi R., Pilati T., Simonetta M., *Acta Cryst.* **B34**, 2157 (1978).
26. Hedberg L., Hedberg K., *J. Amer. Chem. Soc.*, **107**, 7257 (1985).
27. Minkin V. I., Minyaev R. M., *Nonclassical Structures of Organic Compounds*. Mir Publ. Moscow. 1987.
28. Jackson J. E., Allen L. C., *J. Amer. Chem. Soc.*, **106**, 591 (1984).
29. Würthwein E.-U., Chandrasekhar J., Jemmis E. D., Schleyer P.v.R., *Tetrahedron Lett.*, **22**, 843 (1981).
30. Belvile D. J., Pabon R. A., Bauld N. L., *J. Amer. Chem. Soc.*, **107**, 4978 (1985).
31. Barkhash V. A., *Nonclassical Carbocations* (in Russian). Nauka. Sibirian Div. Novosibirsk. 1984; *Topics Curr. Chem.*, **116/117** (1984).
32. Mislow K., *Bull. Soc. Chim. Belge*, **86**, 595 (1977).
33. Nourse J. G., *J. Amer. Chem. Soc.*, **102**, 4883 (1980).
34. Davidson R. A., *J. Amer. Chem. Soc.*, **103**, 312 (1981).
35. Forsen S., Nilsson M., In: *The Chemistry of Carbonyl Group* (Patai S. ed.), vol. 2. J. Wiley. New York. 1970.
36. Hoffmann R., Stohrer W. D., *J. Amer. Chem. Soc.*, **93**, 6941 (1971).
37. Vogel E., Boll E. A., Gunther H., *Tetrahedron Lett.*, 609 (1965).
38. Doering W., *Zh. Vsesoyuzn. D. I. Mendeleev Obschz.*, **7**, 308 (1962).
39. Schröder G., *Chem. Ber.*, **97**, 3140 (1964); Mereny R., Oth J. F. M., Schröder G., *Chem. Ber.*, **97**, 3150 (1964).
40. Saunders M., *Tetrahedron Lett.*, 1699 (1963).
41. Leone R. E. and Schleyer P.v.R., *Angew. Chem. Intern. Ed. Engl.*, **9**, 860 (1970).
42. Dugundji J., Gillespie P., Marquarding D., Ugi I., and Ramirez F., In: *Chemical Applications of Graph Theory* (Balaban A. T. Ed.). Academic Press. New York. 1976., p. 107.
43. Oth J. F. M., *Pure Appl. Chem.*, **25**, 573 (1971).
44. Balaban A. T., Farcasiu D., *J. Amer. Chem. Soc.*, **89**, 618 (1967).

45. Binsch G., Eliel E., and Kessler H., *Angew. Chem.*, **83**, 618 (1971).
46. Mislow K. and Raban M., In: *Selected Problems of Stereochemistry* (in Russian). Mir Publ. Moscow. 1970; *Topics in Streochemistry*, vol. 1. (1967).
47. Eliel E. L., *J. Chem. Educ.*, **48**, 163 (1971).
48. Bersuker I. B. *Electronic Structure and Properties of Coordination Compounds*. Khimiya. Leningrad. 1976., p. 286.
49. Cohen B. and Weiss S., *J. Phys. Chem.*, **87**, 3606 (1983).
50. Muetterties E. L., *Inorg. Chem.*, **4**, 769 (1965).
51. Drago R. S. *Physical Methods in Chemistry*. W. B. Saunders Co., Philadelphia. London. Toronto. 1977.
52. Reutov O. A. *Theoretical Foundations of Organic Chemistry* (in Russian) Moscow University Publ. House. Moscow. 1964. Ch. 10.
53. Minyaev R. M., *Zh. Struct. Khim.*, **27**, 140 (1985).
54. Cao H. Z., Allavena M., Tapia O., and Evleth E. M., *J. Phys. Chem.*, **89**, 1581 (1985).
55. Gunnarson G., Wennerstrom H., Egan W., and Forsen S., *Chem. Phys. Lett.*, **38**, 96 (1976).
56. Baughcum S. L., Duerst R. W., Rowe W. R., Smith Z., and Wilson E. K., *J. Amer. Chem. Soc.*, **103**, 6296 (1981); Baughcum S. L., Smith Z., Wilson E. B., and Duerst R. W., *J. Amer. Chem. Soc.*, **106**, 2260 (1984).
57. Bicerano J., Schaefer H. F., and Miller W. H., *J. Amer. Chem. Soc.*, **105**, 2550 (1983); Frish M. J., Scheiner A. C., Schaefer H. F., and Binkley J. S., *J. Chem. Phys.*, **82**, 4194 (1985).
58. Haddon R. C., *J. Amer. Chem. Soc.*, **102**, 1807 (1980).
59. Brown R. S., Tse A., Nakashima T., and Haddon R. C., *J. Amer. Chem. Soc.*, **101**, 3157 (1979).
60. Kunze K. L. and de la Vega J. R., *J. Amer. Chem. Soc.*, **106**, 6528 (1984).
61. Bondybey V. E., Haddon R. C., and Rentzepis P. M., *J. Amer. Chem. Soc.*, **106**, 5969 (1984).
62. Jackman L. M., Trewela J. C., and Haddon R. C., *J. Amer. Chem. Soc.*, **102**, 2519 (1980).
63. Bell R. P. *The Tunnel Effect in Chemistry*. Chapman and Hall. London. 1980.
64. Goldanskii V. I., Trachtenberg L. I., and Floerov V. I. *The Contemporary Views on Tunneling Heavy Particles in Chemical Transformations*. (in Russian). VINITI Publ. Moscow. 1985.
65. Carpenter B. K., *J. Amer. Chem. Soc.*, **105**, 1701 (1983).
66. Dewar M. J. S., Merz K. M., and Stewart J. J. P., *J. Amer. Chem. Soc.*, **106**, 4040 (1984).
67. Pearson R. G., *J. Mol. Struct. THEOCHEM*, **103**, 25 (1983).
68. Bersuker I. B. *The Jahn—Teller Effect and Vibronic Interactions in Modern Chemistry*. Plenum Press. New York. London. 1984.
69. Dewar M. J. S. and Merz K. M., *J. Phys. Chem.*, **89**, 4739 (1985).
70. Ammeter J. H., *Nouv. J. Chim.*, **4**, 631 (1980).
71. Taube H., *Science* **226**, 1028 (1984).
72. Ingold C. K. *Structure and Mechanism in Organic Chemistry*. Cornell University Press. Ithaca and London. 1969.
73. Olah H. A., *Uspekhi Khimii*, **44**, 79 (1975).
74. Hückel W. *Theoretische Grundlagen der Organischen Chemie*. Band 1. Leipzig. 1952.
75. Ingold C. K., Shoppee C. W., and Thorpe J. F., *J. Chem. Soc.*, 1477 (1926).
76. Balaban A. T., *Rev. Roumaine Chim.*, **12**, 875 (1967).
77. Hendrickson J. B., *Angew. Chem. Intern. Ed. Engl.*, **13**, 47 (1974).
78. Zefirov N. S. and Trach S. S., *Zh. Org. Khim.*, **11**, 225, 1785 (1975).
79. *Chemical Applications of Graph Theory*. (Balaban A. T. Ed.). Academ. Press. New York. London. 1976.
80. Bauer J., Herges R., Fontain E., and Ugi I., *Chimia*, **39**, 43 (1985).
81. Beak P., *Acc. Chem. Res.*, **10**, 186 (1977).
82. Katritzky A. R., *Uspekhi Khimii.*, **41**, 700 (1971).
83. Koltsov A. I. and Kheifetz G. M., *Uspekhi Khimii.*, **40**, 1646 (1971); **41**, 877 (1972).
84. Bell R. P. *The Proton in Chemistry*. Chapman and Hall. London. 1973.
85. Grunwald E. and Ralph E. K. In: *Dynamic Nuclear Magnetic Resonance Spectroscopy* (Jackman L. M., Cotton F. A. Eds.), p. 621. Academic Press. New York. 1975.
86. Hibbert F., *Acc. Chem. Res.*, **17**, 115 (1984).

87. Smith K. K. and Kaufmann K. J., *Phys. Chem.*, **85**, 2895 (1981).
88. Scheiner S., *Acc. Chem. Res.*, **18**, 174 (1985).
89. Wilkinson G. and Piper T. S., *J. Inorg. Nucl. Chem.*, **2**, 32 (1956).
90. Bennett M. A., Cotton F. A., Davison A., Faller J. W., Lippard S. Y., and Morehouse S. M., *J. Amer. Chem. Soc.*, **88**, 4371 (1966).
91. Sergeyev N. M., *Progress in NMR*, **9**, 71 (1973).
92. Childs R. F., *Tetrahedron*, **38**, 567 (1982).
93. Cotton F. A., *Acc. Chem. Res.*, **1**, 257 (1968); *J. Organometal. Chem.*, **100**, 29 (1975).
94. Nesmeyanov A. N. and Kravtsov D. N., *Doklady Akad. Nauk USSR*, **135**, 331 (1960), **137**, 614 (1961); **140**, 1334 (1961).
95. Fedorov L. A., *Uspekhi Khimii*, **42**, 1481 (1973); *NMR Spectroscopy of Organometallic Compounds* (in Russian). Nauka. Moscow, 1984.
96. Kemmitt R. D. W. and Smith M. A. R., *Inorganic Reaction Mechanisms*, **3**, 460 (1974).
97. Fedorov L. A., Kravtsov D. N., and Peregudov A. S., *Uspekhi Khimii*, **50**, 1304 (1981).
98. Minkin V. I., Olekhnovich L. P., Knyazhanskii M. I., Mikhailov I. E., and Lyubarskaya A. E., *Zh. Org. Khim.*, **10**, 817 (1974).
99. Kuzmina L. G., Bokii N. G., Struchkov Yu. T., Minkin V. I., Olekhnovich L. P., and Mikhailiv I. E., *Zh. Struct. Khim.*, **15**, 659 (1974).
100. Pinnavaia T. J., Collins W. T., and Howe J. J., *J. Amer. Chem. Soc.*, **92**, 4544 (1970).
101. Meinema H. A. and Noltes J. G., *J. Organometal. Chem.*, **16**, 257 (1969).
102. Jain V. K., Bohra R., and Mehrotra R. C., *Structure and Bonding*, **52**, 148 (1982).
103. Bally J., Ciornei E., and Balaban A. T., *Rev. Roumaine Chem.*, **13**, 1507 (1965).
104. Kravtsov D. N., *Zh. Vsesoyuzn. D. I. Mendeleev Obschz.*, **12**, 53 (1967).
105. Holm R. H. In: *Dynamic Nuclear Magnetic Resonance Spectroscopy*. (Jackman L. M., Cotton F. A. Eds.) Academic Press. New York. London., p. 317.
106. Schkolnikova L. M. and Poray-Koshitz M. M., *Kristallokhimiya*, **16**, 117 (1982). VINITI. Moscow.
107. Eisch J. J., *Inst. Eng. Chem. Prod. Res. Devel.*, **14**, 11 (1975).
108. Rosenblum M., *J. Amer. Chem. Soc.*, **94**, 1239 (1972).
109. Charkin O. P., *Stability and Structure of Inorganic Molecules, Radicals and Ions in the Gas Phase* (in Russian). Nauka. Moscow. 1980.
110. Pratter S. J., Cooper M. H., Aroney M. J., and Filiupczuk A., *J.C.S. Dalton Trans.*, 1761 (1985).
111. Dewar M. J. S. and Ford G. F., *J. Amer. Chem. Soc.*, **101**, 783 (1979).
112. Dewar M. J. S. and Reynolds C. H., *J. Amer. Chem. Soc.*, **106**, 1744 (1984).
113. Dannenberg J., Goldberg B. J., Barton J. K., Dill K., Weinwuzzei D. H., and Longas M. O., *J. Amer. Chem. Soc.*, **103**, 7764 (1981).
114. Yates K., *Progress Theor. Org. Chem.*, **2**, 269 (1976).
115. Paquette L. A., *Tetrahedron*, **31**, 2855 (1975).
116. Mann K. E., *Progress in NMR Spectroscopy*, **11**, 95 (1977).
117. Ben-Shoshan R. and Pettit R., *J. Amer. Chem. Soc.*, **89**, 2231 (1967).
118. Schwartz J. and Sadler J. E., *J. C. S. Chem. Commun.*, 172 (1973).
119. Anh N. T., Elian M., and Hoffmann R., *J. Amer. Chem. Soc.*, **100**, 110 (1978).
120. Albright T. A., Hofman P., Hoffmann R., Lilliya C. P., and Dobosh P. A., *J. Amer. Chem. Soc.*, **105**, 3396 (1983).
121. Ustyunyuk Yu. A., *Vestn. Moscow Univ., Ser. 2. Khimiya.* **23**, 605 (1982).
122. Nesmeyanov A. N., Ustyunyuk Yu. A., Makarova L. G., Andrae S., Ustyunyuk N. A., Novikova L. M., and Luzikov Yu. N., *J. Organometal. Chem.*, **154**, 451 (1978).
123. Kwart H. and Johnson N., *J. Amer. Chem. Soc.*, **92**, 6064 (1970).
124. Holdschmidt H., *Angew. Chem.*, **80**, 942 (1968).
125. Samarai L. I. and Drach B. S., *Zh. Org. Khim.*, **7**, 2229 (1971).
126. Breslow R., Ryan G., and Groves J. T., *J. Amer. Chem. Soc.*, **92**, 988 (1970).
127. Gloss G. L. and Harrison A. M., *J. Org. Chem.*, **37**, 1051 (1972).
128. Curci R., *J. Org. Chem.*, **38**, 3149 (1973).

129. Kessler H. and Feigel M., *Acc. Chem. Res.*, **15**, 2 (1982).
130. Feigel M., Kessler H., and Walter A., *Chem. Ber.*, **111**, 2947 (1978).
131. Feigel M., Kessler H., Leibfrietz D., and Walter A., *J. Amer. Chem. Soc.*, **101**, 1943 (1979).
132. Anet F. L. A., *J. Amer. Chem. Soc.*, **84**, 671 (1962).
133. Paquette L. A., *Pure Appl. Chem.*, **54**, 987 (1982).
134. Glukhovtsev M. N., Simkin B. Ya., and Minkin V. I., *Zh. Org. Khim.*, **22**, 212 (1986).
135. Günther H., Runsink J., Schmickler H., and Schmidt P., *J. Org. Chem.*, **50**, 289 (1985).
136. Cheng A. K., Anet F. A. L., Mioduski J., and Meinwald J., *J. Am. Chem. Soc.*, **96**, 2887 (1974).
137. Stohrer W. D. and Hoffmann R., *J. Amer. Chem. Soc.*, **94**, 779 (1972).
138. Dewar M. J. S. and Lo D. H., *J. Amer. Chem. Soc.*, **93**, 7201 (1971).
139. Miller L. S., Grohmann K., and Dannenberg J. J., *J. Amer. Chem. Soc.*, **105**, 6862 (1983).
140. Gleiter R. and Gygax R., *Topics Curr. Chem.*, **63**, 49 (1979).
141. Yamamoto Y. and Akiba K., *Heterocycles*, **13**, 297 (1979).
142. Pogortzelec P. J. and Reid D. H., *J. C. S. Chem. Commun.*, 289 (1983).
143. Hansen L. K. and Nordvik A., *J. C. S. Chem., Commun.*, 800 (1974).
144. Luckenbach R. *Dynamic Stereochemistry of Pentacoordinated Phosphorous and Related Elements*. G. Thieme. Stuttgart. 1973.
145. Arbuzov B. A. and Prilezhaeva N. A., *Uspekhi Khimii*, **43**, 933 (1974).
146. Dannenberg J. J., *Angew, Chem. Intern. Ed. Engl.*, **15**, 519 (1976).
147. Newton M. D. In: *Application in Electronic Structure Theory*, p. 223. Modern Theoretical Chemistry, vol. 4. Plenum Press. New York. 1977.
148. Jones G., Reinhardt T. E., and Bergmark W. R., *Solar Energy*, **20**, 241 (1978).
149. Wittig G., *Acc. Chem. Res.*, **7**, 6 (1974).

2. Carbonotropy

1. The Problem of Carbonotropy

1.1. *General Principles*

The section considers in some detail molecular design and research data of carbonotropy, i.e. tautomerism which is characterized by rapid and reversible migrations between individual atomic centers in molecules of groups linked with these centers via a carbon atom. Processes of breaking and making of the carbon—heteroatom bonds are of prime importance in the reactions studied by organic chemistry and biochemistry. The problem of the carbonotropic tautomerism as a permanent reaction modelling the elementary acts of such processes was formulated for the first time in the early seventies [1, 2]. With regard to their intrinsic mechanisms, carbonotropic tautomeric systems can be divided into two types. In systems of the first type, the processes of breaking and making of chemical bonds between the carbon atom of the migrating group and nucleophilic centers of the rest .of the molecule occur which model the typical reactions of bimolecular nucleophilic substitution at the carbon atom. Tautomeric compounds in which there occur intramolecular reactions of electrophilic substitution at the carbon atom of the migrating group belong to the second type of carbonotropic systems. Numerous examples have been found in the extensive and thoroughly-studied domain of carbonium ion rearrangements [3—11].

The main feature distinguishing the nucleophilic from the electrophilic type of carbonotropy is that the former can take place in electrically neutral and thermally stable tautomeric compounds whereas the latter is typical of ions which are usually stable in superacidic media. In carbonotropic tautomeric systems of the nucleophilic type, the migrant must have an electron-deficient center. Among the first obtained systems of this kind were the compounds with an acylic migrant, thus giving rise to the name of acylotropic tautomerism [1]. It was later shown [2, 13] that also the moieties of the CH acids as well as other carbon-containing groups can serve as migrants in the rapid, reversible rearrangement reactions between nucleophilic centers. This warrants employment of a more general term, i.e. carbonotropy, covering all cases considered in this chapter. Three procedures formed the basis for permitting a search for them, as follows:

(1) In each class of tautomeric systems (diad, triad, pentad etc.), acylotropy and carbonotropy were considered to proceed as a degenerate (2.1)—(2.3) type process, in which case thermodynamic nonequivalence of the tautomers **a** and **b** ($\Delta G^0 = 0$) is ruled out automatically.

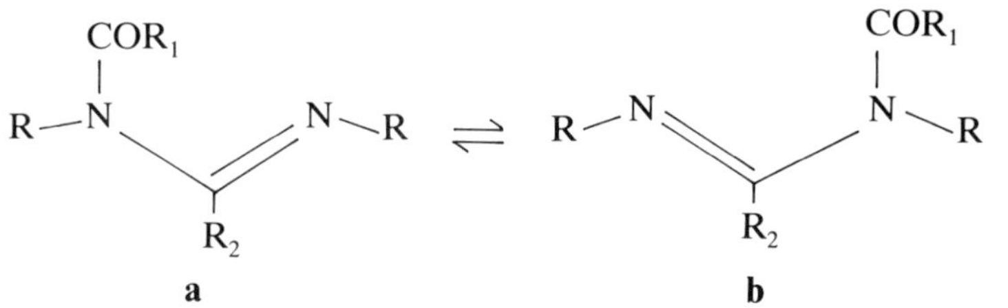

$$(2.1)$$

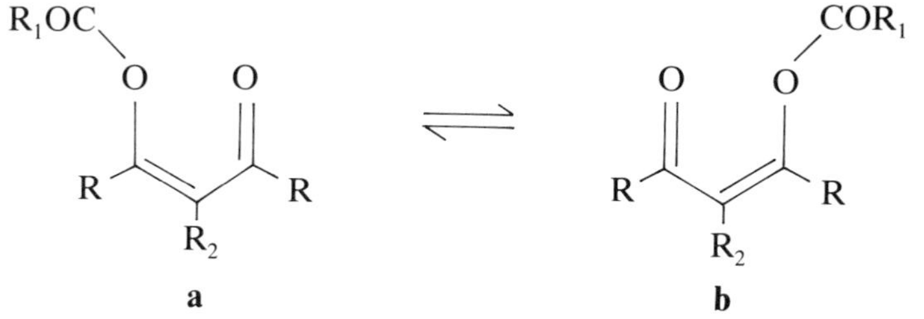

$$(2.2)$$

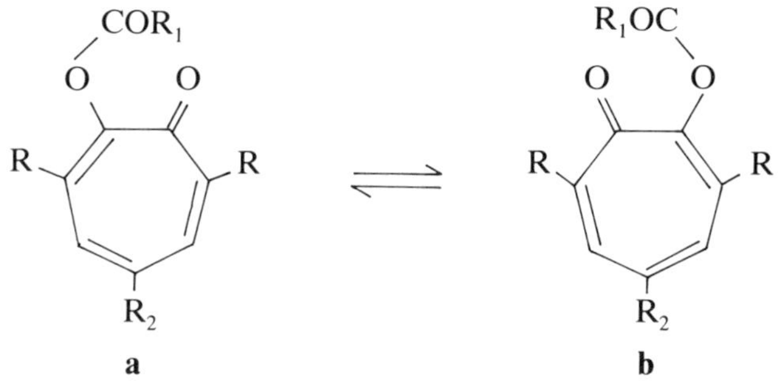

$$(2.3)$$

(2) Through deliberate variation of the reactivity of nucleophilic centers and the migrating group, the activation barrier of the reactions was lowered to the limits consistent with the tautomeric scale ($\Delta G^{\neq} < 25$ kcal/mol).

(3) A directed synthesis of acylotropic and carbonotropic compounds in which intramolecular group transfers are preferable permitted the concentration-dependence of the kinetics, and any appreciable influence of the polar characteristics of the solvent to be neglected.

This strategy enabled the researcher widely to employ the most powerful modern method for studying chemical kinetics, i.e. the method of dynamic nuclear magnetic resonance covering the optimum range of fast chemical transformations (Table 1.I).

The molecular design of carbonotropic reactions required a solution to a number of important theoretical problems. For example, regarding the mechanisms of acylotropic rearrangements, it was fundamentally important to find out whether the act of acyl transfer is a concerted (synchronous, single-step) or nonconcerted, two-step process. What are the roles of steric factors of a reaction site, of stereochemical rigidity and nonrigidity of tautomeric systems in the kinetics of their transformations? What possibilities are there to control the carbon group transfer rates by varying the degree of electron deficiency of a migrating center? Is analysis of mechanisms of carbonotropy possible when applying principles of orbital symmetry retention and stereoselection rules? What factors influence the thermodynamic parameters of the undegenerate tautomeric carbonotropic equilibria?

Variability of tautomeric systems — in particular, of the migrating carbon groups — opens up possibilities for a new approach to the modelling of the key stages of biochemical reactions in which transferase enzymes take part, and to the comparison of the conditions and kinetics of acylotropy with those of the biochemical transacylation.

Analysis of the data obtained in the studies of carbonotropic systems may serve as a paradigm for a directed approach to the development of the tautomeric systems in which the derivatives of the main group elements act as the migrants.

What are the chief distinctive features of the reactions proceeding with the breaking and making of the carbon bonds as compared to the reactions forming the basis of elementotropic transformations? May carbon be regarded as a special element possessing unusually strong chemical bonds or are the reactions directed at the carbon centers of a molecule characterized by especially rigid stereochemistry? The definite answer is that the main specificity of reactions at carbon centers lies, undoubtedly, in an extremely high steric requirement of proper transition state structures, which has to be taken into account when developing the carbonotropic tautomeric systems. This important point may be explained using the following comparison.

Silylotropy of acetylacetone derivatives [14] can be readily achieved requiring fairly low activation energy for the O, O'-transfers of the trialkylsilyl group. On the other hand, substitution of this group by an alkyl migrant in O-alkyl derivatives of analogous structure does not lead to (2.4) type tautomerism

$$\text{Ia} \quad\rightleftharpoons\quad \text{Ic} \quad\rightleftharpoons\quad \text{Ib} \tag{2.4}$$

Thus, the intramolecular process (2.4) is infeasible while the reaction of the isoelectronic silicon derivative proceeds without difficulty. This is explained by extremely stringent requirements of the sp^3-carbon center for the formation of the transition state structure in the nucleophilic substitution reaction possessing the trigonal bipyramidal configuration, with the leaving and the entering group necessarily occupying apical positions

II, D_{3h} **III**, C_s **IV**

Therefore, the transition state of the alkyl group tautomeric transfer reaction (2.4) must have, not the structure **Ic**, but **IV**. However, realization of this structure involves significant energy expenditure for producing deformations of valence angles as well as bond-stretching in the *cis*-enol skeleton of the dicarbonyl compound. Indeed, as is evident from the X-ray structural data [15], the distance $O_1—O_2$ in the acetylacetone *cis*-enol amounts approximately to 2.4 Å. The minimal distance $O_1—O_2$ in **IV** must be greater than 3.5 Å.

If relative position of interacting atomic centers in a molecule permits the trigonal bipyramidal transition state of a substitution reaction of the sp^3-carbon atom to be obtained, then such processes as the one reported by Martin and Basalay [16], i.e. the degenerate intramolecular rearrangement of 1,8-di-(thioaryl)antracene-9-carbinyl cations **V**, may be observed. This rearrangement models the S_N2-substitution reaction at the tetrahedral carbon atom

(2.5)

Va   **Vc**   **Vb**

In the cations **Va, b**, the pyramidality of the sulfonium center leads to diastereotopy of the methyl groups revealed in the ^{1}H NMR spectra at room temperature. A rise in temperature speeds up the process (2.5) passing through the transition state (or intermediate) **Vc** which results in the coalescence of the methyl signals ($\Delta G_{298}^{\neq} \simeq 14—16$ kcal/mol). Reaction (2.5) formally corresponds to the case of the alkylotropic tautomerism. Although the 'migrating' carbon group is rigidly fixed within the molecule, there takes place transfer of the σ-bond from one nucleophilic center to another.

At the same time, the bimolecular nucleophilic substitution reaction at the silicon sp^3 center is much less discriminating in the choice of the transition state structure. As is evidenced by numerous experimental [17—20] as well as sufficiently rigorous calculation [20, 21] data, the energies of the two types **II** and **III** (M = Si) of the transition states in nucleophilic substitution reactions at silicon are fairly close. Hence, depending on the structure of the system to which the silicon center is attached the reaction proceeds readily enough via type **II** or **III** transition state or intermediate structures.

In the degenerate silylotropic reaction (2.6), migration of the silyl group can proceed only through structure **III**. Unlike the potentially carbonotropic system (2.4), this route is not prohibited in reaction of substitution at the silicon center. It should be also noted that the two reaction paths under consideration produce different stereochemical results, *viz.*, inversion when passing through the transition state **II** (**IV**) and retention of configuration at the silicon when passing through **III**. That the silylotropic tautomerism in the processes (2.6) includes the transition state (or intermediate) of type **III** was proved experimentally by the dynamic NMR method with diastereotopically substituted silyl groups [22].

VIa   VIb   (2.6)

VIIa   VIIb

The example considered exposes the inadequacy of the conventional approach to the prediction of a tautomeric reaction using formal replacement of one migrant by another. Not only the structures of the initial and final products of the transformation should be taken into account, but also the mechanism of reaction (structure of the transition state and reaction trajectory) is to be specified. But we have seen that this mechanism may differ a great deal even for analogous migrants (CR_3, SiR_3) so that their inclusion in the same category because of the identical schemes of the process (sigmatropic shifts) becomes purely formal. Evidently, the mechanism of an expected tautomeric transformation has to be analyzed not simply as regards the topology of redistribution of single and multiple bonds in the system carrying a migrant, but in the light of the steric and electron requirements of the process occurring at the migrating center whose chief characteristics are trajectory of the reaction path and structure of the transition state.

For a search of tautomeric systems it is fruitful to address oneself to the wide and varied domain of molecular rearrangements. Many of them represent limiting elementary stages of important chemical reactions. As was noted in Section 1.3, Chapter 1, rearrangements differ from tautomerizations not in the mechanism of the reaction, but only in the energy difference between isomers (irreversible rearrangements) and in the higher energy barrier for their interconversion. Thus, the problem is reduced to the modification of the known rearrangement process whose thermodynamic and activation parameters have to be fitted in with the scale of tautomeric processes, Equation (1.5). The reversibility condition does not emerge for degenerate rearrangements in symmetrical compounds (virtual reaction) [23]. The second condition, that of the sufficiently high rate of interconversion of isomers, is ensured through structural variation of the steric and electronic factors of reactivity.

This approach, based on target-oriented selection, absolves the researcher of the necessity to empirically sort out all conceivable variants while still retaining sufficient challenge to chemical imagination.

1.2. *Choice of a Migrant*

Consideration of the possibility of tautomeric sigmatropic shifts of the silyl and alkyl groups is based on a comparison of the structures of transition states of the appropriate nucleophilic substitution reactions. This geometrical approach should be supplemented by an analysis of electron factors influencing the reactivity of atomic centers taking part in the key stage of the tautomeric transformation.

Thus, among various carbon compounds the alkyl derivatives, as compared to the carbonyl derivatives, are characterized by low reactivity with respect to most nucleophilic reagents. To achieve reasonable rates of substitution of the group X at the sp^3-carbon atom in **VIII**, a considerable activation of **VIII** is required.

VIII   **IX**

The alkyl groups which undergo transfer in biological systems are usually linked with the atom carrying a considerable positive charge as in the case of the sulfonium ion of *S*-adenosyl methionine **X**, thiazolium ion of thiamine **XI** and protonated 5-methyl tetrahydrofoleic acid **XII** [24].

X

XI

XII

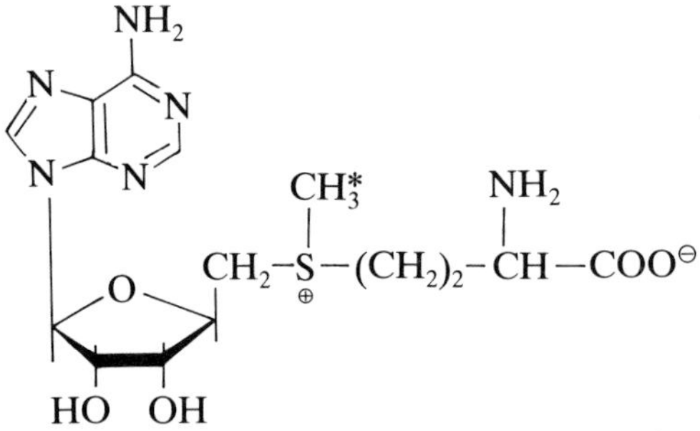

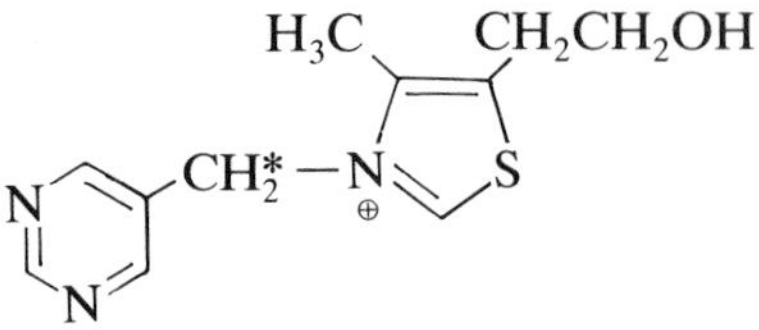

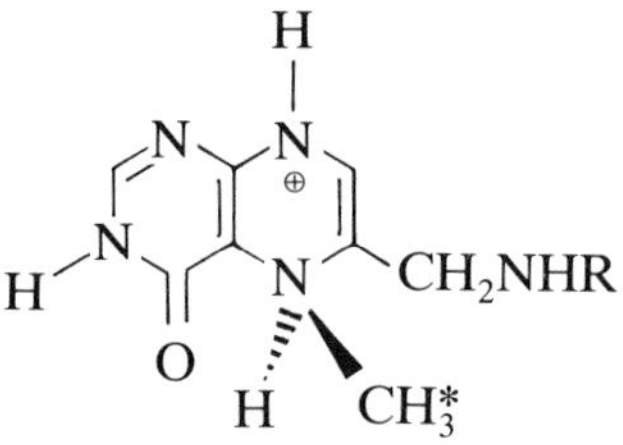

In the typical alkylating agents utilized in chemical practice (**XIII–XVI**), the principle of activation is also implemented through conjunction of the alkyl being transferred with a strong electrophile.

$H_3C{-}X$

$Alk_3O^{\oplus}BF_4$

$Alk_3S^{\oplus}OH^{\ominus}$

XIII	**XIV**	**XV**	**XVI**
X = Cl, Br, J	R = OCH_3, Ph, F		

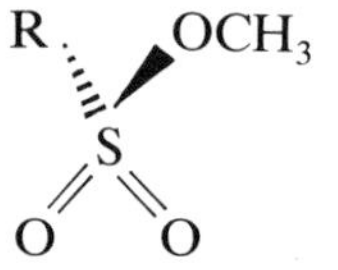

However, this activation (polarization) of the alkyl carbon bond with the adjoining electronegative center inevitably worsens the steric conditions for the development of the S_N2 reaction which is fairly sensitive to the bulkness and electron characteristics of the substituents linked with the alkyl carbon. Hence, an accumulation of electronegative substituents at the carbon center may lead to its total screening and inaccessibility to nucleophilic attack. For example, perfluorohydrocarbons are not susceptible to the reactions of nucleophilic substitution [25]. The geometrical demands of the sp^2-carbon atom differ from those of sp^3-carbon in that they are the same both in nucleophilic and electrophilic substitution reactions, accompanied in each case by formation of tetrahedral adducts. Hence, no steric hindrance is expected in the reversible reaction of the acyl group transfer, for instance, in the system of 1,3-diketone enol analogous to (2.4).

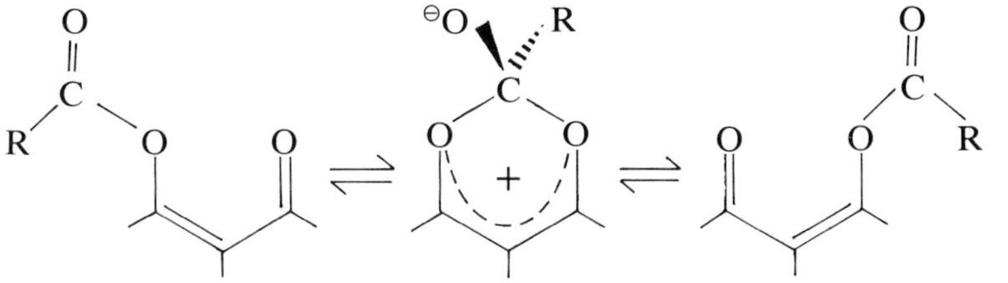

$$\qquad\qquad (2.7)$$

XVIIa   **XVIIc**   **XVIIb**

An additional factor favoring enhanced reactivity of the sp^2-carbon center in carbonyl compounds **XVIII** is the shift of the valence-electrons from carbon to the electronegative atom.

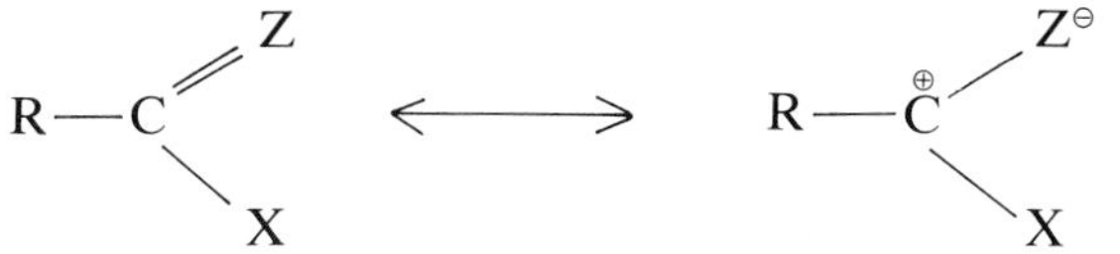

XVIII   **XIX**

Polarization of the **XVIII** ↔ **XIX** type facilitates the nucleophilic attack on the electron deficient carbon atom, so that nucleophilic substitution preceded by an addition to the carbonyl group proceeds, usually, under fairly mild conditions.

Acyl transfer, i.e. transacylation, reactions develop at particularly high rates in biochemical processes of synthesis and degradation of higher carbonic acids, lipids, carbohydrates and steroids [26]. Biosynthesis of protein molecules effected by the reaction of aminoacidic group transfer from the transport RNA onto the growing polypeptide chain in the corresponding enzyme systems (ribosomes) may proceed at extremely high rates. A typical example, illustrating the evaluation of the low limit of acceleration in bioorganic acyl reactions, is provided by the comparison between the rates of the urea hydrolysis reaction

$$(NH_2)_2C{=}O = H_2O \rightarrow 2\,NH_3 + CO_2$$

in the presence of an enzyme (urease) ($k_{21} \sim 3 \times 10^4\,s^{-1}$) and without it ($k_{21} \sim 3 \times 10^{-10}\,s^{-1}$) [28, 29]. Such a striking difference (10^8—10^{15} times) between the rates of the same chemical reaction is one of the main motives for studying the enzymes, as well as various simpler molecular systems which model enzymic action [30—32].

Consider the following sequence of reactions of esterification [33] as one example of oriented approach to such systems

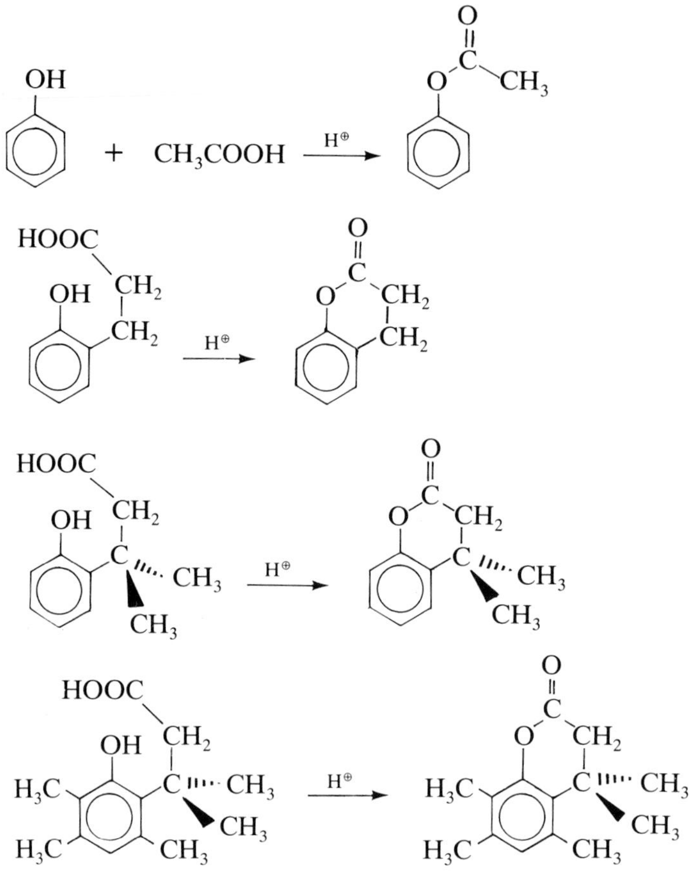

$$k = 10^{-10}\,\text{s}^{-1}*$$

$$k = 5.94 \times 10^{-6}\,\text{s}^{-1}$$

$$k = 2.62 \times 10^{-2}\,\text{s}^{-1}$$

$$k = 5.9 \times 10^{5}\,\text{s}^{-1}$$

* Reduced to the dimension of the pseudomonomolecular reaction constant.

Transition from intermolecular to intramolecular reaction and restriction of the intramolecular mobility of reacting groups through introduction of bulky methyl groups (adjusting the reaction site to a suitable conformation) results in a more than 10^{15} times increase in the reaction rate! The magnitude of this effect is of the same order as the enzymatic acceleration of reactions [30—32]. According to the widespread view, adopted largely as a result of Koshland's work [34], the role of an enzyme consists in (a) bringing together of reacting groups, (b) their mutual orientation, and (c) optimizing the directions of the forming and rupturing bonds between all pairs of the reacting atomic centers of the substrates (steering). This role may be played not only by an enzyme's adjusting its conformation so as to satisfy the above conditions in forming an enzyme—substrate complex. In principle, it may be assumed by any molecular system carrying the reacting centers in an intramolecular reaction (rearrangement) which is congruent with the enzymatic one. The more the conformation of this system corresponds to the proximity and optimum orientation of the reacting centers, the lower is the free activation energy of the process. Lowering of the activation barrier is, naturally, due, in the first place, to the entropy factor, and even insignificant structural variations may bring about extraordinary changes in the rate of intramolecular reaction, thus rivalling enzymes in their efficiency [35, 36]. Thus, the problem of the search for tautomeric carbonotropic systems is directed, first of all, towards fulfillment of the stereo-chemical requirements of the limiting stage of transfer of the carbon-centered migrant. If they are satisfied, enormous accelerations of an intramolecular reaction (10^{15}—10^{20} orders of magnitude) may be achieved by varying the electronic characteristics of the reaction centers, e.g., by increasing the electrophilicity of the migrant in nucleophilic carbonotropic systems.

2. Sigmatropic Acyl and Aryl Rearrangements

Whereas the first acylotropic and arylotropic tautomeric systems were documented only in the early seventies, the rearrangements of this type had become much earlier a subject for extensive studies. Analysis presented below is aimed at revealing such rearrangements which might be suitable candidates for modification into tautomeric systems.

2.1. *Acyl 1,3-rearrangements*

Apparently, the first example of the acyl 1,3-migrations was the observation [37] of unusual behavior of *N*-benzoyl-*N*-phenylbenzamidine **XX**.

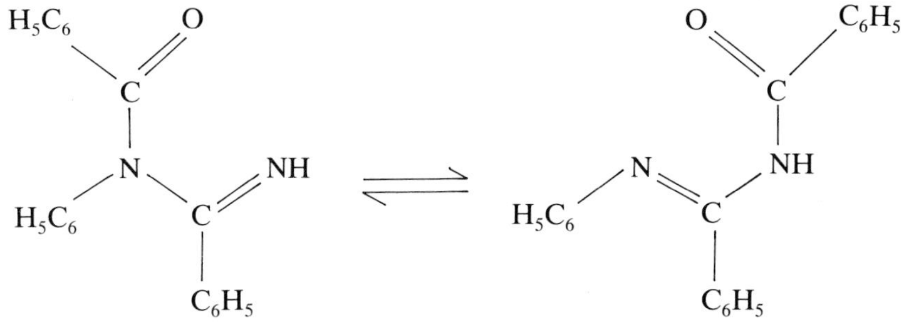

$$(2.8)$$

      XXa   **XXb**

    Upon standing for some time, or when an attempt at recrystallization from ethanol is made, **XXa** (m.p. 97°) converts to the isomer **XXb** (m.p. 135°). A similar $N \rightleftharpoons N'$-rearrangement was later discovered [38] for the triazene derivative **XXI** after recrystallization from acetone, pyridine and ethanol.

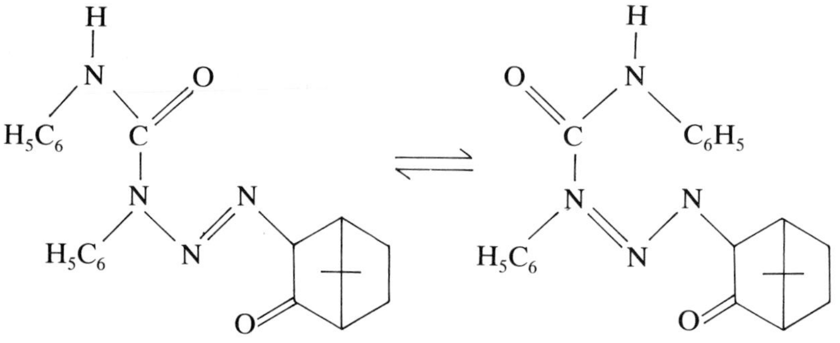

XXIa   **XXIb**

(2.9)

Upon heating, yellow phenylazotribenzoylmethane **XXIIa** rearranges into red
α-phenylazo-β-benzoylhydroxybenzalacetophenone **XXIIb** and colorless benzoyl-
phenylhydrazone **XXIIc** through competing C → O and C → N 1,3-shifts of the
benzoyl group [39].

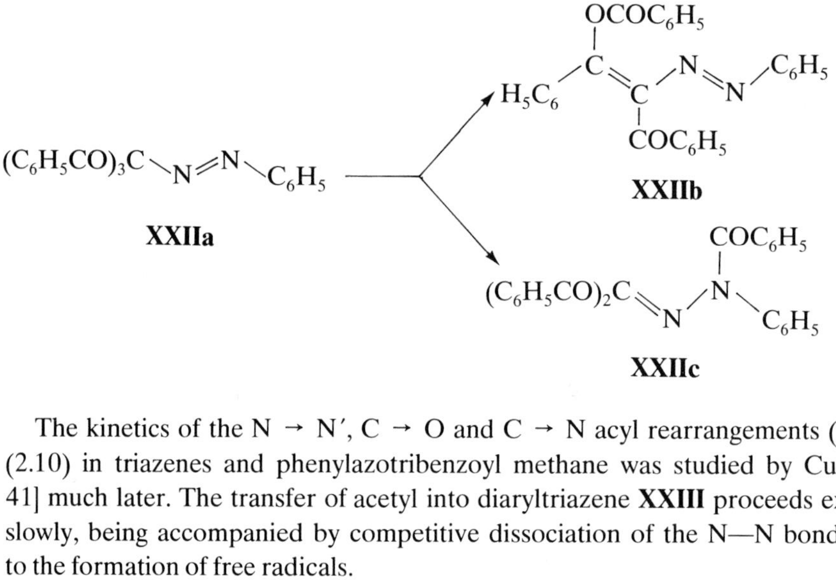

$(C_6H_5CO)_3C\diagdown_{N}\diagup^{N}\diagdown_{C_6H_5}$

XXIIa

$\overset{OCOC_6H_5}{\underset{COC_6H_5}{H_5C_6\diagup C = C \diagup N \searrow_N \diagup C_6H_5}}$

XXIIb

(2.10)

$(C_6H_5CO)_2C \diagup \diagdown N \diagup \overset{COC_6H_5}{N} \diagdown C_6H_5$

XXIIc

The kinetics of the N → N′, C → O and C → N acyl rearrangements (2.9) and
(2.10) in triazenes and phenylazotribenzoyl methane was studied by Curtin [40,
41] much later. The transfer of acetyl into diaryltriazene **XXIII** proceeds extremely
slowly, being accompanied by competitive dissociation of the N—N bond leading
to the formation of free radicals.

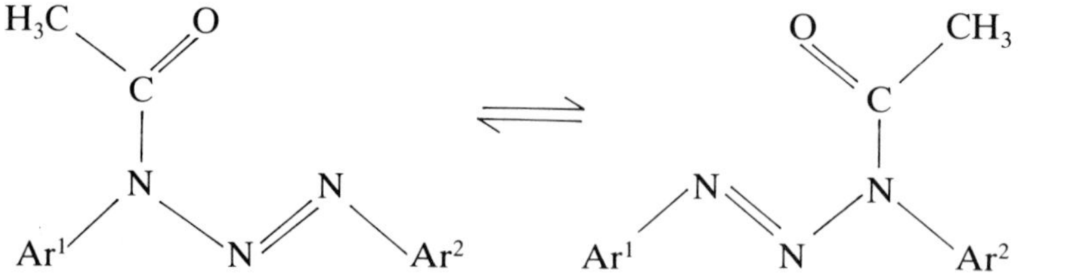

XXIIIa   **XXIIIb**

(2.11)

For instance, transfer of benzoyl groups in triazene **XXIV** requires an activation energy of 27—28 kcal/mol [42] whereas in the case of radical homolysis this energy is only 1—3 kcal/mol higher (28—30 kcal/mol).

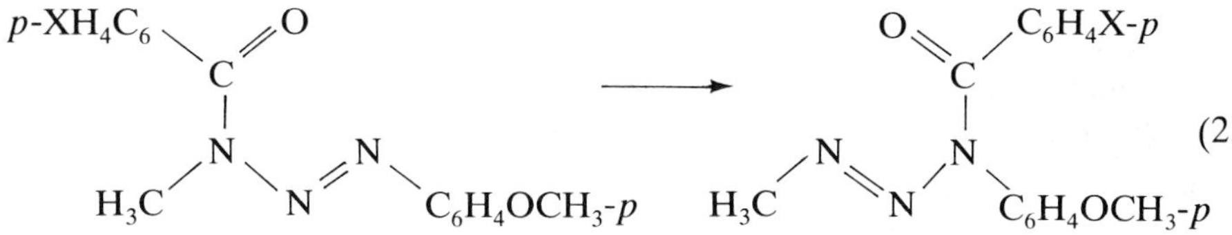

XXIVa   **XXIVb**

$$(2.12)$$

Phenylazotribenzoylmethane **XXII** rearranges somewhat faster [40]. C → O and C → N transfers proceed at almost equal rates in benzene ($2.3 \times 10^{-5} \, s^{-1}$) and in dioxane ($2.35 \times 10^{-5} \, s^{-1}$) at 43°C ($\Delta G^{\neq} = 26.2$ kcal/mol). On passing from triazene **XXIII** to benzimidine systems, the activation energy of the benzoyl group transfer falls to 25.2 kcal/mol for compound **XX** [40].

Rearrangement of isopropenyl acetate [43] is one of the few examples of the 1,3-O → C acetyl transfer, which is the reverse of the C → O shift (2.10). It occurs under extremely severe conditions of pyrolysis (450°C) and proceeds, apparently, according to the homolytical mechanism:

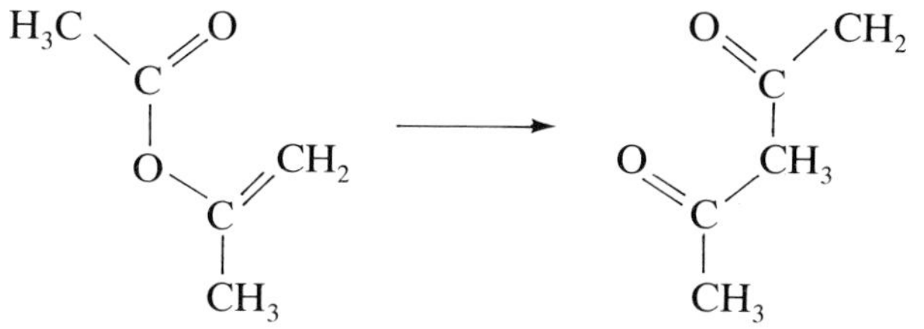

XXVa   **XXVb**

$$(2.13)$$

Benzoyl O → N 1,3-migrations in the *O*-derivatives of isoamides **XXVI** have been studied in detail [44—46]. Free activation energies of the aroyl group transfer reaction (2.14)

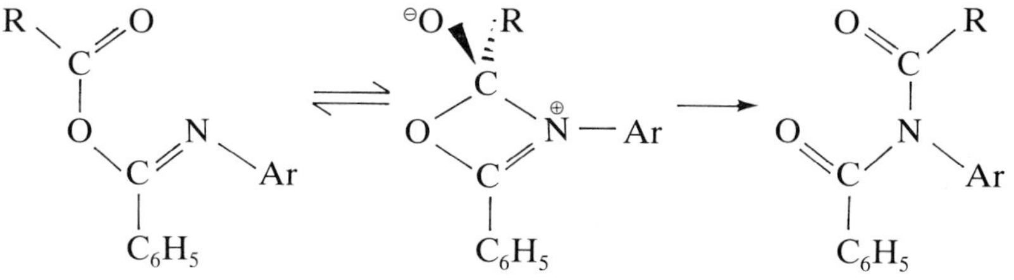

XXVIa   **XXVIc**   **XXVIb**

$$(2.14)$$

were determined from the integral intensities of the $\nu_{C=N}$ and $\nu_{C=O}$ bands in

IR-spectra, at 40—50°C they amount to 21—25 kcal/mol [47]. The rate of the O → N shifts rises upon introduction of acceptor *p*-substituents; to a lesser degree it is influenced by the polarity of the solvent. The fairly high activation barrier of the isomerization (2.14) is accounted for by the strained four-center cyclic transition state structure **XXVIc**. Hindered rotation (or planar inversion) about the C=N bond in initial isoamides **XXVIa** is not the limiting stage for the acyl transfer [45]. Curtin [40] reviewed all the then known data on the 1,3-acyl rearrangements described by the symbolic Equation (2.15). Equilibria are shifted to the left or to the right depending on the type of X, Z and Y centers as well as on that of substituents R.

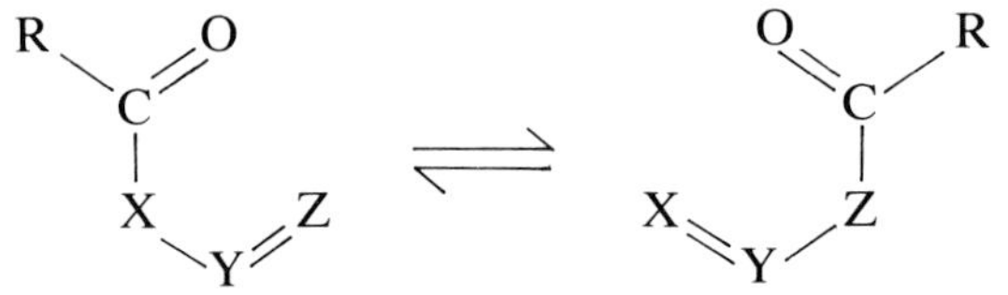

$$(2.15)$$

XXVIIa　　　　　　　**XXVIIb**

X, Z = C, O, N;　　Y = C, N;　　R = Ar, Alk

A similar type of the 1,3-acyl shifts can be observed in the molecules in which the triad **XXVII** serves as a fragment of a heterocyclic system.

The *O*-acyl isomers of 3-hydroxy-1,2-thiazole **XXVIIa** rearrange into *N*-isomers upon prolonged standing or on brief heating [48]. Acyl derivatives of 3-hydroxy-benz-1,2-oxazole behave in a similar manner [49].

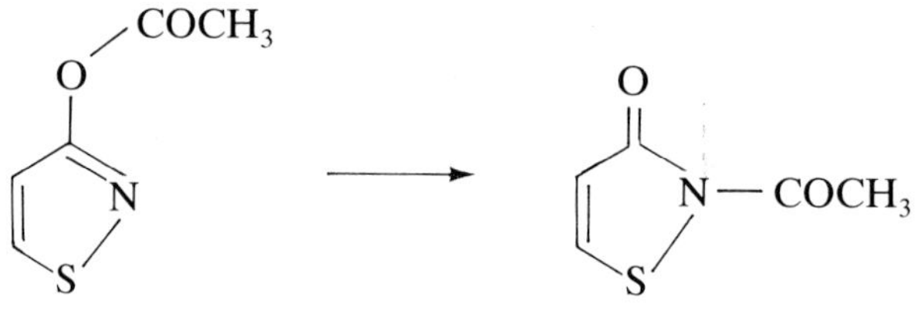

$$(2.16)$$

XXVIIIa　　　　　　　**XXVIIIb**

Reversible O ⇌ N acyl migrations were found only in two cases. Acylation of the sodium salt of 6-hydroxyphenanthridine leads at −20°C to the *O*-isomer **XXIXa** which, upon melting (m.p. 119°C), was converted to the *N*-isomer **XXIXb** (m.p. 191°C). With the aid of IR-spectroscopy, the rate constants and activation parameters of the direct and reverse reactions (2.17) were found [40]. Free activation energy amounts for the benzoyl derivative (X = H) to 24.5 kcal/mol. The rate of the O ⇌ N exchange falls for *p*-methoxybenzoyl and grows in the case of *p*-chlorobenzoyl derivatives. 17% Content of **XXIXa** (X = N) and 83% of **XXIXb** in the mixture corresponds to the state of dynamic equilibrium at ambient temperature.

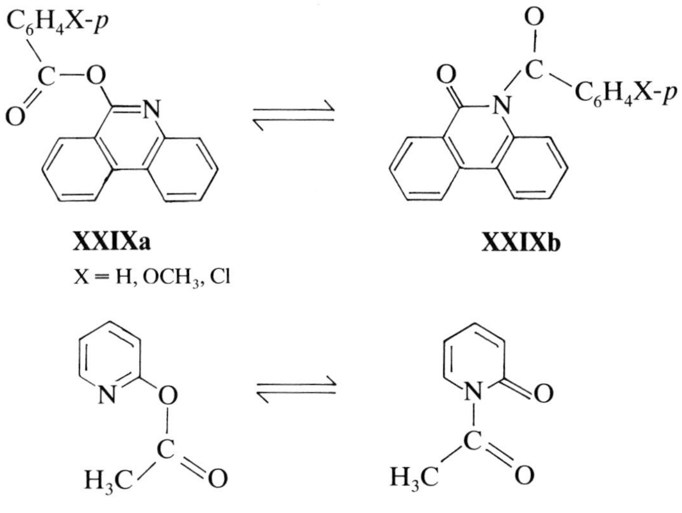

XXIXa

X = H, OCH$_3$, Cl

XXIXb

(2.17)

XXXa

XXXb

Likewise, the *O*-acetyl derivative **XXXa** formed in the acylation of the Na- or Tl-salt of 2-pyridone is stable at lower temperatures, but, when heated, it rearranges into an *N*-isomer **XXIXb** [50]. In methylene chloride, the equilibrium mixture is composed of 10% **XXXb** and 90% **XXXa** [51].

2.2. *Aryl 1,3-rearrangements*

Elaborating upon earlier work by Mumm [52, 53], Chapman [54] carefully studied the rearrangements of *O*-arylisomides resulting from the 1,3-transfers of aryl groups.

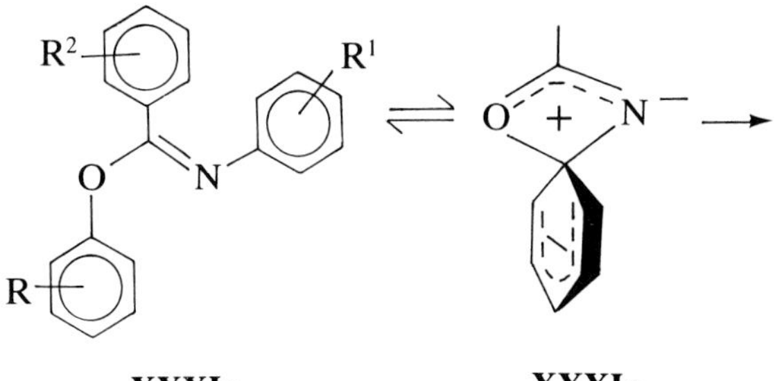

XXXIa

XXXIc

XXXIb

(2.18)

Chapman's rearrangement proceeds under rather severe conditions, at 150—300°C: it is frequently employed as a preparative method for obtaining diarylamines [55].

Chapman suggested that the reaction (2.18) proceeded intramolecularly via transition structure **XXXIc**. The intramolecular character of the reaction was demonstrated by a crossover experiment. Upon heating of the mixture of compounds **XXXIa** ($R = R_1 = R_2 = H$) and **XX** ($R = R_1 = CH_3$-p, $R_2 = H$) and subsequent hydrolysis, only symmetrical diarylamines were isolated. These results and conclusions were later fully confirmed [56]. Electron-acceptor substituents in the O-aryl group facilitate the O → N rearrangement, as do electron donor substituents with regard to the reverse, i.e. N → O, process.

The nitrogen atom is less nucleophilic in arylimines than in alkylimines, and N-alkyl imidates might have been expected to rearrange more readily than the compounds **XXXI**. It turned out, however, that they did not undergo isomerization [54, 57].

In the case of amidines, Chapman's rearrangement proved to be reversible. Thus, heating (at 250°C) of the pure isomers **XXXIIa** and **XXXIIb** produces an equilibrium mixture of identical composition [58].

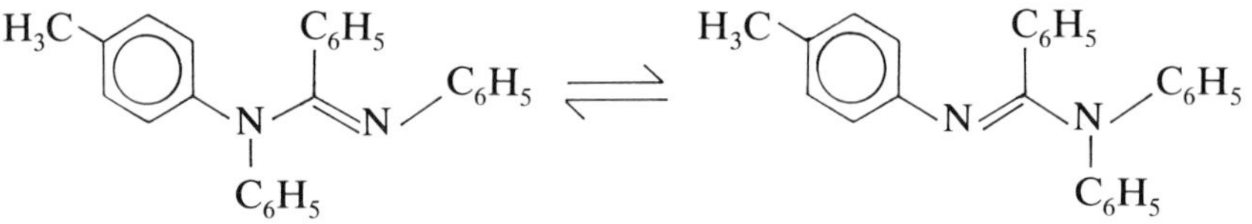

$$(2.19)$$

XXXIIa (60—65%)   **XXXIIb** (35—40%)

Crossover experiments on the mixtures of N-phenylamidines have also shown the rearrangements of the (2.19) type to be intramolecular.

1,3-aryl rearrangements in isoimides **XXXIa** proceed 10^4—10^5 times faster than in amidines **XXXII**. Activation barriers for aryl shifts amount to 39.5—40.5 kcal/mol in isoimides and 44.5—46.5 kcal/mol in amidines.

Chapman also studied rearrangements of isothioimidates **XXXIIIa**

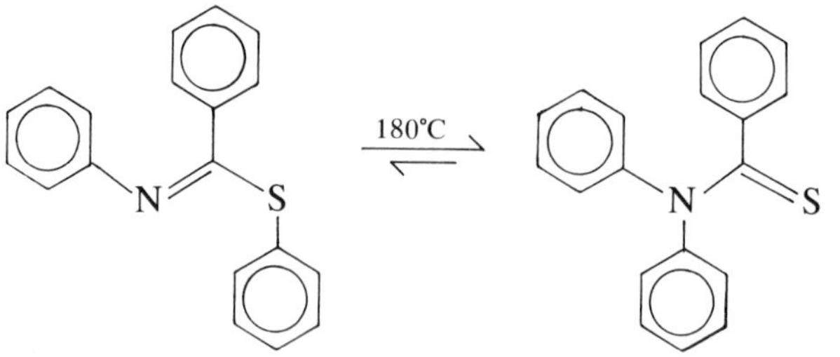

$$(2.20)$$

XXXIIIa   **XXXIIIb**

In this case, the aryl group migrates towards nitrogen approximately 10 times slower than in the oxygen analog **XXXIa**. This may be accounted for by enhanced nucleophilicity of sulfur in comparison with oxygen and by greater stability of **XXXIIIa**.

Stevens' rearrangement [59] also belongs to analogous 1,3-aryl transfers.

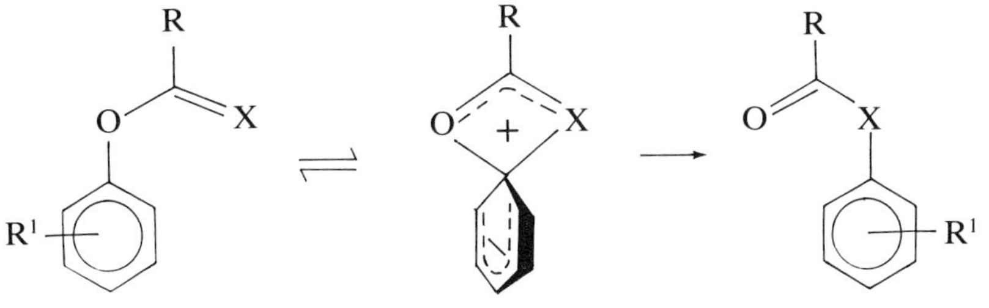

| **XXXIVa** | **XXXIVc** | **XXXIVb** | X = S, Se |

$$(2.21)$$

This reaction is an O → S (Se) transfer, it readily progresses upon heating of starting compounds to 100—180°C and may serve as a convenient preparative method enabling transition from phenols to thio- and selenophenoles. Stevens' rearrangement can be catalyzed by acids and is accelerated when electron-acceptor groups R_1 are present in the *O*-aryl nucleus or when polarity of the solvent is enhanced.

Besides Chapman and Stevens rearrangements, a number of other 1,3-aryl transfers are described by Eq. (2.21), including those of Schönberg [59] and Newman—Kwart [60] (R = ArO, X = S, $\Delta H^{\neq}$ ~ 38 kcal/mol and R = R_2N, X = S, $\Delta H^{\neq}$ ~ 38 kcal/mol, respectively) which in their activation energies are close to the Chapman rearrangement of *O*-aryl isoimides.

The data presented above on 1,3-acyl and aryl rearrangements bear witness to the following:

(1) Only nondegenerate, hence, almost exclusively, irreversible rearrangements have been studied enabling one to employ for detecting the content of the isomers the methods of electron and vibrational spectroscopy.
(2) Solely the (2.17) general type processes can be assigned to equilibrium isomerism and, therefore, tautomerism of acyl derivatives **XXVII**.
(3) Rate constants and free activation energies indicate rather low rates of the 1,3-acyl transfer. Even for the fastest of the reversible reactions (2.17), the free activation energy actually exceeds the value that would correspond to the range of tautomeric reactions.
(4) Reactions of the 1,3-aryl transfer predominantly follow the intramolecular mechanism.
(5) The rate of the 1,3-acyl and 1,3-aryl transfer reactions is fairly sensitive to structural variations in the system.

2.3. *1,5- and 1,7-acyl Rearrangements*

Interaction of *S*-acetyl-*p*-mercaptobenzaldehyde with aromatic amines readily affords azomethines **XXXV**. However, in the case of more basic alkylamines, there is observed under broadly varied conditions a transfer of the acetyl group from the sulfur atom to amine nitrogen leading to the formation of *N*-substituted acetamide.

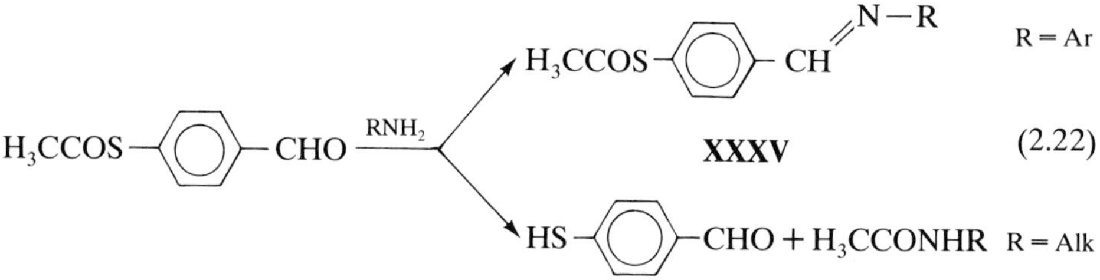

$$H_3CCOS \text{—} \langle \bigcirc \rangle \text{—} CHO \xrightarrow{RNH_2}$$

$$H_3CCOS \text{—} \langle \bigcirc \rangle \text{—} CH \overset{N-R}{\diagup} \qquad R = Ar$$

$$\textbf{XXXV} \qquad (2.22)$$

$$HS \text{—} \langle \bigcirc \rangle \text{—} CHO + H_3CCONHR \quad R = Alk$$

As regards *S*-acetyl-*o*-mercaptobenzaldehyde, its reaction with amines results in the S → N transfer in the case of both alkyl and aromatic derivatives [1]. The reason for the observed transformations lies in the labilization of the *S*-acetyl bond in acetyl mercaptobenzaldehydes and their imines, which serve as carriers of the acetyl group. Later Kemp and Velaccio [61] made use of the idea to employ acylated aromatic hydroxyaldehydes in the role of acyl carriers and pointed out that they were highly promising as peptide binding reagents. They were able to show that the reaction of *ortho-* and *peri*-derivatives had an intramolecular character at the acyl transfer stage. This was evidenced by the interaction of 8-acetoxy-1-naphthaldehyde with amines in which the formation of carbinolamine was reliably demonstrated.

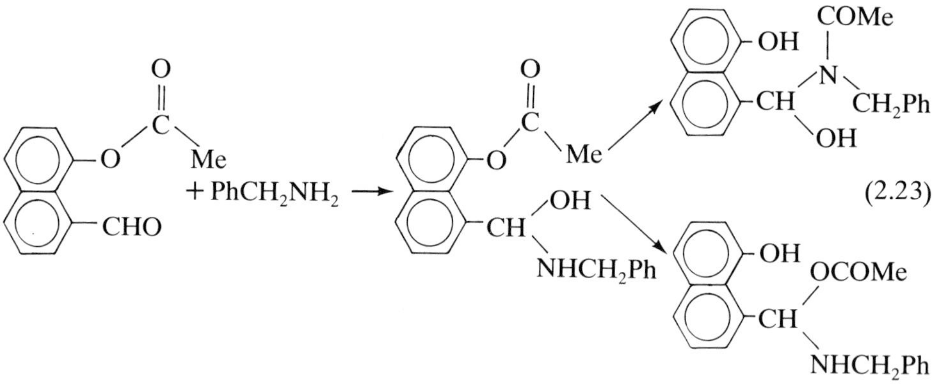

$$(2.23)$$

It is not known if a similar intramolecular mechanism could be operative in the (2.22) reaction, although the corresponding carbinolamine can attain the conformation of **XXXVI** which, in principle, allows intramolecular 'gliding' of the acyl group over the plane of the aromatic nucleus, as required by the mechanism of the Dewar rearrangements [62]

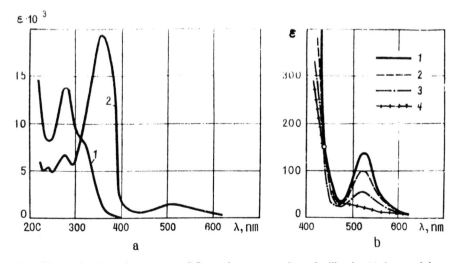

XXXVI

It is more likely that in this case the reaction proceeds in the appropriately formed bimolecular associate. The same is, apparently, true of the first example of acylotropic tautomerism in the series of *p-S*-acetylmercaptobenzaldimines detected by us [1, 2].

$$(2.24)$$

XXXVIIa   **XXXVIIb**

The tautomeric equilibrium (2.24) was studied by means of electron spectra in various solvents. As Fig. 2.1 demonstrates, the spectrum of compound **XXXVII** in

Fig. 2.1 Electronic absorption spectra of *S*-acetyl-*p*-mercaptobenzylaniline in: (a) 1 — cyclohexane, 2 — (Me₂N)₃PO; (b) in the solvent mixture of benzene —(Me₂N)₃PO: (1) 0, (2) 1 : 4 (3) 1 : 1 (4) ∞ .

the nonpolar solvent is completely analogous to the spectrum of the *S*-methyl derivative. In polar solvents, however, *p*-acetylmercaptobenzaldimines show two more weak bands at 410 and 520 nm than are revealed in the spectra of the quinonoid forms of mercaptoazomethines [63], indicating a certain amount of the structure **XXXVIIb**, totalling 5—10% ($\Delta G° \geqslant 1$ kcal/mol).

The mobile character of the equilibrium (2.24) can be seen from the data of Fig. 2.1, in which spectra in a mixed solvent are given. The $S \to N$-acetyl transfer reactions belong to the fast and reversible class since, after varying the polarity of the solvent, the equilibrium is established practically instantaneously. This is in accord with the kinetic data [61] which indicate an extremely sharp acceleration of the acyl transfer in the reaction (2.23) in the polar solvents, such as dimethylsulfoxide and acetonitryl.

Unlike the tautomerism of acetyl mercaptobenzaldimines, the equilibrium isomerism of acetyl-4-pyridones [51] — studied at almost the same time as the former — belongs to the slow rearrangements. (The reactions (2.25) and (2.23) were studied only in nonpolar methylene chloride)

$$\text{XXXVIIIa (53\%)} \rightleftharpoons \text{XXXVIIIb (47\%)} \tag{2.25}$$

Interesting intramolecular 1,5-rearrangements of acyl moieties were identified by Woodward and Olofson when studying the mechanism of the reactions (2.26) between the isoxazolium cation and nucleophilic agents, whose investigation had been initiated by Mumm's work [52. 65]. It has been found that under the action of a strong base the isoxazolium ring is opened with the formation of the corresponding ketene imine, which then adds carbonic acid to give rise to the enol tautomer **XXXIX** of isoimide, which subsequently rearranges via cyclic intermediate compound **XXXIXd** into the *N,N*-diacylamide form **XL** through the 1,5-transfer of acyl.

Thus, the transformation **XXXIX** → **XL** proceeds not according to the 1,3-$O \to N$ type of acyl shift, as Mumm assumed, but via a series of successive 1,5-sigmatropic proton and acyl rearrangements. Woodward suggested utilization of the high intramolecular mobility of compounds **XXXIX** (free activation energy of the slowest stage **XXXIXc** → **XL** $\Delta G_{30}^{\neq} = 21.7$ kcal/mol) for peptide synthesis, since the intermediates **XXXIX** are good acyl donors in the intermolecular $O \to N$-transacylation.

Further substantial progress in the studies of acyl rearrangement was made after one had started investigating degenerate reactions and employing the technique of dynamic nuclear magnetic resonance (DNMR).

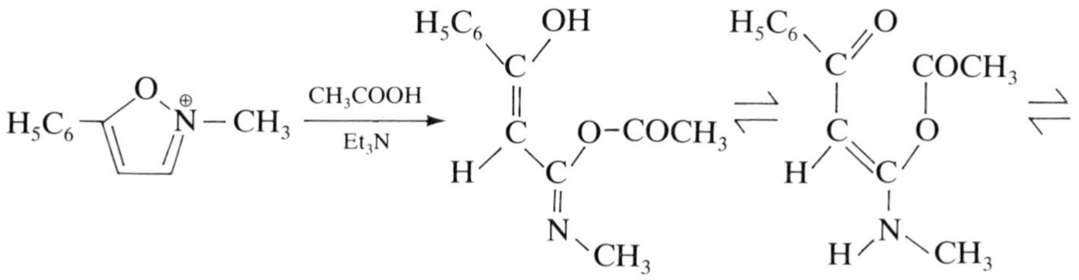

XXXIXa   XXXIXb   (2.26)

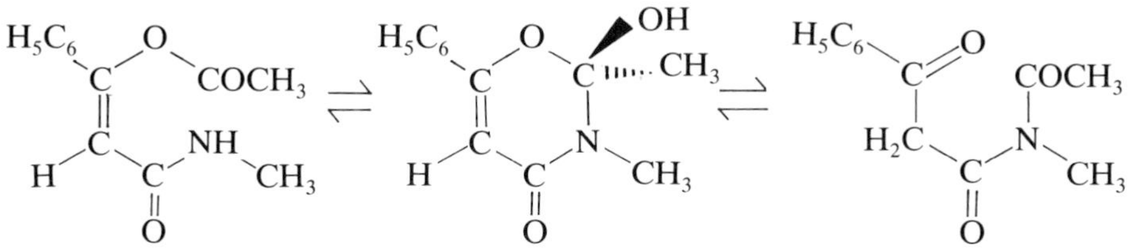

XXXIXc   XXXIXd   XL

Calder and Cameron [66] applied the DNMR method to studying kinetics of reversible rearrangements of naphtazarine acetates.

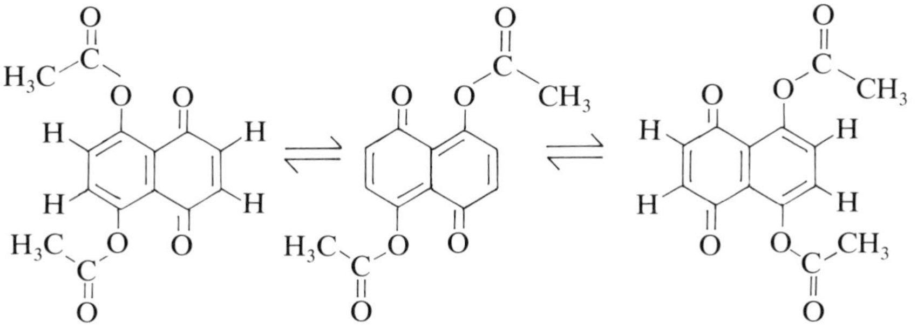

XLIa   XLIc   XLIb   (2.27)

When the temperature of a deuteronitrobenzene solution of compounds **XLIa, b** is raised above 170°C, there occurs a broadening of the upfield (quinonoid) and downfield (benzoid) proton peaks, brought about by the chemical exchange reaction (2.27). Coalescence of signals does not take place even at 200°. The magnitude of the free activation energy ($\Delta G^{\neq}_{298}$ = 22.3 kcal/mol) indicates moderate rates of the $O \rightleftarrows O'$-acetyl shifts (k_{298} – 10^{-4} s^{-1}) at room temperature. The authors [66] did not observe any appreciable amounts of the symmetrical isomer **XLIc** in equilibrium with **XLLa, b** which was expected to appear in the ^{1}H NMR spectrum as a characteristic four-proton AB-quartet at temperatures below the coalescence point. From this it can be inferred that either the thermodynamic difference between **XLIa, b** and **XLIc** exceeds 4 kcal/mol or else the reaction (2.27) proceeds as a concerted process. The observed rearrangement does not contradict the latter possibility.

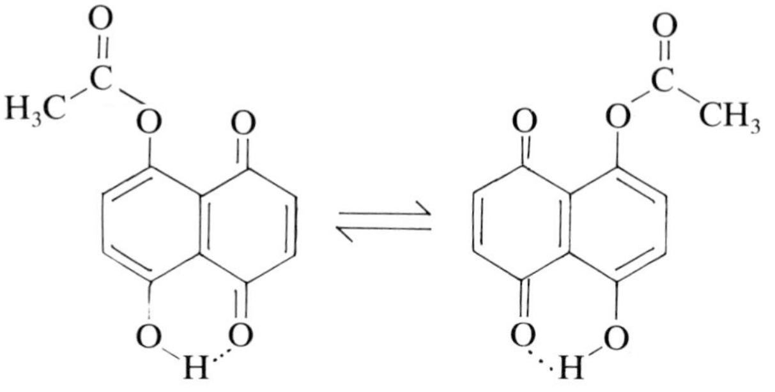

XLIIa   **XLIIb**

(2.28)

In contrast with (2.27), a substantial lowering of the activation barrier could have been expected in the process (2.28) since a proton transfer cannot require large activation energies. However, the value of $\Delta G^{\neq}_{298}$ obtained equal 24.5 kcal/mol, which is even higher than in the case of diacetate **XLI**. Hence, it is precisely the acyl group transfer which determines the kinetics of the overall reaction (2.28). Slow exchange processes in the series of *N*-acyl derivatives of pyrazoles observed by means of the DNMR method [67] belong to acyl rearrangements of the 1,5-sigmatropic shift type, initiated by temperature variations. Heating of solutions of compounds **XLIII** (1,5—7) above 170°C leads to the isochronicity of signals of the indicator groups in the 3 and 5 positions of the pyrazole cycle. This can be accounted for by the rearrangement (2.29)

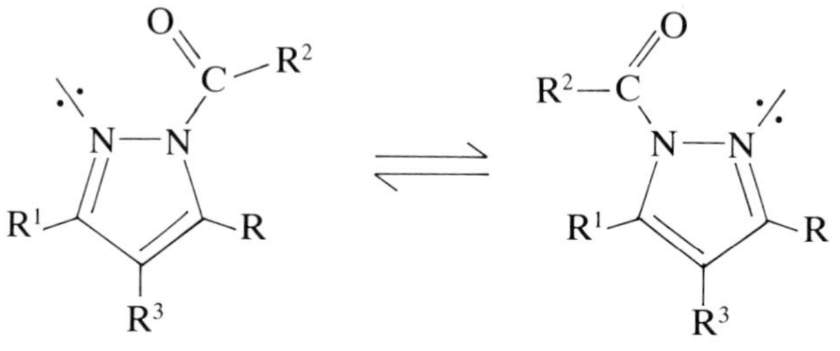

XLIIIa   **XLIIIb**

(2.29)

	R_1	R_2	R_3
1.	Me	NHPh	H
2.	Me	$C_6H_4NO_2$-n	H
3.	Me	$C_6H_3(NO_2)_2$-(2,4)	H
4.	Br	Br	H
5.	Me	NHC_6H_4Me-n	H
6.	Me	NHPh	Me
7.	Me	NHC_6H_4Me-n	Me

A crossover experiment answered the question if the mechanism of acyl shifts was intra- or intermolecular. Upon heating of equimolar mixtures of compounds **XLIII** (1 + *) and (5 + 6) to 200°C in 1-chloronaphthalene, there is in every case a mixture formed of all four possible *N*-acylpyrazoles which is indicative of intermolecular exchange. This process is believed to occur via reversible dissociation of carbamoyl pyrazoles into pyrazole and isocyanates followed by recombination to the initial and the crossover products.

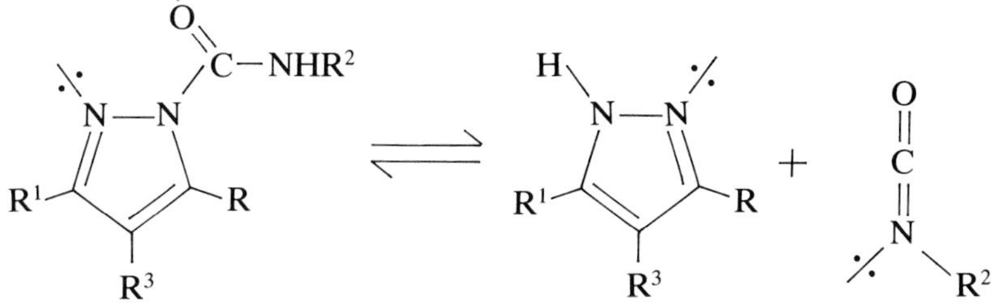

$$
\text{(2.30)}
$$

In the case of the *N*-acyl derivatives of pyrazoles **XLIII** (2—4), no exchange process was observed [67, 68].

Apparently, the intramolecular 1,5-sigmatropic reaction cannot at all be accomplished via the N $\rightleftharpoons$ N′ shift of the acyl groups owing to the fact that the orientation of the lone electron pair axis of the neighbouring nitrogen atom is unfavorable for the nucleophilic attack on carbonyl carbon.

The pyrazole system illustrates particularly clearly the difference between the steric requirements of the metallotropic and carbonotropic tautomerism.

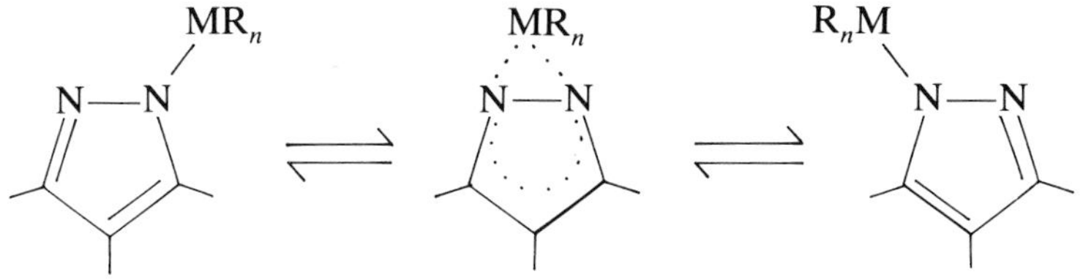

$$
\text{(2.31)}
$$

MR$_n$ = SiMe$_3$, GeMe$_3$, SnMe$_3$, HgPh

Metallotropic reversible intramolecular rearrangements (2.31) occur readily enough in the series of silicon-, germanium-, and tin-derivatives of pyrazole [69] with the activation barrier in this series consecutively dropping from 24 to 8 kcal/mol. Moreover, in the case of the phenylmercuric derivative of 3,5-dimethylpyrazole, rapid reversible shifts of the HgPh-group with the frequency of over 50 s^{-1} are observed even at $-100°$C.

3. Acylotropic tautomerism

Some of the rearrangements considered above may, in regard to their energy parameters, be assigned to tautomeric transformations. Such are, for example, the acyl 1,3-rearrangements (2.17) and 1,5-rearrangements (2.27). However, these isolated cases can give us no idea either of the degree of generality characterizing this new type of the *i,j*-sigmatropic tautomerism, nor of whether it is possible to achieve, through structural modifications, a substantial increase in the frequency of acyl migrations which in these systems is rather low ($k < 10^{-4}$ s^{-1}).

This section presents new experimental data on fast and reversible intramolecular *i,j*-acyl migrations (*j* = 3, 5, 7, 9) of nucleophilic type which have been obtained mainly in the present authors' laboratory over the last 10—12 years. They show that if optimal structural conditions are provided, the rates of acyl migrations become comparable with the proton migration rates.

Considering that the fit of the molecular reaction site is particularly important for producing a sufficiently low activation energy barrier of rearrangement, the emphasis was laid on studying degenerate or nearly degenerate reactions and symmetrical or nearly symmetrical molecules. In studying reactions of this type, a remarkable role belongs to the dynamic NMR method by which all important kinetic data have been obtained.

3.1. *Tautomeric 1,3-acyl rearrangements of the* N,N'*-diarylamidine derivatives*

Degenerate 1,3-acyl rearrangements of the (2.15) type can be carried out in the amidine systems **XXVII** (X = Z = NR). The observation by Wheeler [37] that introduction of the 2-aryl substituent accelerates the acyl migration, as, for example, in (2.8), should prove important for structural regulation of the magnitude of the activation energy barrier. Thus, it could be expected that the *N*-acyl derivatives of benzamidines would be susceptible to a thermally initiated rearrangement caused by the 1,3-transfer of the acyl groups that would be fast enough to be studied by the dynamic NMR method.

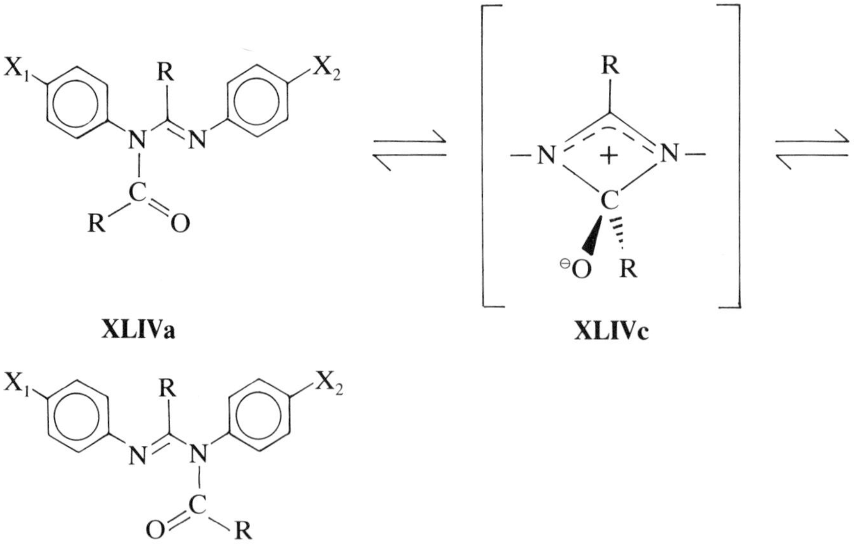

XLIVa	**XLIVc**	(2.32)

XLIVb

Indeed, it was found [70, 71] that in the ^{1}H-NMR spectra of solutions of amidines **XLIV** ($X_1 = X_2 = CH_3$ and $X_1 = H$, $X_2 = CH_3$, $R = C_6H_5$) there are observed at room temperature well separated signals of nonequivalent methyl groups (Fig. 2.2) which with the rising temperature become broader coalescing finally into common peak. Such behavior of indicator signals is consistent only

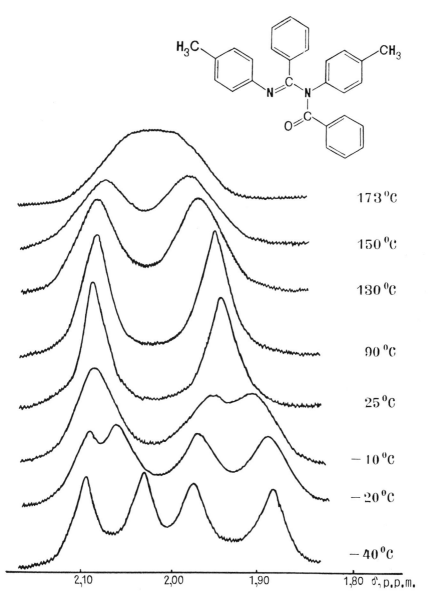

Fig. 2.2 Acylotropy and stereodynamics observed in the ^{1}H—NMR spectra in *N*-benzoyl *N,N'*-di-(*p*-tolyl) benzamidine (No. 9 in Table 2.I).

with the transformations of the type **XLIVa** ⇌ **XLIVb**, since rotations about amidine and amide C—N bonds do not average the spectral picture (Fig. 2.2), generally, these internal rotations are characterized by lower activation barriers [72, 73].

An analogous spectral pattern is also observed in the case of the nondegenerate tautomeric rearrangement (2.32) for two different indicator groups ($X_1 \neq X_2$) with the nearly equal population of the tautomers **XLIVa, XLIVb** (Fig. 2.3).

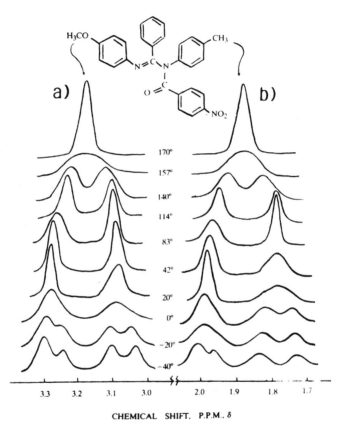

Fig. 2.3. Temperature dependence of the ^{1}H-NMR spectra of amidine **XLV** (No. 18 in Table 2.I): (a) in the region of methoxy-group signals; (b) in the region of methyl-group signals. Solvents: o-dichlorobenzene (ablve 40°C), chlorobenzene (below 40°C) [75].

In order to elucidate the influence of various structural factors on the rate of the 1,3-acyl migrations, the parameters of rearrangements were determined for a wide range of amidine derivatives by successive variation of the substituents R, R_1, X_1 and X_2. Table 2.I contains rate constants and activation parameters for the type (2.32) acylotropic rearrangements obtained from the analysis of the total line shape of indicator signals of the ^{1}H-NMR spectra. These data permit us to draw

Table 2.I. Kinetic and Activation Parameters of Acylotropic Rearrangements of Amidines (2.32) (Solvent: *o*-Dichlorobenzene) [70, 71, 75]

No.	Compound XLIV				$k_{25} \times 10^5$, s^{-1}	$\Delta H^{\neq}$, kcal/mol	$\Delta S^{\neq}$, e.u.	$\Delta G^{\neq}_{25}$ kcal/mol
	R	R_1	X_1	X_2				
1	$C_6H_4NO_2$-p	$C_6H_4OCH_3$-p	CH_3	CH_3	17.4	23.2 ± 0.3	2.3 ± 0.7	22.5
2	$C_6H_4NO_2$-p	C_6H_5	CH_3	CH_3	2.18	26.0 ± 0.3	7.8 ± 0.8	23.7
3	$C_6H_4NO_2$-p	C_6H_4Br-4	CH_3	CH_3	0.87	26.1 ± 0.4	6.0 ± 1.0	24.3
4	$C_6H_4NO_2$-p	$C_6H_4NO_2$-p	CH_3	CH_3	0.44	25.6 ± 0.5	3.0 ± 1.2	24.7
5	$C_6H_4NO_2$-p	$C_{10}H_7$-1	CH_3	CH_3	~0	—	—	> 27.0
6	$C_6H_4NO_2$-p	H	CH_3	CH_3	0.02	26.7 ± 1.0	0 ± 1.5	26.7
7	C_6H_4Br-p	C_6H_5	CH_3	CH_3	0.55	26.6 ± 0.2	6.7 ± 0.5	24.6
8	C_6H_4Cl-p	C_6H_5	CH_3	CH_3	0.87	26.3 ± 0.2	6.7 ± 0.5	24.3
9	C_6H_5	C_6H_5	CH_3	CH_3	0.17	25.0 ± 0.4	2.9 ± 1.0	24.1
10	$C_6H_4OCH_3$-p	C_6H_5	CH_3	CH_3	0.14	27.2 ± 0.3	6.0 ± 0.7	25.4
11	C_6H_4Br-p	$C_6H_4OCH_3$-p	CH_3	CH_3	0.44	26.1 ± 0.2	4.8 ± 0.6	24.7
12	C_6H_4Br-p	C_6H_4Br-p	CH_3	CH_3	0.13	27.0 ± 0.3	5.4 ± 0.7	25.4
13	$C_6H_4NO_2$-p	C_6H_5	OCH_3	OCH_3	0.69	25.3 ± 0.2	6.5 ± 0.5	23.4
14	CH_3	C_6H_5	OCH_3	OCH_3	0.12	27.3 ± 0.4	5.9 ± 0.7	25.5
15	$C_6H_4NO_2$-p^a	C_6H_5	CH_3	H	3.27	25.7 ± 0.1	7.4 ± 0.3	23.5
16	$C_6H_4OCH_3$-p^a	C_6H_5	CH_3	H	0.14	27.0 ± 0.5	5.3 ± 0.7	25.4
17	CH_3	H	CH_3	CH_3	~0	—	—	> 27.0
18	$C_6H_4NO_2$-p^b	C_6H_5	CH_3	OCH_3	1.24	26.5 ± 0.3	8.0 ± 0.6	24.1
19	$C_6H_4NO_2$-p^c	C_6H_5	CH_3	Br	4.72	25.3 ± 0.5	7.2 ± 0.8	22.3
20	$C_6H_4NO_2$-p^d	C_6H_5	CH_3	NO_2	39.3	24.0 ± 0.5	7.0 ± 1.1	21.0

[a] At 25°, the content of the tautomer **XLIVa** is 45%;
[b] 55% of **XLIVa** at 150°;
[c] 35% of **XLIVa** at 150°;
[d] 13% of **XLIVa** at 150°.

certain conclusions as to the influence of the structural factors of the acylamidine system on its ability to undergo reversible isomerizations.

The rate of acyl migrations is greatly influenced by the substituent at the central carbon atom of acylamidines. As seen from the data of Table 2.I (compounds 1—6), the acylotropic transformations proceed sufficiently fast on the ^{1}H-NMR time scale only if there is an aryl group in position 2 which is conjugated with the π-system of the amidine fragment. This can be accounted for by the structure **XLIVc** of the tetrahedral transition state (or intermediate), typical of reactions of nucleophilic substitution at the carbonyl center. Upon stabilization of **XLIVc**, one may expect a lowering in the activation energy of the rearrangement (2.32) facilitated by delocalization of the positive charge in the conjugated amidine cation. In accordance with this, the electron-donating substituents at the aryl nucleus R_1 increase the rate of acyl migrations whereas the electron-accepting substituents lower if (*cf.* Nos 1 and 4 in Table 2.I). A considerable increase in the energy barrier of the *N,N'*-acyl transfers in the case of the 2-(1-naphthyl) amidine derivative (No 5 in Table 2.I) is, apparently, caused by the steric demands of the 1-naphthyl group, which force it out of the plane of the amidine triad and break the conjugation. In Table 2.II, correlation equations are given which quantitatively describe the relationships governing the influence of the substituents R and R_1. In contrast to the substituents in the 2-aryl group, the electron-accepting substituents in the migrating *N*-aryl group accelerate the exchange and the electron-donating substituents slow it down (Nos 2, 7—10 in Table 2.I). The best correlation is attained when electrophilic constants of substituents are used (Table 2.II). this shows that concentration of the positive charge at the carbon atom of the migrating group is conducive to the increase in the rate of acyl migrations.

Table 2.II. Structural Correlations for Reaction (2.32)

Reaction Series	Changeable Substituent	Nos of Compounds in Table 2.I	Correlation Equation	$r(s)$
$C_6H_4NO_2$-*p* = R $X_1 = X_2 = CH_3$	R_1	1—4	$\log k/k_0 = -1.15\sigma°$	0.993, 0.103
C_6H_5 = R $X_1 = X_2 = CH_3$	R	2.7—10	$\log k/k_0 = 0.64\sigma^+$	0.996, 0.060

Since reactions of the 1,3-acyl transfer in amidines proceed at sufficiently high rates only at elevated temperatures (+150—200°C), the possibility of an inter-molecular process should be examined. It was noted earlier that the *N,N'*-acyl exchange of *N*-carbamoylpyrazoles in 1-chloronaphthalene at 200° (2.30) was brought about by intermolecular reactions. On the other hand, reactions of the 1,3-acyl transfer in amidines are genuinely intramolecular, as has been confirmed by experiments on the *N,N'*-acyl migrations in solutions containing two different *N*-acyl-*N,N'*-diarylamidines, e.g. **XLIV**-9 and **XLIV**-13 (the second index denotes the number of the compound in Table 2.I).

The temperature dependence of the signals of methyl and methoxyl groups in this mixture is identical to that of the line shape of these signals in the solutions containing amidines (**XLIV**-9) and (**XLIV**-13) separately. The methyl doublet signal is coalesced at + 170°C, and the doublet signal of the methoxyl groups at +150°C. After returning to room temperature, the original spectrum is fully reproduced. Such spectral behavior of the mixture of amidines in solution allows one to rule out intermolecular acyl exchange, which would have led to the formation of new compounds. After 15—30 min of standing at +170°C and subsequent cooling, no new signals of the methyl and methoxyl groups of amidines (**XLIV**-2) and (**XLIV**-21) could be detected.

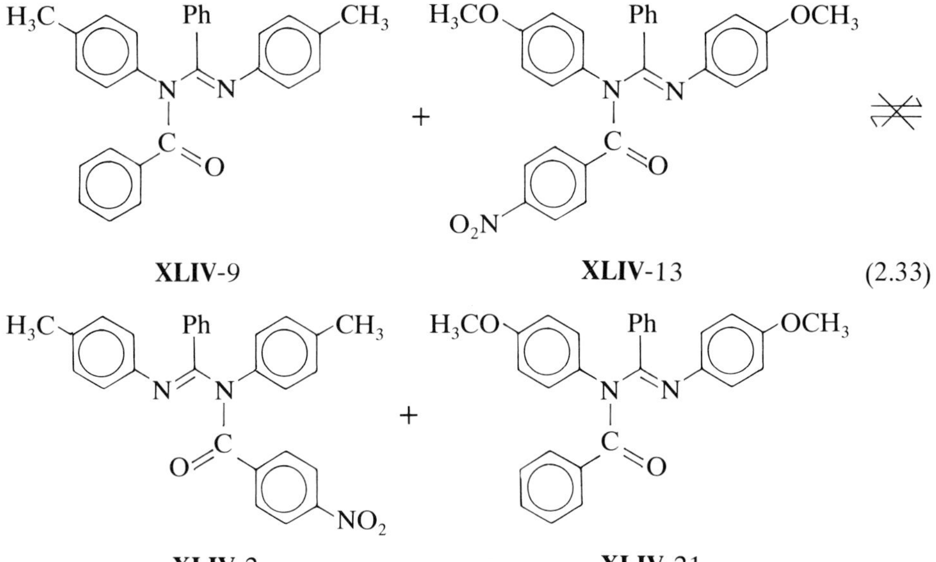

XLIV-9   **XLIV**-13   (2.33)

XLIV-2   **XLIV**-21

Tentatively assuming that the sensitivity of the method permits detection of compounds (**XLIV**-2) and (**XLIV**-21) when their concentration attains 1—2%, and bearing in mind that during 15—30 min at +170°C there occur in amidine (**XLIV**-9) 3.6×10^4 and in amidine (**XLIV**-13) 1.78×10^5 intramolecular *N,N'*-migrations, it can be calculated that for 10^3 intramolecular *N,N'*-migrations there takes place only one intermolecular migration — if any at all.

Analysis of data on nondegenerate tautomeric rearrangements (2.32) $X_1 = X_2$ (compounds Nos. 15, 16, 18—20 in Table 2.I) shows that the electron-acceptor substituents X lower the equilibrium content of tautomers **XLIVa** in which the migrating group is linked with the nitrogen atom closest to the varied substituent. There is an order of 30 difference between the rate constants of the forward and reverse reactions [75, 76]. Unlike the influence of the substituent R in an acyl group correlated by σ^+-constants of the substituents (Table 2.II), the influence of the substituent X on the rate of acyl transfers **XLIVa** → **XLIVb** is correlated by the σ-constants (Table 2.III).

Table 2.III. Structural Correlations for Nondegenerate Reactions (2.32) [75]. Variable X_1 Substituent with $X_2 = CH_3$.

Correlation Equation	r	s
$\Delta G_{25}^{\neq} = 0.12 + 1.21\sigma$	0.986	0.16
$\log k_{150}$ (a ⇌ b) $= 1.26 + 1.08\sigma$	0.988	0.12
$\log k_{150}$ (a → b) $= 1.00 + 2.35\sigma$	0.992	0.06
$\log k_{150}$ (b → a) $= 0.91 + 0.44\sigma$	0.980	0.19

From the data of Tables 2.II and 2.III, it follows that the substituent X_1 (with constant $X_2 = CH_3$) exerts the greatest influence on kinetics of the forward **XLIVa** → **XLIVb** reaction.

As seen from Figures 2.2 and 2.3, the NMR spectra of *N*-acylamidines **XLIV** show quite a complex picture of the appearance of different stereodynamic effects. These are caused by hindered amide (3) and amidine (2) rotations and by a planar inversion of the nitrogen at the azomethine bond (1). Fig. 2.3 illustrates the temperature dependence of the ^{1}H-NMR spectrum of the amidine derivative **XLV** (No 18 in Table 2.I).

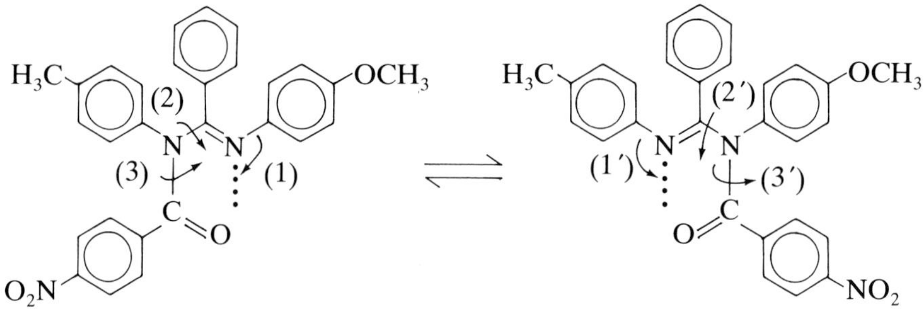

XLVa   XLVb

The amide rotation (3) appears, in agreement with literature data [72, 73, 77—79], to be the most probable mechanism responsible for the splitting of signals of the indicator groups in the temperature range observed. The amidine rotation (2) cannot be observed in the temperature range under investigation since in *N*-arylamidines it is characterized by lower activation barriers than those in the amide rotation (3), namely, $\Delta G^{\neq} < 15$ kcal/mol [71]. Only the presence of alkyl substituents on the nitrogen atoms of amidines and guanidines [79] raises the activation barrier of the amidine rotation (2) to 14—16 kcal/mol. Additional broadening of proton signals of the methol and methoxyl groups which grows appreciably with the rising temperature is thought to be associated with the topomerization (1) contributed to the stereodynamic process $Z \rightleftharpoons E$. Computer modelling of the NMR spectra [75] has made possible an evaluation of the activation barriers of these processes, the values of which are shown in Table 2.IV.

Table 2.IV. Kinetic and Activation Parameters of the Dynamic Effects Recorded in the NMR Spectra of the Amidine **XLV** [75].

Dynamic Process	Tautomeric Form	k_{150}, s^{-1}	$\Delta G^{\neq}_{25}$, kcal/mol
Migration	a → b	5.2	24.5
Migration	b → a	4.2	24.6
(1)	a	1.7×10^3	18.9
(1′)	b	1.3×10^3	19.0
(2), (2′)	a, b	$> 3.5 \times 10^5$	~ 12
(3)	a	3.7×10^4	14.5
(3′)	b	2.0×10^4	14.9

As follows from the data of Table 2.IV, the rotation and inversion processes in the amidines **XLIV**, **XLV** proceed at much higher rates than the acyl rearrangements. Their role consists in the preparation of a molecule for transition to a conformation sterically favorable for the occurence of an intramolecular rearrangement, i.e. 1,3-acyl migration.

3.2. *Stereochemistry of Acyl Migrations*

Judging from their mechanism, the acyl rearrangements and tautomerizations considered above represent reactions of nucleophilic substitution at the sp^2 hybridized carbon atom of the carbonyl group. The stereochemistry of these reactions depends, in the first place, on the direction of approach of the attacking nucleophile to the carbonyl group plane. The pattern of the MERP for a nucleophilic attack can be determined from data obtained from quantum mechanical calculations and structural mapping.

The pathway of a reaction of the simplest nucleophilic reagent, a hydride-ion, with the simplest carbonyl compound, formaldehyde, calculated by *ab initio* as well as semiempirical MINDO/3 methods is shown in Fig. 2.4.

When the hydride-ion is far away from the electrophilic center, the electrostatic

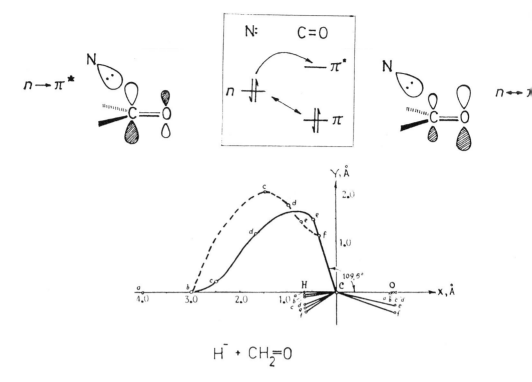

Fig. 2.4. The MERP for nucleophilic addition of a hydride ion to formaldehyde according to *ab initio* [80] (dashed curve) and MINDO/3 [81] (solid curve) calculations. The points a,b,c,d,e,f, denote the positions of the H, C, O atoms of the molecule being attacked and of the H$^-$ atom with the distances between the nucleophile and the carbon atom equalling, respectively, 4 Å, 3 Å, 2.5 Å, 2.0 Å, 1.5 Å and 1.12 Å. In the initial state, the molecule of formaldehyde lies in the *X0Z* plane.

interaction between the reagents prevails, and the nucleophile occupies a position collinear with the C=O bond. For distances less than 3 Å the orbital interaction effects are predominant, *viz.*, the stabilizing *n-π** interaction between the electron lone pair orbital of the nucleophile and the lowest unoccupied MO of the substrate as well as the destabilizing effect of the exchange repulsion of filled orbitals of the reagents. The attacking nucleophile leaves the plane of the carbonyl group and approaches the carbonyl fragment at an angle of about 110°. Fairly close values (± 10°) were obtained in the MERP calculations for the nucleophilic addition reactions involving other carbonyl substrates and other nucleophiles [82—84]. A MERP of the type presented in Fig. 2.4 is also confirmed by the data from the mapping of the reaction path of amino-(hydroxy-) and carbonyl-containing compounds [80, 85, 86]. This reaction path can be traced by analyzing a series of structures with the intramolecular N····C=O interaction ordered according to the diminishing N----C distance. Some of these compounds are given below.

XLVIa
$d_l = 2.910$ Å

XLVIb
$d_l = 2.760$ Å

XLVIc
$d_l = 2.581$ Å

XLVId
$d_l = 2.457$ Å

XLVIe
$d_l = 1.993$ Å

XLVIf
$d_l = 1.640$ Å

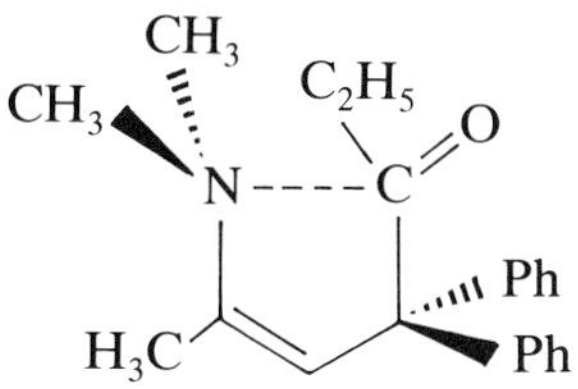

For all the compounds presented above, the N·····C distance is appreciably smaller than the sum of the van der Waals radii, indicating a strong attraction between amine and carbonyl fragments. In the series of compounds **XLVI**, there exists a correlation between the distance N·····C, the length of the C═O bond and the angle of pyramidalization. These dependences are shown in a snapshot series in Fig. 2.5. The angle of approach of the nucleophile to the plane of the carbonyl group, computed from the structural mapping data is 107°.

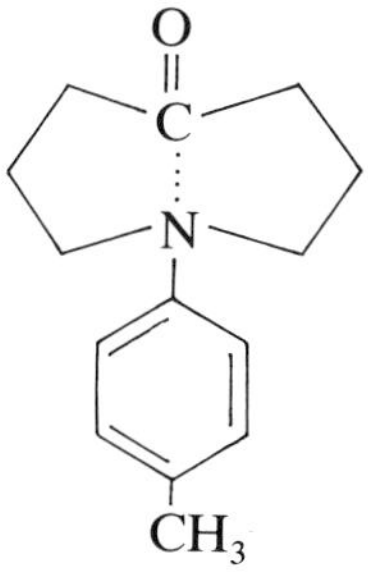

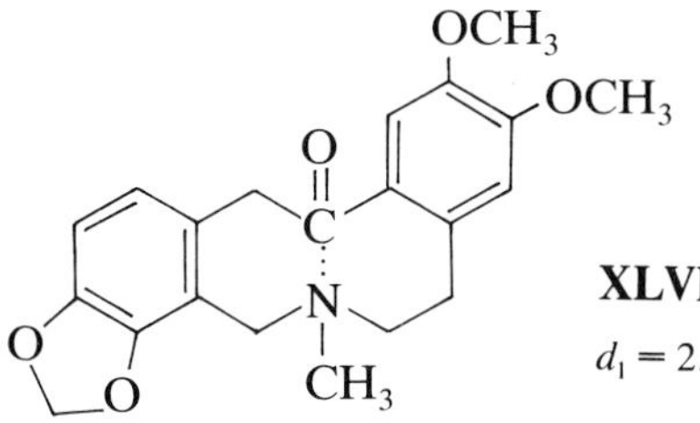

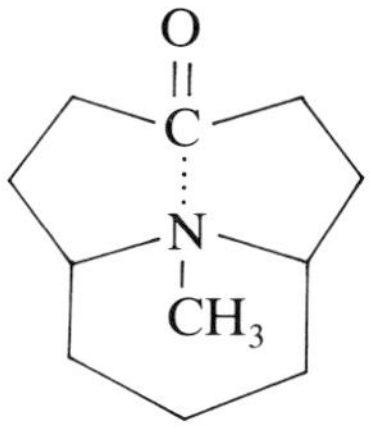

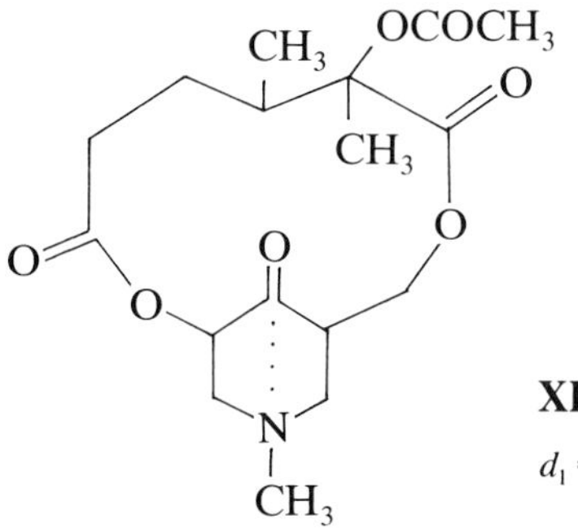

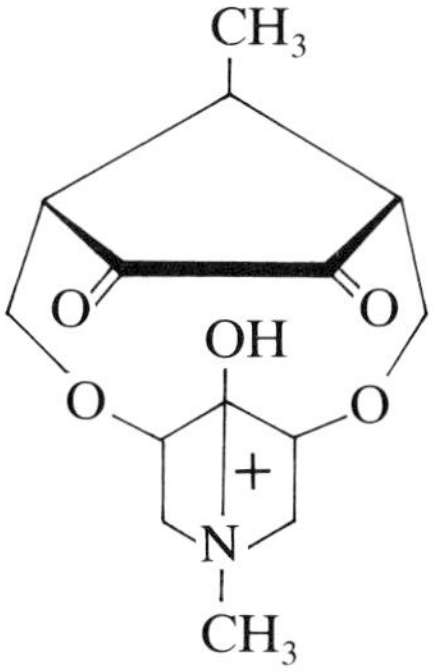

An important question is how stringent is the requirement for optimum directionality of a nucleophilic attack for the reaction to occur. How significant should be the deviations from the MERP so as to inhibit the formation of a tetrahedral intermediate (in the case of an ion—molecular reaction) or a transition state? Calculations by Lipscomb *et al.* [84] gave a theoretical answer to this question. As has been exemplified by the ion—molecular reaction of addition of the methoxyl-anion to formamide, which models reactions between the enzyme trypsine and various substrates, the region of the attractive potential on the

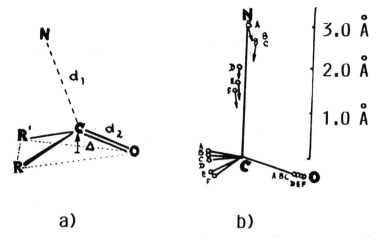

a)   b)

Fig. 2.5. (a) Structural parameters of the reaction site of nucleophilic addition to a carbonyl group according to [80, 87]. When $d_1 \rightarrow \infty$, $\Delta = 0$ and d_2 corresponds to the length of the double bond C=O. (b) Projection of the reaction coordinate onto the NCO plane. The points A, B, C, D, E, F denote the structures **XLIa–f**. The arrows indicate the direction of the orbital axis of the electron lone pair of the nucleophile.

corresponding PES, in which the reaction proceeds without surmounting the energy barrier giving rise to an adduct, is large enough. Making use of the distance O⋯⋯C=O as a reaction coordinate, the authors [84] have found that the reaction funnel representing the region of attractive potential occupies about 20% of the whole hemisphere with its center at the carbonyl carbon (Fig. 2.6).

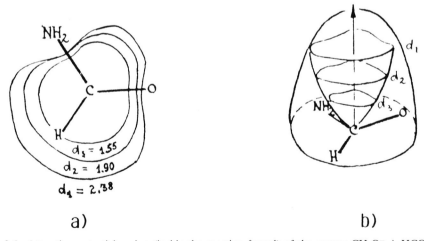

a)   b)

Fig. 2.6. Attractive potential region (inside the reaction funnel) of the system $CH_3O^- + HCONH_2$ constructed for three distances O=C⋯⋯O (in Å) by the PRDDO method from the data of calculations [84]. (a) Projections of null isoenergy curves onto equatorial plane. (b) Reaction funnel showing cross-sections at various d_1, d_2, d_3 values. Outside the reaction funnel is the region of repulsive potential.

Thus, although the optimal angle for a nucleophilic attack is approximately 110°, the reaction channel for the primary stage of nucleophilic substitution is wide enough. Hence, even fairly large deviations from the optimal path along which the reactive centers draw together do not inhibit the reaction. This conclusion is particularly important for the design of intramolecular reactions in which there may occur considerable steric hindrances to optimal angularity in the approach of the attacking group to the carbonyl function. As an example of such reactions, there may be cited the 1,3-acyl transfer in amidines **XLIV** which was examined in the previous Section. Fig. 2.7 shows how the steric requirements of a nucleophilic attack on the carbonyl group are fulfilled in a model intramolecular transacylation reaction (2.34).

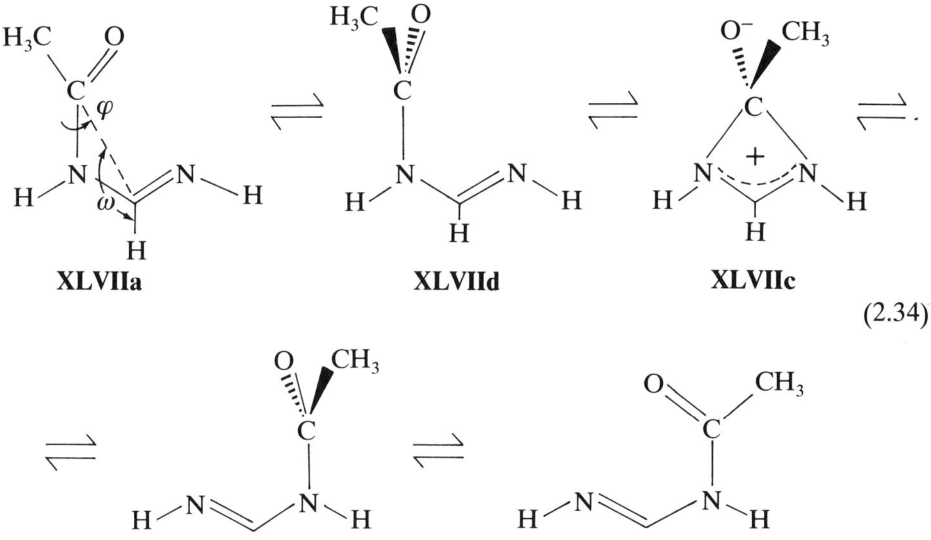

XLVIIa   XLVIId   XLVIIc

(2.34)

XLVIIe   XLVIIb

Calculations [88] of the rearrangement (2.34) carried out by the MINDO/3 method show that the 1,3-transfer of an acetyl group in the amidine system starts with the rotation of the acetyl fragment in a stable planar conformation **XLVIIa** about the C—N bond at an angle $\varphi \simeq 90°$ (conformations **XLVIId, e**). Approximately the same arrangement of the fragments of the molecule — acetyl and amidine — is also retained in the transition state of the reaction ($\omega = 168°$). In these structures, requirements of the MERP for a nucleophilic attack are closely reproduced also in the absence of steric constraints (Fig. 2.4). It should be noted that although in the perpendicular conformation **XLVIId** the angle HN····C=O (plane) equalling 125° considerably exceeds the optimum value, the calculated activation energy of 25 kcal/mol, which is close to the experimental value (Table 2.I), still corresponds to the limit of the tautomeric scale.

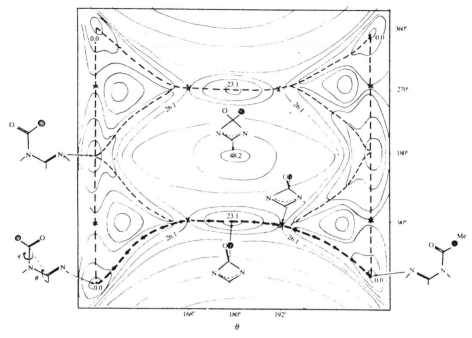

Fig. 2.7. Potential energy surface of the degenerate reaction of 1,3-acetyl transfer (2.34) in the coordinates φ, ω from the data of calculations [88] by the MINDO/3 method. The arrows indicate the minimal energy reaction path and the crosses denote the transition states.

3.3.   Tautomeric 1,5-Acyl Rearrangements of O-Acylenols of 1,3-Diketones

It has been already noted that the *O*-acyl derivatives of the classical tautomeric system of 1,3-diketones (2.7) provide favorable possibilities for the occurence of the acylotropic tautomerism. This fragment is present in the molecules of *O*-acetyl derivatives of naphthazarine **XLI**, which are characterized by slow intramolecular $O \rightleftharpoons O'$-acetyl transfers (2.27) corresponding to a tautomeric equilibrium of very low mobility. A considerable reduction in the activation barrier of degenerate acyl migrations was achieved [80—94] is *cis*-(*Z*)-acylenols of 1,3-diketones **XLVIII** possessing the same structural moiety for the 1,5-shift of an acyl group as **XLI**.

$$\text{XLVIIIa} \rightleftharpoons \text{XLVIIIc} \rightleftharpoons \text{XLVIIIb} \tag{2.35}$$

XLVIIIa   **XLVIIIc**   **XLVIIIb**

Upon acylation of 1,3-diketones with acetyl chloride, an approximately 1 : 1 mixture of the *Z* (**XLIX**) and *E* (**L**) isomers is formed in almost all cases [90—96] so that the ¹H-NMR spectrum reflects, at low temperature, the signals of both isomers (Fig. 2.8). They are readily separable chromatographically [92], but the

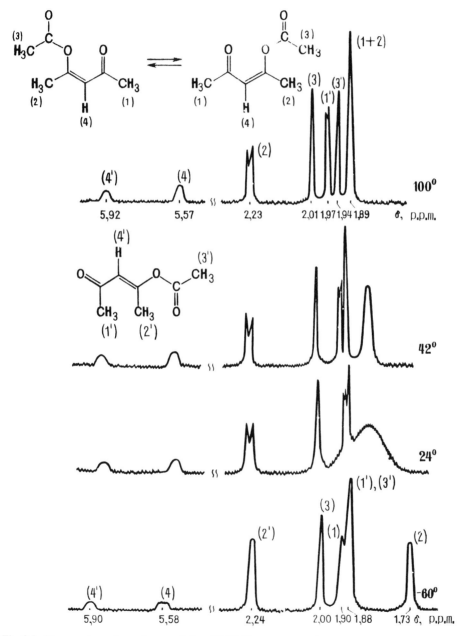

Fig. 2.8. Temperature dependence of the ¹H-NMR spectrum of acetylacetone *Z*- and *E*-enol acetates **XLIX** and **L** in chlorobenzene solution [89, 90].

tautomerization (2.35) can be observed even in the presence of the unreactive
E-isomeric form **L**.

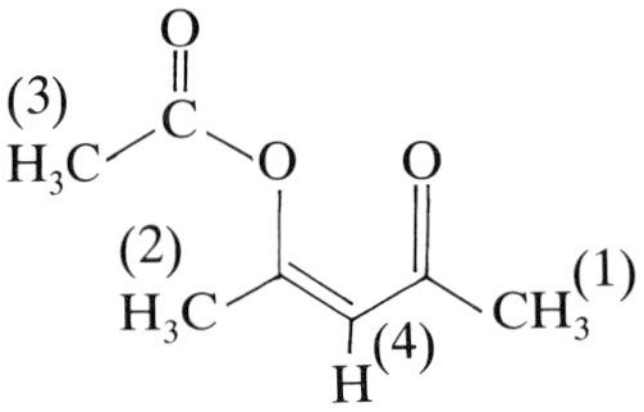

XLIX (*Z*) 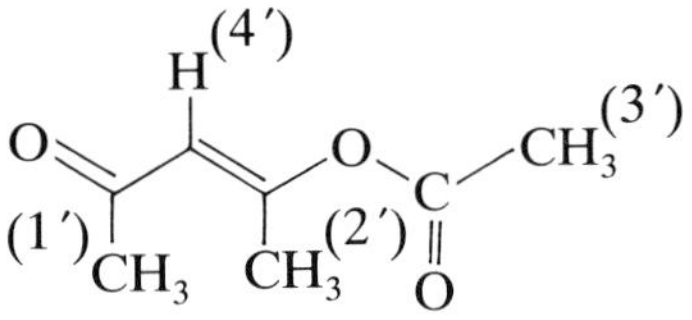   **L** (*E*) 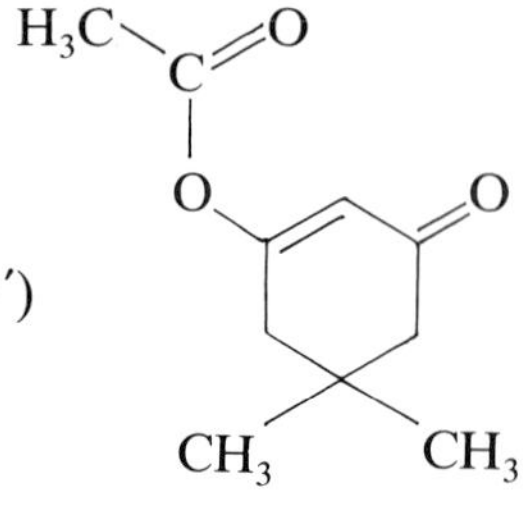   **La**

When the solution temperature is increased, the spectrum of the *E*-isomer
shows no changes similar to the spectrum of *O*-acetyldimedone **La** in which the
E-configuration is rigidly fixed.

On the other hand, the NMR spectra of the *Z*-isomer of **XLIX** with increasing
temperature show an averaging of the signals of the methyl groups of R_1 and R_3,
indicating a rapid (k_{25} = 35 sec^{-1}) *O,O′*-transfer of the acetyl group. The rate of
the reactions (2.35) does not depend on the solution concentration or on the ratio
of *E* and *Z* forms, pointing to intramolecular nature of the reaction. This is also
demonstrated by the fact that, upon mixing and heating (to 100°C) of solutions of
two different acylenols **XLVIIId, e** even after several hours no indications of the
cross-products **XLVIIIf, g** confirming an intermolecular transacylation are found in
the ^{1}H-NMR spectra.

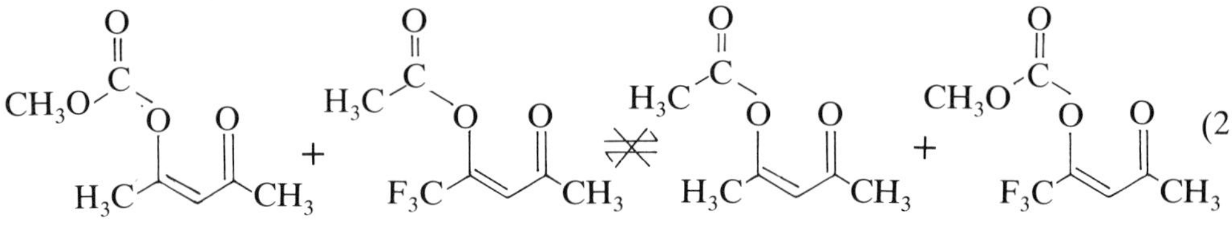

XLVIIId   **XLVIIIe**   **XLVIIIf**   **XLVIIIg**   (2.36)

The simplest spectral picture of the degenerate tautomerization is observed in
the case of the *O*-methoxycarbonyl derivative of acetylacetone **XLVIIId** which is
obtained in the form of the pure *Z*-isomer. Fig. 2.9 shows the ^{1}H-NMR spectra of
compounds **XLVIIId** at 0°C (slow exchange), at 130°C (fast exchange) and at
intermediate exchange rates in the region of the indicator methyl groups. At 0°C,
assignment of the signals of the methyl groups (1) and (2) is facilitated by
observing allyl spin-coupling ($^4J_{HH}$ = 1.1 Hz) of the —CH$_3$ protons (2). Comparison between Figures 2.8 and 2.9 shows that the rate of exchange of the
methoxycarbonyl group positions is significantly smaller than that in the case of
the acetyl group.

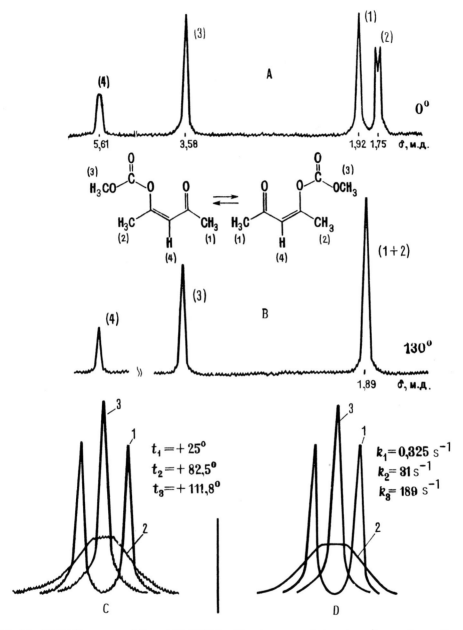

Fig 2.9. ^{1}H-NMR spectra of Z-enol **XLVIIId** in chlorobenzene solution: A — slow exchange, B — very rapid exchange, experimental, and theoretically calculated line shape of indicator C-methyl groups (with H(4) spin-decoupling).

Table 2.V contains data on kinetics of both degenerate and non-degenerate tautomeric acyl rearrangements (2.35) obtained by the dynamic NMR method.

Table 2.V. Rate Constants and Free Activation Energy of the Acyl Rearrangements (2.35). (Solvent: chlorobenzene) [90, 91, 93].

| No. | Compound **XLVIII** | | | | | $\Delta G_{25}^{\neq}$, |
	Migrant C(O)R	R_1	R_2	R_3	k_{25}, s^{-1}	kcal/mol
1	CH_3	CH_3	H	CH_3	34.7	15.3
2	$C(CH_3)_3$	CH_3	H	CH_3	43.2	15.2
3	OCH_3	CH_3	H	CH_3	0.3	18.1
4	NEt_2	CH_3	H	CH_3	6.5×10^{-5}	23.1
5	$C_6H_4OCH_3$-p	CH_3	H	CH_3	0.8	17.6
6	$C_6H_4CH_3$-p	CH_3	H	CH_3	2.68	16.8
7	$C_6H_4OCH_3$-m	CH_3	H	CH_3	5.31	16.4
8	C_6H_5	CH_3	H	CH_3	6.8	16.3
9	C_6H_4Cl-p	CH_3	H	CH_3	11.1	16.0
10	C_6H_4Br-m	CH_3	H	CH_3	23.0	15.6
11	$C_6H_4NO_2$-m	CH_3	H	CH_3	58.5	15.0
12	$C_6H_4NO_2$-p	CH_3	H	CH_3	90	14.7
13	OCH_3	CH_3	CH_3	CH_3	0.8	17.4
14	OCH_3	CH_3	COOEt	CH_3	3.3	16.7
15	OCH_3	CH_3	SEt	CH_3	0.6	17.7
16	OCH_3	CH_3	Cl	CH_3	7×10^{-2}	19.0
17	CH_3	C_6H_5	H	CH_3	15.3	15.8
18	CH_3	CF_3	H	CH_3	1.8	17.1
19	CH_3	C_6H_5	H	CF_3	4.8	16.7
20	OCH_3	C_6H_5	H	CH_3	0.9	17.5
21	$C(CH_3)_3$	C_6H_5	H	CH_3	47.2	15.1

The rate of the O,O'-migration of acyl groups in the compounds **XLVIII** depends on the degree of electron deficiency of the carbonyl carbon atom of the migrant. As can be seen from the data of Table 2.V, the migration frequency increases by six orders of magnitude when going from a diethylcarbamoyl deriva-tiver (No. 4) to an acetyl derivative (No. 1). Electron donor substituents in the phenyl ring also significantly retard the migration of aroyl groups in **XLVIII** (R = Aryl). Fig. 2.10 shows the correlation dependence of the (2.35) rearrangement rates on the σ^+-constants of the substituents obtained.

The positive sign of the reaction constant ρ is unusual in the correlation equation which includes this type of substituent constants. It points to the decisive role which the positive charge at the carbonyl carbon has in facilitating the stabilizing $n \to \pi^*$ interaction between the reacting groups (Fig. 2.4).

This, in fact, accords with the CNDO/2 and MINDO/3 calculations of electron disributions in the ground and transition states of the rearranging compounds **XLVIII**. Calculation of reaction pathways of the 1,5-acyl O,O'- and $N,N,'$-transfer in the model compounds **LI** reveals the same types of intramolecular reorganiza-tions leading to formation of a transition state structure [88]. The initial stage of an intramolecular reaction is represented by the low-barrier *s-cis—trans-*isomerization of the stable conformation **LIa** with simultaneous 90° rotation of the

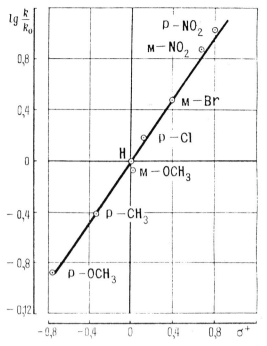

Fig. 2.10. Correlation of the rates of the reversible rearrangements (2.35) ($R_1 = R_3 = CH_3$; $R_2 =$ H = Aryl) of Z-O-aroylacetylacetones **XLVIII** at 25°C, $\rho = 1.34 \pm 0.12$.

acyl group plane. The transition state of the **LIc** acyl transfer reaction retaining only one symmetry plane is formed from the sterically adjusted metastable (local minimum on the PES) conformer **LId**.

(2.36)

The effect of substituents in the 1(3) and 2 positions of the Z-acylenols **XLVIII** is less pronounced. π-Acceptor substituents R_2 somewhat increase the frequency of acyl exchange and π-donors retard it. Acceptors in the 1(3) position also retard the migration process [91, 93].

In undegenerate tautomeric systems of type **XLVIII** (Nos. 17—21 in Table 2.V), the position of the tautomeric equilibrium remains practically unchanged over the temperature range of −50 to 150°C, which indicates a value of the entropy factors close to zero. For benzoylindenyl acylates, however, whose thermal rearrangements were investigated by Hartke [94, 95, 97—99], the temperature has a significant effect on the ratio of O-acyl tautomers, which may be seen from the thermodynamic parameters of the following tautomeric rearrangement [99].

$$(2.37)$$

LIIa	**LIIb**

$\Delta H^{\neq} = 14.4 \pm 0.4 \text{ kcal/mol}$ $\Delta H° = 0.44 \pm 0.08 \text{ kcal/mol}$
$\Delta S^{\neq} = -10.5 \pm 1.2 \text{ e.u.}$ $\Delta S° = 3.7 \pm 0.3 \text{ e.u.}$
$\Delta G^{\neq}_{25} = 17.6 \pm 0.6 \text{ kcal/mol}$ $\Delta G°_{25} = -0.66 \text{ kcal/mol}$

As there is a considerable separation of charges (*cf.* **LIc**) in the transition state of the acyl migration reaction, it could be assumed that the reaction rate would depend on polarity of the solvent. A specific investigation of the dependence of the methoxycarbonyl O,O'-migration rates in Z-enol **XLVIIId** has shown the solvent effect on the frequency of intramolecular O,O'-migrations to be insignificant. Data obtained are given in Table 2.VI.

A much stronger influence upon the rate of acyl migrations is exerted by general acid catalysis, however, only when large amounts of sufficiently strong acids are applied.

A considerable acceleration of the O,O'-acyl migrations in **XLVIIIg** is also achieved in the presence of lithium perchlorate as well as europium tris(dipivaloylmethane).* Clearly, the effect of a catalyst will depend on the position of the equilibrium.

For the acceleration of intramolecular O,O'-acyl migrations to occur, it is necessary that the equilibrium (2.38) be shifted to the left, whereas the stabilization of the complex **LIIIb** in which the nucleophilic center is blocked would rather

---

* In the work [100] an erroneous interpretation of the Eu(DPM)₃ effect is given, resulting from the erroneous assumption of the existence of a stable type **XLVIIIc** cyclic structure.

Table 2.VI. The Solvent Influence on Kinetics of O,O'-Migration of Methoxycarbonyl Group (2.35) in Compounds **XLVIII** (R = OCH$_3$, R$_1$ = R$_3$ = CH$_3$, R$_2$ = H).

No.	Solvent	k_{25}, s^{-1}	$\Delta H^{\neq}$, kcal/mol	$\Delta S^{\neq}$, e.u.	$\Delta G_{25}^{\neq}$, kcal/mol
1	None	0.44	12.9 ± 0.3	−16.8 ± 0.7	17.9
2	Methoxyethanol	0.10	18.8 ± 0.3	0.0 ± 1.2	18.8
3	Pyridine	0.16	19.0 ± 0.2	1.7 ± 1.0	18.5
4	Perfluorotoluene	0.22	17.6 ± 0.2	−2.4 ± 0.6	18.3
5	Hexachlorobutadiene	0.26	17.7 ± 0.2	−1.9 ± 0.5	18.2
6	Chlorobenzene	0.33	15.7 ± 0.1	−7.8 ± 0.3	18.1
7	*o*-Dichlorobenzene	0.53	13.9 ± 0.2	−12.2 ± 0.4	17.8
8	Dioxane	0.75	13.5 ± 0.3	−13.8 ± 0.5	17.6
9	Dimethylsulfoxide	1.19	14.4 ± 0.3	−10.0 ± 0.6	17.4
10	Nitrobenzene	1.49	14.0 ± 0.2	−11.2 ± 0.4	17.3
11	Ethanol	1.76	15.8 ± 0.3	−4.3 ± 0.6	17.1
12	Benzonitrile	2.11	13.8 ± 0.3	−10.6 ± 0.5	17.0

Table 2.VII. General Acid Catalysis of O,O'-Migration of Methoxycarbonyl Group in Compound **XLVIII** (R = OCH$_3$, R$_1$ = R$_3$ = CH$_3$, R$_2$ = H) [2].

Catalyst	Solvent	k_{25}, s^{-1}	$\Delta H^{\neq}$, kcal/mol	$\Delta S^{\neq}$, e.u.	$\Delta G_{25}^{\neq}$, kcal/mol
—	Chlorobenzene	0.33	15.7 ± 0.1	−7.8 ± 0.3	18.1
CH$_3$COOH (1 : 1)	Chlorobenzene	70.0	15.9 ± 0.2	3.1 ± 0.6	15.0
—	Hexachlorobutadiene	0.26	17.7 ± 0.2	−1.9 ± 0.5	18.2
(CH$_3$)$_3$CCOOH(1 : 1)	Hexachlorobutadiene	0.88	17.5 ± 0.2	0.0 ± 0.8	17.5

$$\text{(2.38)}$$

LIIIa　　　　　**LIIIb**　　　　K = H, Li, Eu(DPM)$_3$

inhibit a tautomeric rearrangement. On the other hand, coordination to the carbonyl oxygen atom in **LIIIa** substantially enhances the electrophility of the neighboring carbon which, as was shown earlier (Fig. 2.10), leads to a lowering of the activation energy barrier against acyl migrations. Acceleration of acyl migrations can also be easily explained in terms of orbital interactions by taking into account the mechanism of this intramolecular reaction (Fig. 2.11).

Addition of a proton or metal ion to a carbonyl group results in the lowering of its π_{CO} as well as π_{CO}^* energy levels which leads to narrowing of the energy gap

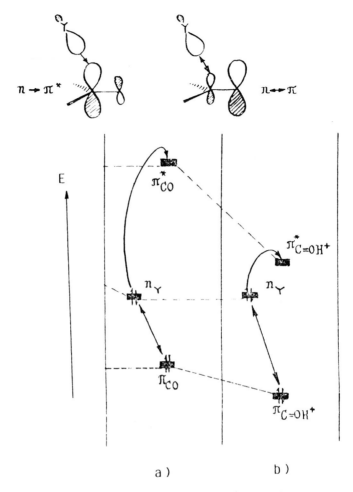

Fig. 2.11. Orbital interactions determining the formation of the Y—CO bond in the nucleophilic addition of Y to a carbonyl group: (a) uncatalyzed intramolecular nucleophilic attack in **XLVIII**; (b) reaction catalyzed by the coordination of a proton or a Lewis acid to the oxygen atom of the migrating acyl group.

between π and π^* orbitals (attractive interaction). In contrast, the repulsive n-π^* interaction diminishes due to moving apart of these energy levels. Both these effects strongly favor, in accordance with the theory of orbital interactions [101], reactions of intramolecular nucleophilic substitution in the type **LIIIa** complexes.

3.4. *Acylotropy of 9-Acyloxyphenalen-1-Ones*

9-Acyloxyphenalen-1-ones **LIV** contain the same reaction site as the acyl derivatives of naphthazarin and acetylacetone.

$$ \text{(2.39)} $$

LIVa   LIVc   LIVb

A specific feature of the molecular system **LIV** is its stereochemical rigidity due to the linking of the reaction site to the phenalene moiety inhibiting all possible rotations in the acyloxyenone moiety except for the low-barrier C—O rotations. Table 2.VIII contains kinetic characteristics of the degenerate tautomeric rearrangement (2.39) obtained by the dynamic ^{1}H-NMR method. They are compared with the data on analogous derivatives of acetylacetone *cis*-enols. Fig. 2.12 shows variation of the ^{1}H-NMR spectrum of 9-methoxycarbonylphenalen-1-one when the solution temperature is raised under conditions of rapid exchange (2.39).

Table 2.VIII. Kinetic and Activation Parameters of O,O'-Migration of Acyl Groups in Compounds **XLVIII** ($R_1 = R_3 = CH_3$, $R_2 = H$) and **LIV** according to ^{1}H-NMR investigation [102]. (Solvent: tetrahydrofuran).

Compound	k_{25}, s^{-1}	$\Delta H^{\neq}$, kcal/mol	$\Delta S^{\neq}$, e.u.	$\Delta G^{\neq}_{25}$, kcal/mol
XLVIII, R = CH$_3$	35	14.3 ± 0.2	−3.3 ± 0.4	15.3
LIV, R = CH$_3$	49	9.1 ± 0.1	−20.2 ± 0.3	15.1
XLVIII, R = OCH$_3$	0.22	17.6 ± 0.1	−2.4 ± 0.3	18.3
LIV, R = OCH$_3$	0.50	11.8 ± 0.3	−20.0 ± 1.0	17.8

It can be seen that the rigid fixation of the reaction site in compounds **LIV** does not lead to any appreciable changes in the rates and the free energy of activation of the reaction (2.39) as compared to (2.35) even though very significant negative values of the entropy of activation have been calculated for the former. This might have been explained by the assumption that the achievement of the intermediate structure **LIVc** of C_s-symmetry on the reaction path (2.39) should require a substantial structural reorganization during averaging of the bond lengths in the hydrocarbon fragment. Admittedly, in the formal sense the reaction (2.39) corresponds to a 1,13-sigmatropic shift thus requiring realternation of six single and double bonds. It appears, however, that the structural reorganization of this fragment resulting from the acyl transfer **LIVa** $\rightleftharpoons$ **LIVb** should not be that

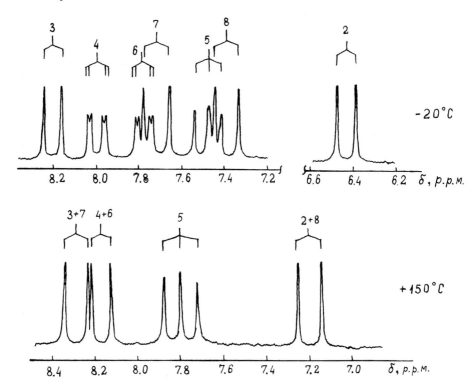

Fig. 2.12. ^{1}H-NMR spectrum of compound **LIV** (R = OCH$_3$) at slow (in tetrahydrofuran solution) and rapid (in dimethylacetamide solution) acyl migration [102]. $J_{23} = J_{78} = 10$ Hz, $J_{45} = J_{56} = 7.8$ Hz, $J_{46} = 2.2$ Hz

significant. This may be corroborated by a comparison with the reaction of an intramolecular proton transfer in 9-hydroxyphenalen-1-one. Though the double-minimum potential originating from an unsymmetrical hydrogen bond was proved for this compound by a variety of experimental techniques including X-ray crystallography at ambient temperature [103] and vibronically resolved laser-excited fluorescence in an argon matrix at 4 K [104] and by *ab initio* (STO-3G) calculations [105] as well, the interminimal distance is very small (in the range of 0.4—0.64 Å). As may be seen from the calculated positions of the atoms in the carefully optimized geometries relative to the center of the mass, only those appertaining to the upper hydroxyenone moiety of the rearranged molecule are affected by the proton shift. The symmetrical structure is thus neither a minimum nor a transition state, and this is possibly also true for the intramolecular acyl transfer reaction (2.39).

3.5. *Tautomerism of 2-Acyloxytropones*

The *O*-acyl derivatives of tropolone **LV** as well as of 9-acyloxyphenalen-1-ones

LIV possess a rigid molecular framework holding two nucleophilic centers in a fixed position. Only two nonrigid modes of rotation about the C—O bonds are retained which fit the reaction site to an optimal structure for the transition state of the acyl group transfer reaction.

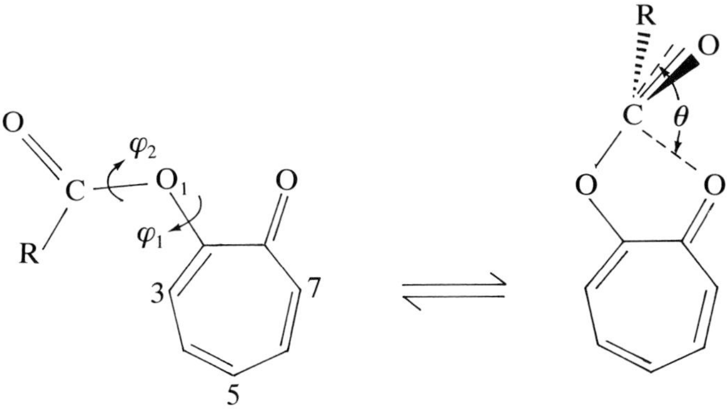

$$(2.40)$$

LVa$_1$ **LVa**

According to an X-ray study [106], the conformation **LVa**$_1$ is realized in the case of the *p*-chlorobenzoyl derivative of tropolone (R = C_6H_4Cl-*p*). Assuming the band lengths and valence angles in the conformation **LVa** to be identical to those in **LVa**$_1$, one may calculate that the distance =O····C=O equals 2.2 Å while the angles between the direction =O····C and the C(O)R plane in the conformation **LVa** are $\theta = 99°$, $\varphi_1 = 0°$.

In the case of 3,7-disubstituted 2-acyl-oxytropones, the conformation **LVa**$_1$ is sterically inhibited, and the equilibrium (2.40) must be shifted toward **LVa**. The data of an X-ray study [107] of 3,5,7-trimethyl and 3,5,7-tribromo derivatives of 2-acetoxytropone presented in Fig. 2.13 show that, even though the distances O····C=O are actually greater than 2.2 Å, they are still smaller than the sum of the van der Waals radii of oxygen and carbon (3.0 Å) and the angle $\theta \simeq 110°$ is close to the optimal value for an intramolecular nucleophilic attack.

Thus, in the series of the tropolone derivatives, one may expect to find the most rapid intramolecular O,O'-migrations of the acyl groups. Formally, they may be ranked as 1,9-sigmatropic rearrangements.

In full accordance with expectations based on stereochemical preperties of the tropolone derivatives, the most mobile acylotropic systems of those known at the present time, with a frequency of acyl migrations of up to 10^6 sec^{-1} and higher, even at room temperature, were indeed found in this series of compounds [68, 108—112].

The results of kinetic investigations of the rearrangements (2.41) by the method of the dynamic ^{1}H and ^{13}C-NMR are shown in Table 2.IX.

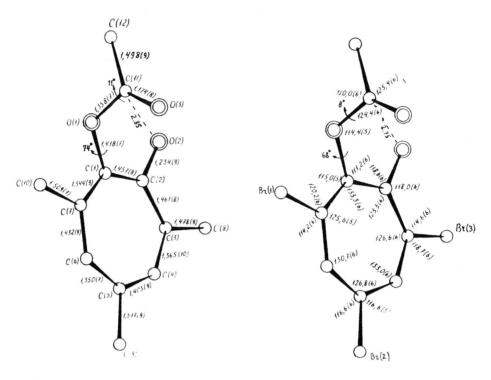

Fig. 2.13. Molecular structures of 2-acetoxy-3,5,7-trimethyltropone and 2-acetoxy-3,5,7-tribromot-ropone from the data of an X-ray study [107].

$$\text{LVa} \quad \rightleftharpoons \quad \text{LVc} \quad \rightleftharpoons \quad \text{LVb} \quad (2.41)$$

LVa   **LVc**   **LVb**

In the series of tropolone derivatives **LV** the dependence of the rate of acyl migrations on the electronic effect of the R group in the acyl migrant is expressed even more strongly than in the case of the compounds **XLIV** and **XLVIII**. The migration frequency of the acyl group is reduced by eight orders of magnitude upon transition from an acetyl ($R = CH_3$) to a carbamoyl ($R = N(CH_3)_2$) migrant, which is related to the decrease in electrophilicity of the carbonyl carbon. On the other hand, for the trifluoroacetyl derivatives **LV** ($R = CF_3$), the *O,O*-migration of

Table 2.IX. Kinetic and Activation Parameters of *O,O'*-Rearrangement of Acyl Groups (2.41) [110, 112, 113].

No.	Compound **LV**				k_{25}, s^{-1}	$\Delta H^{\neq}$, kcal/mol	$\Delta S^{\neq}$, e.u.	$\Delta G^{\neq}_{25}$, kcal/mol
	Migrant **COR**	R_1	R_2	R_3				
1	CH$_3$[a]	H	H	H	2.4×10^5	10.1 ± 0.4	0.0 ± 0.9	10.1
2	OCH$_3$	H	H	H	2.1×10^3	13.8 ± 0.5	-0.3 ± 1.3	13.9
3	N(CH$_3$)$_2$	H	H	H	2.1×10^{-2}	19.6 ± 0.5	-0.4 ± 1.2	19.7
4	OCH$_3$	CH$_3$	H	H	8.2×10^3	—	—	12.2
5	N(CH$_3$)$_2$	CH$_3$	H	H	5.9×10^{-2}	19.1 ± 0.3	0.0 ± 1.0	19.1
6	CH$_3$	CH$_3$	CH$_3$	CH$_3$	1.2×10^6	9.2 ± 0.2	0.0 ± 1.1	9.2
7	OCH$_3$	CH$_3$	CH$_3$	CH$_3$	5.6×10^4	11.0 ± 0.2	0.1 ± 1.2	11.0
8	N(CH$_3$)$_2$	CH$_3$	CH$_3$	CH$_3$	3.0×10^{-2}	19.5 ± 0.2	0.0 ± 1.1	19.5
9	C$_6$H$_4$OCH$_3$-*p*	CH$_3$	CH$_3$	CH$_3$	1.0×10^4	12.0 ± 0.3	-0.3 ± 1.2	12.0
10	OCH$_3$	H	Br	H	72	14.8 ± 0.2	-0.3 ± 1.3	14.9
11	OCH$_3$	Br	H	Br	18	15.7 ± 0.3	0.1 ± 0.8	15.7
12	OCH$_3$	Br	Br	Br	22	15.6 ± 0.2	0.0 ± 0.6	15.6
13	CH$_3$	CH$_2$C$_6$H$_5$	H	CH$_2$C$_6$H$_5$	3.1×10^6	8.3 ± 0.5	-1.1 ± 1.6	8.6
14	N(CH$_3$)$_2$	CH$_2$C$_6$H$_5$	H	CH$_2$C$_6$H$_5$	3.8×10^{-1}	18.0 ± 0.3	0.0 ± 1.0	18.0
15	OCH$_3$	CH$_2$C$_6$H$_5$	H	CH$_2$C$_6$H$_5$	9.7×10^3	12.0 ± 0.3	-0.1 ± 0.7	12.0
16	CH$_3$	CH$_2$C$_6$H$_5$	H	CH$_2$C$_6$H$_5$	5×10^7	—	—	< 7.0

[a] According to the data of [114]

the acyl cannot be 'frozen' even at −120°C (compound 16 in Table 2.IX). The characteristic v_{CO} band of the trifluoroacetyl group is not detected in the IR spectrum of *O*-trifluoroacetyltropolone, which might be isolated by crystallization only in the form of a stable H-complex of the intermediate form **LVc** with a tetrahedral carbon atom. The number of compounds of this type known at the present time is very small [74].

The effect of substituents in the seven-membered ring shows up most strongly in an acceleration of the *O,O'*-acyl migrations of the bulky 3- and 7-substituents. The reason for this is evidently the fact that 3(7)-substituents stabilize an *s-cis* conformation **LVa**, in which the closest approach of the carbonyl carbon to the tropone oxygen is achieved. This readiness of the reactive center of the tropolone derivatives for rearrangement determines the small contribution of the entropy term to the free activation energy of acyl transfer (Table 2.IX).

As follows from the preceding analysis, transfer of the acyl group is preceded by formation of an orthogonal conformation of the type **LVa**. Whereas, for acetyl (R = CH$_3$) and aroyl (R = Ar) derivatives, transition to an orthogonal conformation requires the overcoming of only small torsion barriers, for carbomethoxy (R = OCH$_3$) and carbamoyl (R = N(CH$_3$)$_2$) derivatives, these barriers are quite high. In the spectra of carbomoyl derivatives, besides the acyl migration, there is also observed the dynamic process of a hindered amide rotation (Fig. 2.14).

LVIa   **LVIb**

From the comparison of rate constants of the above processes observed separately in the NMR spectra, it can be estimated that during one O → O′ shift of a carbonyl group (at 25°C) there occur no fewer than 200 rotations of the dimethylamino group. It can be asserted that the withdrawal of this group out of the plane of the amide fragment is an important factor facilitating the act of acyl migration. Since in the nonplanar conformation **LVIb** the electron lone pair of nitrogen does not participate in the conjugation with the carbonyl group, the electron deficiency of the carbonyl carbon atom is enhanced, which accelerates the migration of the carbonyl group. Apparently, this conformational effect plays an important part in the acceleration of nucleophilic reactions of amides, including the biochemical reactions of peptide hydrolysis and the transacylation.

The most complete information on the stereodynamic processes preceding the acyl migration **LVa** → **LVb** has been derived by studying the temperature-variable ^{1}H-NMR spectra of 2-acyloxy-3,7-dibenzyltropones (R$_1$ = R$_3$ = CH$_2$Ph, R$_2$ = H) [109, 112].

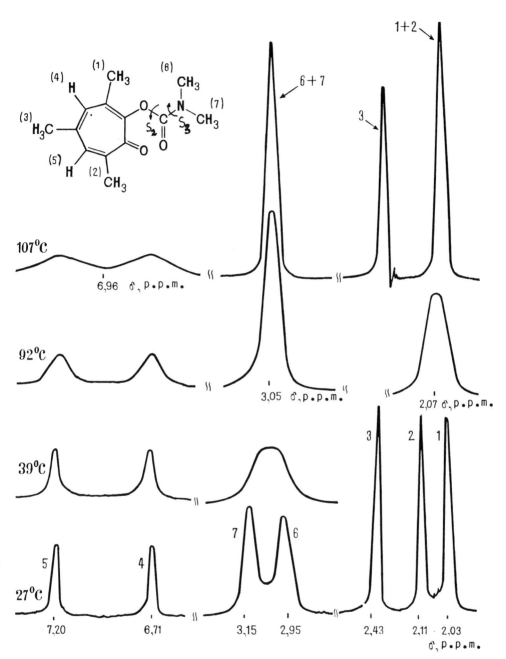

Fig. 2.14. Temperature-variable ¹H-NMR spectrum of *O*-carbamoyl-3,5,7-trimethyltropolone **LV**
(R = NMe₂, R₁ = R₂ = R₃ = Me) in perfluorotoluene solution [110].

The observation of diastereotopy of both methylene groups in the ¹H-NMR
spectra of these compounds at low temperatures has made it possible to carry
out — using a computer simulation of their spectra — an evaluation of the

stereodynamic processes associated with rotations about all single bonds in the acyloxy moiety. The results presented in Table 2.X show that the fast migrations of an acyl group take place only under conditions of complete conformational mobility of the acyloxy fragment, ensuring an adjustment of the reactive center of the system to the steric demands of the transition state structure.

Table 2.X. Kinetic and Activation Parameters of the Dynamic Processes Recorded in the NMR Spectra of the 2-Acyloxy-3,7-Dibenzyltropones **LV** ($R_1 = R_3 = CH_2C_6H_5$; $R_2 = H$) [112].

R	Type of Dynamics	k_{25}, s^{-1}	$\Delta H^{\neq}$, kcal/mol	$\Delta S^{\neq}$, e.u.	$\Delta G_{25}^{\neq}$, kcal/mol
	Acylotropy	0.38	18.0 ± 0.2	0.0 ± 0.9	18.0
$N(CH_3)_2$	Amide Rotation (S_3)	18.3	12.1 ± 0.3	−12.0 ± 0.8	15.7
	Acyloxy Rotation (S_1)	2.16×10^3	11.4 ± 0.4	−5.1 ± 1.2	12.9
	Acylotropy	0.97×10^4	12.0 ± 0.3	−0.1 ± 0.7	12.0
OCH_3	S_1 rotation	2.24×10^4	10.1 ± 0.3	−4.8 ± 0.8	11.5
	S_2 rotation	8.58×10^4	10.0 ± 0.3	−2.4 ± 0.7	10.7
CH_3	Acylotropy	3.1×10^6	8.3 ± 0.5	−1.1 ± 1.6	8.6
	S_1, S_2 rotation	3.1×10^6	−	−	8.4

4. Tautomeric Rearrangements of Aryl Groups

An analysis of *i,j*-rearrangements of the acyl and aryl groups (Section 2 of this Chapter) reveals the common character of their stereochemistry and mechanism. The migration of an aryl group between two nucleophilic centers corresponds to the intramolecular substitution at the sp^2-hybridized carbon atom of the aromatic ring. The trajectory of a nucleophilic attack on this atom is analogous to the trajectory for a nucleophile approaching the sp^2-hybridized carbon atom of a carbonyl group (Fig. 2.4). When the aryl group contains strong electron-accepting substituents, the intermediate tetrahedral structures emerging upon addition of the nucleophile will be highly stabilized. If the attacking nucleophile is an anion, such intermediate structures, known as σ-complexes or Meisenheimer—Jackson complexes, are quite stable and can be isolated preparatively [115—117]. Thus, it may be expected that when acyl groups in type **XLIV**, **XLVIII**, **LV** acylotropic systems are substituted by aryl groups activated with electron-accepting substituents, it will lead to another type of carbonotropic tautomeric compound in which there may be observed rapid rearrangements of aryl groups, i.e. moieties of CH-acids.

4.1. 1,3-Rearrangements of Aryl Groups in the Derivatives of N,N'-diarylamidines

The first examples of rapid and reversible intramolecular rearrangements of aryl groups were found in the series of *N*-(2,4,6-trinitrophenyl) amidines **LVII** [118—119].

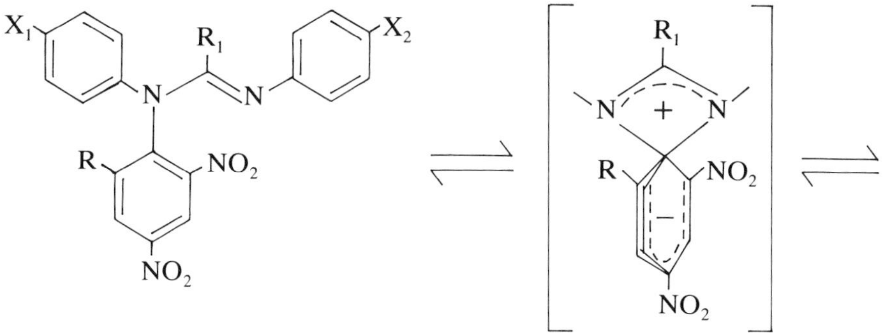

| **LVIIa** | **LVIIc** | (2.42) |

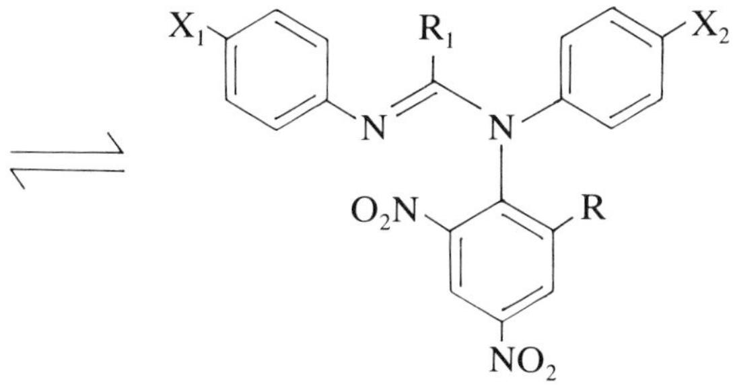

LVIIb

The results of a kinetic investigation of the degenerate and nondegenerate rearrangements (2.42) are presented in Table 2.XI. As can be seen from Fig. 2.15, *N,N'*-diarylamidines exist in solution as *E*-isomers. The (2.42) migrations are easily detected by the line shape of the indicator methyl group.

By comparison with the data for *N,N'*-acyl migrations in the amidines **XLIV** it can be seen that the activation energies of the migrations of picryl groups in **LVII** are lower by 4—6 kcal/mol. Upon transition from a trinitrophenyl migrant in **LVII** (R = NO$_2$) to a 2,4-dinitrophenyl migrant (R = H), the rate of the rearrangements **LVIIa** ⇌ **LVIIb** decreases by approximately six orders of magnitude (compounds

Table 2.XI. Kinetic and Activation Parameters of the Carbonotropic Rearrangements (2.43). (Solvent: Chlorobenzene) [75, 119].

No.	Compound LVII				k_{25}, s^{-1}	$\Delta H^{\neq}$, kcal/mol	$\Delta S^{\neq}$, e.u.	$\Delta G^{\neq}_{25}$, kcal/mol	P_a	P_b
	R_1	X_1	X_2	R						
1	$C_6H_4OCH_3$-p	CH_3	CH_3	NO_2	1.95	15.7 ± 0.1	-4.3 ± 0.3	17.0	0.5	0.5
2	C_6H_5	CH_3	CH_3	NO_2	0.69	16.3 ± 0.1	-4.4 ± 0.2	17.6	0.5	0.5
3	C_6H_4Br-p	CH_3	CH_3	NO_2	0.31	16.2 ± 0.2	-6.4 ± 0.4	18.1	0.5	0.5
4	$C_6H_4NO_2$-p	CH_3	CH_3	NO_2	0.08	17.3 ± 0.3	-5.5 ± 0.9	18.9	0.5	0.5
5	$C_{10}H_7$-1	CH_3	CH_3	NO_2	0.69	16.2 ± 0.1	-4.6 ± 0.4	17.6	0.5	0.5
6	OCH_3	OCH_3	OCH_3	NO_2	1.25	15.5 ± 0.2	-6.0 ± 0.5	17.3	0.5	0.5
7	C_6H_5	CH_3	CH_3	H	1.9×10^{-6}	25.2 ± 0.6	0.2 ± 1.2	25.2	0.5	0.5
8	H	OCH_3	OCH_3	NO_2	10^{-7}	—	—	>27.0	0.5	0.5
9	$C_6H_4OCH_3$-p	CH_3	H	NO_2	1.23	16.7 ± 0.1	-2.1 ± 0.4	17.4	0.6	0.4
10	C_6H_5	CH_3	H	NO_2	0.69	15.6 ± 0.2	-6.8 ± 0.6	17.6	0.6	0.4
11	C_6H_4Cl-p	CH_3	H	NO_2	0.27	16.8 ± 0.2	-4.7 ± 0.4	18.2	0.6	0.4
12	$C_6H_4NO_2$-p	CH_3	H	NO_2	0.01	20.0 ± 0.3	-0.4 ± 0.7	19.9	0.6	0.4
13	$C_{10}H_7$-1	CH_3	H	NO_2	0.87	15.7 ± 0.1	-6.2 ± 0.4	17.5	0.7	0.3
14	C_6H_5	CH_3	H	H	4×10^{-7}	26.0 ± 0.4	0.3 ± 1.7	25.9	0.5	0.5
15	OCH_3	OCH_3	CH_3	NO_2	1.76	15.8 ± 0.3	-4.5 ± 0.9	17.1	0.6	0.4
16	C_6H_5	Br	CH_3	NO_2	0.25	16.9 ± 0.8	-4.0 ± 0.8	18.4	0.25	0.25
17	C_6H_5	NO_2	CH_3	NO_2	1.5×10^{-2}	18.1 ± 0.4	-6.0 ± 1.0	19.9	0.1	0.9

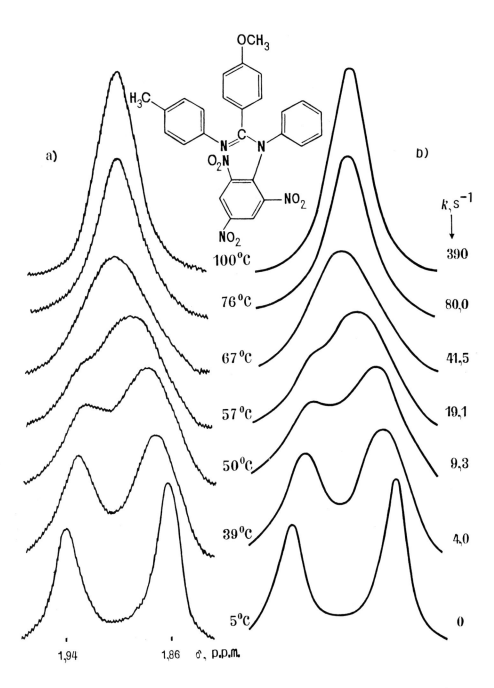

Fig. 2.15. Experimental (a) and theoretically calculated (b) line shape of the ^{1}H-NMR spectrum of the amidine derivative **LVII** ($X_1 = CH_3$, $X_2 = H$, $R = NO_2$, $Y = C_6H_4OCH_3$-p), compound No. 9 in Table 2.XI, in the spectral region of the methyl group X_1 [75].

2 and 7, 10 and 14 in Table 2.XI). Rearrangements of the *p*-nitrophenyl group in *N*-*p*-nitrophenyldiarylamidines are not detected by the dynamic NMR method even at temperatures of approximately 200°C.

Particularly rapid are the 1,3-migrations of the 2,4,6-tris(trifluoromethylsulfonyl)phenyl group in the amidine derivatives **LVIII** [120].

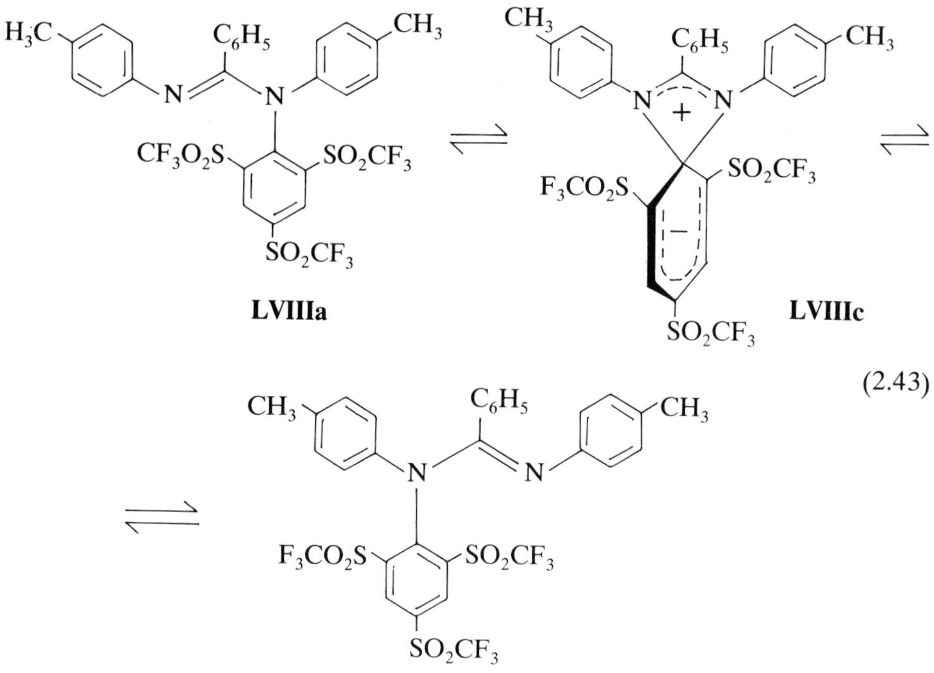

LVIIIa   **LVIIIc**

(2.43)

LVIIIb

The frequency of *N,N'*-migrations of an aryl group at 25°C in compound **LVIII** is 22 s^{-1} (*o*-dichlorobenzene) while the free activation energy $\Delta G^{\neq}_{25} = 15$ kcal/mol which is lower by 2.1 kcal/mol than in the analogous 2,4,6-trinitrophenyl derivative **LVII**. This is due to the exceptionally high electrophilicity of the 2,4,6-tris (trifluoromethylsulfonyl)phenyl group as well as to an increased stabilization of the **LVIIIc** intermediate. The anionic σ-complexes of 1,3,5-tris(trifluoromethylsulfonyl)benzene are known to be more stable and to form more readily than the corresponding σ-complexes of 1,3,5-trinitrobenzene [117]. As in the case of the *N*-acylamidines, activation of the migration process requires the presence of an aryl substituent R$_1$ in the 2 position. Donor substituents in the aryl group R$_1$ accelerate the reaction: (log k/k_0)$_{80°C}$ = −1.35σ.

An enhancement of the nucleophilicity of the nitrogen atoms between which there takes place the migration of the activated aryl group in *N,N*-dialkylamidines **LIX** leads to a competition between two processes, i.e., the 1,3-migration and the irreversible ring closure to form a benzimidazolium salt **LX** [121].

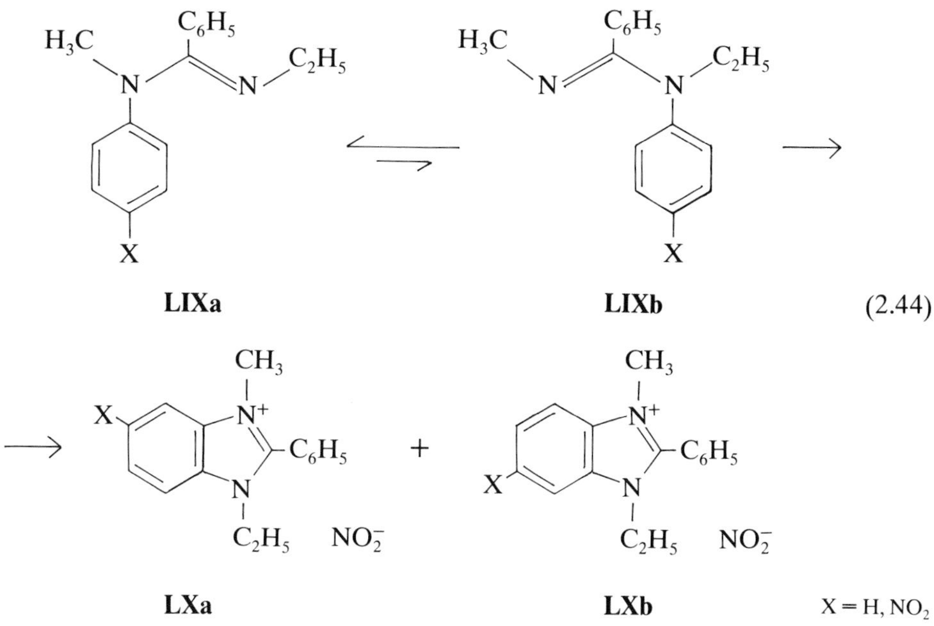

$$\text{LIXa} \qquad \longleftrightarrow \qquad \text{LIXb} \qquad \longrightarrow \qquad (2.44)$$

$$\longrightarrow \qquad \text{LXa} \qquad + \qquad \text{LXb} \qquad X = H, NO_2$$

The successful cyclization of benzamidines **LX** was achieved on heating up to 100—150°C. It depends on the presence of two N,N'-alkyl groups and a nitro group in an *ortho*-position on the aryl moiety.

4.2. O,O′-Migrations of the Aryl Groups in O-Aryl Derivatives of Tropolone and the Synthesis of Bipolar Spiro σ-Complexes

The bipolar spiro complexes **LVIIc, LVIIIc** are short-lived intermediates in the reactions (2.42—2.43) which are not detected by the methods of NMR and flash photolysis (Time resolution 10^{-6} s). Similar structures are postulated as intermediates in many 1,3-aryl shift reactions, for example, in the Chapman, Stevens, Schönberg and Newman—Kwart rearrangements. However, as has been correctly noted [122], no specific experimental proof of the existence of bipolar spiro complexes of the Meisenheimer type has been obtained.

The first reliably characterized bipolar spiro complexes of the Meisenheimer type **LXIc** were obtained in an investigation of the rearrangement of tropolone O-aryl ethers [110, 118].

In the reaction of picryl chloride with sodium salts of 3,7-dibenzyltropolone (**LXI**; $R_1 = R_3 = CH_2C_6H_5$; $R_2 = H$) and 3,5,7-trimethyltropolone ($R_1 = R_2 = R_3 = CH_3$), the products are not O-picryl ethers **LXIa, b** but the stable spiro complexes **LXIc** with high dipole moments (5—6 D) and the deep color (Fig. 2.16) characteristic of Meisenheimer complexes [115]. The quantum mechanical calculations

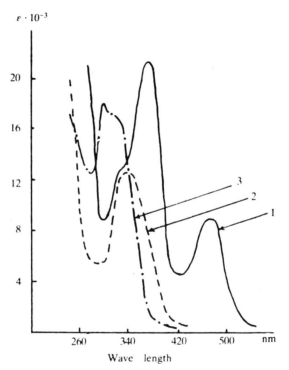

(2.45)

LXIa   LXIc   LXIb

Fig. 2.16. Electronic absorption spectra (in methylene chloride) of the compounds **LXI** ($R_1 = R_2 = R_3 = CH_3$) [110]: (1) Spiro-σ-complex **LXIc**, Z = 2,4,6-$(NO_2)_2$; (2) O-(2,4-dinitrophe-nyl)ether **LXIa**, Z = 2,6-$(NO_2)_2$.

of the electron distribution in the molecules **LXIc** made by the CNDO/2 method show that the total charge of the tropolone ring equals approximately $+0.5$.

The structure of **LXIc** has been proved by the results of X-ray structural investigations [123, 124] shown in Fig. 2.17. Also in this Figure are given molecular structures of the bipolar spiro-complexes of the mono- and diaza-derivatives of compounds **LXIc**

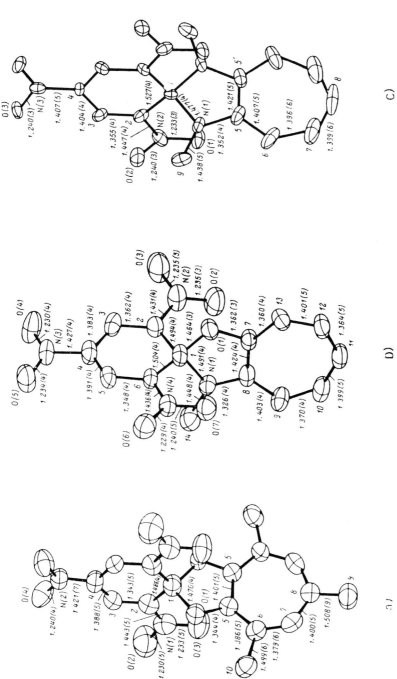

Fig. 2.17. The structure of molecules of bipolar spiro-σ-complexes: (a) **LXIc** (Z = 2,4,6-(NO$_2$)$_3$, R$_1$ = R$_2$ = R$_3$ = CH$_3$); (b) **LXII**; (c) **LXIII** from the data of X-ray structural investigations [123, 124].

LXII   LXIII

A number of other bipolar spiro-complexes of the **LXIc** type obtained so far are shown in Table 2.XIII.

As follows from the electronic spectral data (Fig. 2.16), the O-dinitrophenyl ethers **LXI** ($CR_1 = R_2 = R_3 = CH_3$), in contrast to the trinitrophenyl derivative, show the usual structure of the type **LXIa, b**. The O-(2,6-dinitrophenyl) derivative is of particular interest. Its ^1H and ^{13}C-NMR spectra show an effective C_{2v} symmetry, like that of the O-(2,4,6-dinitrophenyl) derivative, down to $-100°C$, indicating a very fast O,O'-migration of the aryl group ($\Delta G^{\neq} < 7$ kcal/mol). An analogous behavior is characteristic for O-(2,4,6-trinitrophenyl) derivatives of tropolone **LXI** ($R_1 = R_2 = R_3 = H$) and of 3-methyltropolone ($R_1 = CH_3, R_2 = R_3 = H$). The reason for such a low-barrier rearrangement of the heavy aryl residue is made clear by the data of an X-ray analysis of the molecular structure of these compounds. The plane of the aryl fragment in the O-(2,6-dinitrophenyl)ether [123] of 3,5,7-trimethyltropolone is orthogonal to the plane of the seven-membered ring (84.9°), and the *ipso* carbon atom approaches the oxygen atom to a distance (2.45 Å) that is significantly smaller than the sum of the van der Waals radii (3.1 Å). The angle θ at which the latter is oriented with respect to the aryl plane has a value of 107°, i.e., in the molecule there takes place an effective coordination interaction stabilizing the reactive center in the optimal conformation for closing a five-membered ring with the formation of a spiran structure of the type **LXIc**. The O-(2,4,6-trinitrophenyl)ether of tropolone (yellow crystal form) possesses an analogous molecular structure [125] shown in Fig. 2.18.

Thus, the steric role of the 2,6-nitro groups in compounds **LXIa, b** may be seen, consisting of the fine adjustment of the reactive site of the molecules which produces an optimal conformation for the nucleophilic attack, i.e. the conformation with the strongest attractive $C^{1'} \ldots O^2$ interaction.

In Table 2.XII are summarized the results obtained by means of the dynamic ^1H and ^{13}C-NMR spectroscopy characterizing the kinetics of the O,O'-aryl transfer reaction (2.45) as well as that of O,O'-migrations of heteroaryl groups in O-heteroaryl derivatives of 3,5,7-trimethyltropolone **LXIV**.

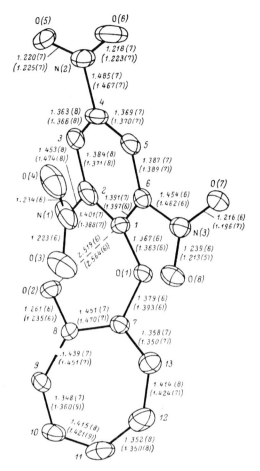

Fig. 2.18. The molecular structure of the *O*-(2,4,6-trinitrophenyl) derivative of tropolone — two crystallographically independent molecular forms from the data of an X-ray structural investigation [125].

$$(2.46)$$

LXIVa   LXIVc   LXIVb

The data of Table 2.XII clearly show that an increase in the electron deficiency of the carbon center linked with the oxygen center sharply enhances the migratory

Table 2.XII. Kinetics of O,O'-Migration of Aryl and Heteroaryl Groups in Derivatives of 3,5,7-Trimethyltropolones **LXI** and **LXIV** According to the Dynamic ^{1}H-NMR Investigation [110, 125, 126].

No.	Type	Compound — Migratory Group	k_{25}, s^{-1}	$\Delta H^{\neq}$, kcal/mol	$\Delta S^{\neq}$, e.u.	$\Delta G^{\neq}_{25}$, kcal/mol
1	**LXI**	Z = 2,6-(NO$_2$)$_2$	5×10^8	—	—	7.0
2		Z = 2,4-(NO$_2$)$_2$	1.1×10^4	8.5 ± 0.2	-10.8 ± 0.5	11.7
3		Z = 2-NO$_2$,4-SO$_2$CH$_3$	3.7×10^4	8.6 ± 0.3	-8.6 ± 0.8	11.2
4		Z = 4-NO$_2$	$<5 \times 10^{-8}$	—	—	>27
5		Het 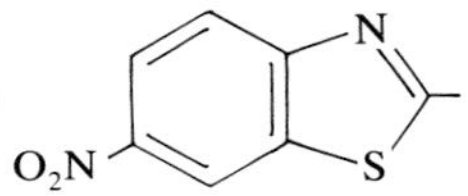 = (6-nitrobenzothiazol-2-yl)	3.5	12.0 ± 0.5	-11.7 ± 1.0	16.7
6		Het 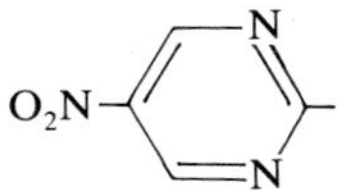 = (O$_2$N-pyrimidinyl)	1.5×10^3	6.2 ± 0.3	-22.9 ± 0.7	13.1
7	**LXIV**	Het 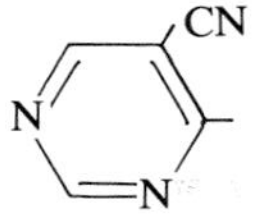 = (CN-pyrazinyl)	2.3×10^2	7.6 ± 0.3	-21.5 ± 0.7	14.3
8		Het 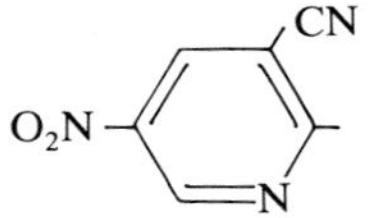 = (O$_2$N-CN-pyridinyl)	9.0×10^2	7.6 ± 0.5	-18.9 ± 0.7	13.4
9		Het 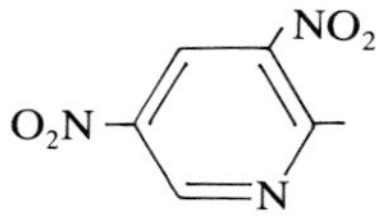 = (O$_2$N-NO$_2$-pyridinyl)	1.2×10^4	6.9 ± 0.5	-14.6 ± 0.8	12.0
10		Het 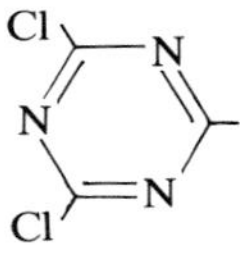 = (dichloro-triazinyl)	$>3 \times 10^8$	—	—	< 7.0

aptitude of both the aryl and the hetaryl group. This trend is illustrated by Fig. 2.19 which shows schematically the dependence of the energy barriers of the aryl migrations (of compounds **LXI**) on the number and location of electron-acceptor nitro groups in the migrant. It shows that not only the whole activation scale of tautomerism (Equation (1.5)), but also the region of mesomerism is involved in a transition from a mononitro (Z = 4-NO_2) to a trinitro derivative (Z = 2,4,6-$(NO_2)_3$).

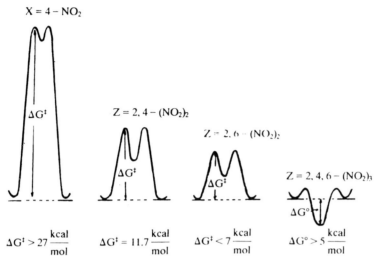

Fig. 2.19. Evolution of the free energy profile of the reaction **LXIa** ⇌ **LXIc** ⇌ **LXb** as a function of the number of nitro groups in the aryl migrant [110].

Fig. 2.20 which includes some data of X-ray diffraction investigations of the compounds **LXI**, **LXV** also reflects this tendency and makes possible a clear presentation of a channel of intramolecular aryl-group transfer at the stage of formation of spiro complexes of the type **LXIc, LXVc**.

The least favorable geometrical characteristics for rearrangement are shown by compound **LXVa** (X = S, Y = O) (A). Compound B is structurally more prepared for O,O'-aryl migration, with a value $\Delta G_{25}^{\neq} = 11.2$ kcal/mol (Table 2.XII). The most favorable conditions are found for compounds C and D, which are characterized by the highest values of the frequencies of O,O'-aryl shifts ($\Delta G^{\neq} < 7$ kcal/mol).

As for the compounds **LV** alkyl and aralkyl substituents in the 3 and 7 positions of the compounds **LXI** show a significant butressing effect, promoting an acceleration in the rearrangement of aryl groups and a shift in the equilibria **LXVa** ⇌ **LXVc** in the direction of the spiro complexes (Table 2.XIII). It is interesting to note that in contrast to the O-(2,4,6-trinitrophenyl) derivative of 3,5,7-trimethyl-tropolone, which exists only in the form **LXIc**, for the analogous derivatives of

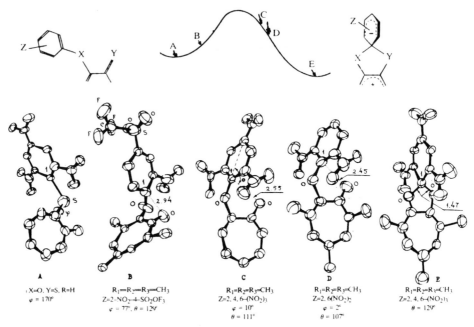

Fig. 2.20. Structural mapping of the reaction channel of the intramolecular transfer of polynitroaryl groups in the tropolone derivatives from the data of [123—125]. The structures are arranged in the order of decreasing distance C_1····O and angle φ (0° corresponds to an *s-cis* conformation) and increasing angle θ (shown in structure C). The saddle point is assumed to lie in the region between the structures B and C where the *s-cis* conformation of the aryl nucleus is formed. After the passage of this region, there occurs strong attractive C_1····O interaction.

3-methyltropolone and unsubstituted tropolone, isomers of both types **LXIc** and **LXIa** can be isolated. The structural characteristics of the latter are shown in Fig. 2.18.

A strong stabilization of type **LXIc** Meisenheimer spiro-complexes gives rise to an interesting example of the valence tautomerism occurring in solutions of compounds **LXI** and **LXV**.

$$\text{LXVa} \rightleftharpoons \text{LXVc} \tag{2.47}$$

LXVa   **LXVc**

Table 2.XIII. The Molecular Structure of Compounds **LXV** in the Crystalline State and in Solution [123—125].

No.	Z	X	Y	R	Structure in the Crystal	% of **LXVc** at 25°C	
						Benzene	DMF
1	2,4,6-(NO$_2$)$_3$	O	O	H	**LXVa**(X-ray), **LXVc**(IR)[a]	0	20
2	2,4,6-(NO$_2$)$_3$	O	O	3-CH$_3$	**LXVa, LXVc**(IR)[a]	0	55
3	2,4,6-(NO$_2$)$_3$	O	O	3,5,7-(CH$_3$)$_3$	**LXVc**(X-ray)	100	100
4	2,4,6-(NO$_2$)$_3$	O	S	H	**LXVc**(X-ray)	0	0
5	2,4,6-(NO$_2$)$_3$	O	NCH$_3$	H	**LXVc**(X-ray)	100	100
6	2,4,6-(NO$_2$)$_3$	S	NC$_3$H$_7$-i	H	**LXVa**(X-ray)	100	100
7	2,4,6-(NO$_2$)$_3$	NCH$_3$	NCH$_3$	H	**LXVc**(X-ray)	100	100
8	2,4-(NO$_2$)$_2$	NCH$_3$	NCH$_3$	H	**LXVc**(X-ray)	100	100
9	2,4,6-(SO$_2$CH$_3$)$_3$	O	O	H	**LXVc**(X-ray)	0	62[b]

[a] Two crystalline forms.
[b] In acetone solution. 0% of **LXVc** in dioxane and 100% of **LXVc** in dimethylsulfoxide.

The position of the equilibrium (2.47) strongly depends on the nature of the nucleophilic centers X, Y and on the polarity of the solvent. It can be easily determined by a measurement of the intensity of the long-wavelength (approximately 500 nm) absorption band of the cyclic tautomeric form **LXVc** in the electronic absorption spectra (see Fig. 2.16). Some pertinent results are shown in Table 2.XIII.

4.3. Valence Tautomerism of O-Aryl Ethers of o-Hydroxyaraldehydes and their Imines

As in the case of aryl ethers of tropolones and their amino and thio derivatives, *O*-aryl ethers of *o*-hydroxyaraldehydes **LXVIa** and their imines **LXVIIa** exist in solution in an equilibrium with their valence tautomers, the bipolar spiro σ-complexes **LXVIc**, **LXVIIc** [127, 128].

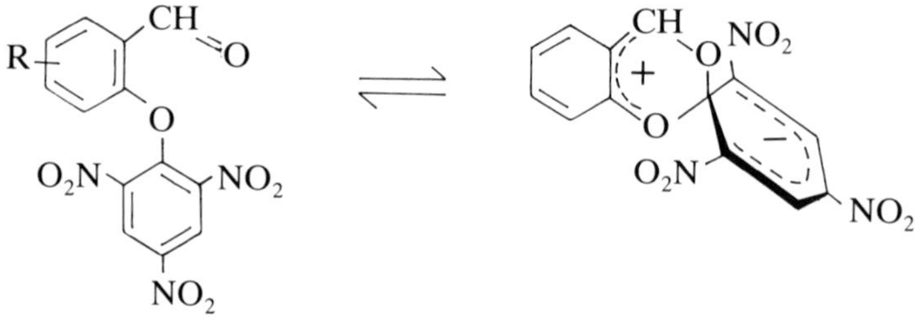

(2.48)

LXVIa   **LXVIc**

(2.49)

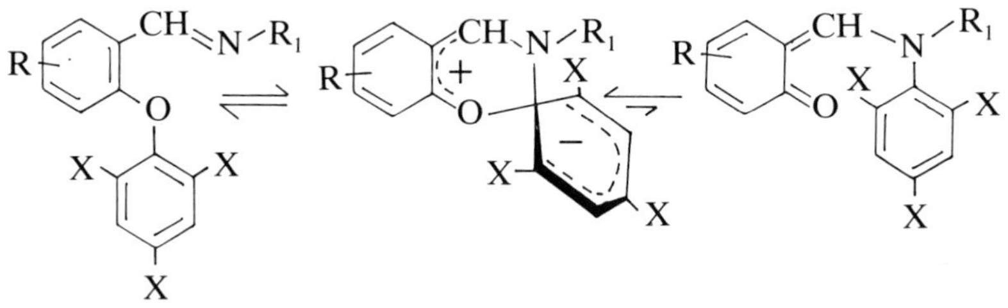

LXVIIa   **LXVIIc**   **LXVIIb**

X = NO$_2$, SO$_2$CF$_3$;   R$_1$ = Alkyl, Ar

In electronic absorption spectra of 2,4,6-trinitrophenyl derivatives of salicyclic and 2-hydroxy-1-naphthaldehyde (**LXVI**, R = 5,6-C$_4$H$_4$) in nonpolar and weakly polar solvents, the absorption bands in the visible region are absent. In polar solvents (acetone, dimethyl sulfoxide) in the 400 and 500 nm regions, there emerge two new absorption bands with a ratio of intensities of 2.5 : 1, which is typical of the trinitrocyclohexadienate form of the Meisenheimer anionic σ-complexes [115]. Such a spectral behavior serves as evidence in favor of the (2.48) equilibrium shifting to the right with the increase in the polarity of the solvent.

Unlike compounds **LXVI**, the aryl ethers of aldimines **LXVII** even in nonpolar solvents have long-wave absorption bands at 400 and 510—540 nm (Fig. 2.21), which points to a strong shift in the equilibrium (2.49) toward the spirocyclic valence tautomers. Table 2.XIV contains some results of studies of the equilibria (2.48) and (2.49) as obtained from the electron absorption and ^{1}H-NMR spectra.

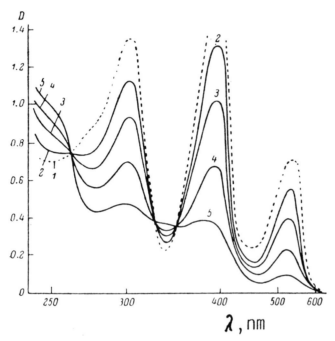

Fig. 2.21. Thermochromism of compound **LXVII** ($X = NO_2$, $R = H$, $R_1 = i$-Pr) as revealed in electronic absorption spectra of its dimethoxyethane solution ($C = 5 \times 10^{-5}$ M) [127] at: (1) −20°C, (2) 22°C, (3) 40°C, (4) 60°C, (5) 80°C.

Table 2.XIV. Tautomeric Equilibria of (2.48), (2.49) according to Electronic Absorption and ^{1}H-NMR Spectral Data [127, 128].

Tautomeric Equilibrium	Solvent	Equilibrium content of spiro-σ-complex	ΔG°_{25}, kcal/mol
LXVIa ⇌ **LXVIc** (R = H)	Dioxane	7% of **LXVIc**	2.6
	DMSO	80% of **LXVIc**	−0.8
LXVIIa ⇌ **LXVIIc** ($R = H, R_1 = i$-Pr, $X = NO_2$)	Dimethoxyethane	70% of **LXVIIc**	−0.6
LXVIIa ⇌ **LXVIIc** ($R = 5,6$-benzo, $R_1 = i$-Pr, $X = NO_2$)	Dimethoxyethane	83% of **LXVIIc**	−0.9
LXVIIa ⇌ **LXVIIc** ($R = 5,6$-benzo, $R_1 = i$-Pr, $X = SO_2CF_3$)	Dioxane	100% of **LXVIIc**	—
	DMSO	100% of **LXVIIc**	—

In the crystalline state, the compounds **LXVII** (R_1 = Alkyl) exist in the form of spiro σ-complexes. An X-ray diffraction study of the molecular structure of two compounds of type **LXVII**, R = 5,6-benzo (X = NO_2 and X = SO_2CH_3) has been carried out. The results for one of these are given in Fig. 2.22.

5. Carbonotropic Tautomerism in Ions, Radicals and Cyclopolyenes

This section considers briefly some data on rearrangements of the carbon-containing groups in anions, cations and stable radicals. As regards the first case, the

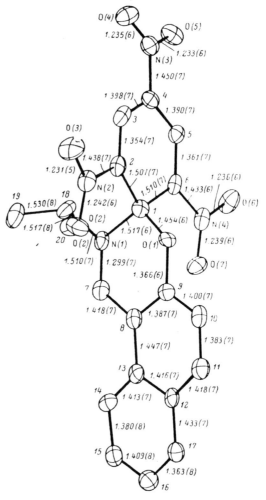

Fig. 2.22. Molecular structure of the spiro σ-complex **LXVIIc** (X = NO_2, R = 5,6-benzo, R_1 = *i*-Pr) according to an X-ray structural investigation [127]. The angle between the planes of nearly orthogonal fragments is equal to 84.5°.

intrinsic mechanism of migration of a carbon-containing group to the anion center parallels that of the associative nucleophilic substitution dealt with in the foregoing sections, with steric conditions for the realization of an intramolecular anion rearrangement being similar to those for the nucleophilic rearrangement in a neutral molecule.

A different situation occurs in the case of rearrangements in the cation and radical systems or in the event of a rearrangement of carbon-containing migrants linked to the polyene systems. Various aspects of such reactions, particularly in the case of carbocations, have been examined in numerous books and review articles; here they will be considered with the specific view of comparing their stereochemical and kinetical characteristics with those of the carbonotropic tautomeric transformations described in the preceding sections.

5.1. *Acylotropic and Arylotropic Rearrangements in Anions*

As the anion groups are the strongest nucleophiles, the formation within the molecule of a negatively charged center facilitates the motion of an electrophilic migrant toward it. This, in fact, constitutes the origin of the base catalysis in many rearrangement reactions of the acyl and aryl groups.

Regarding the nonconjugated systems, there are reliable data suggesting that precisely in the anion of *O*-acetyluridine (aqueous phosphate buffer) there occurs a very rapid transfer of an acetyl group between the hydroxyl oxygen atoms in positions 2 and 3 [32].

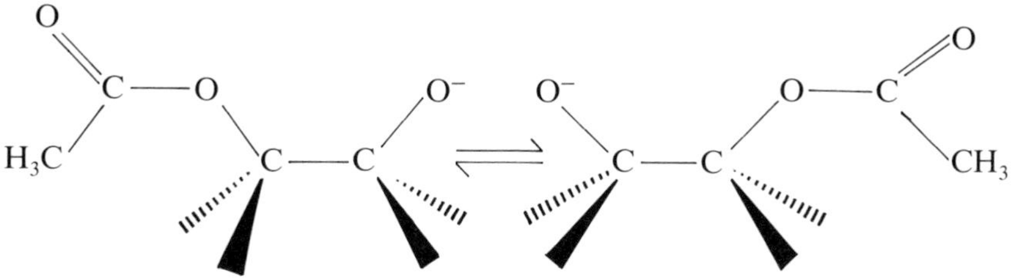

$$(2.50)$$

LXVIIIa **LXVIIIb**

The rate of the intramolecular reaction of an acetyl transfer is higher by five orders of magnitude than that of the intramolecular transfer of an acetyl group to the solvent (water) molecules.

The rapid intramolecular transfer reaction of an acyl group in the anion is an important stage in the base hydrolysis of *o*-carboxylate substituted aryl esters. It has, for example, been shown that in the hydrolysis of labelled 3,5-dinitro-substituted acetylsalicylic acid, the isotopically labelled oxygen is found to be included in the carboxylate group, which can be explained by the following reaction mechanism [129, 130]:

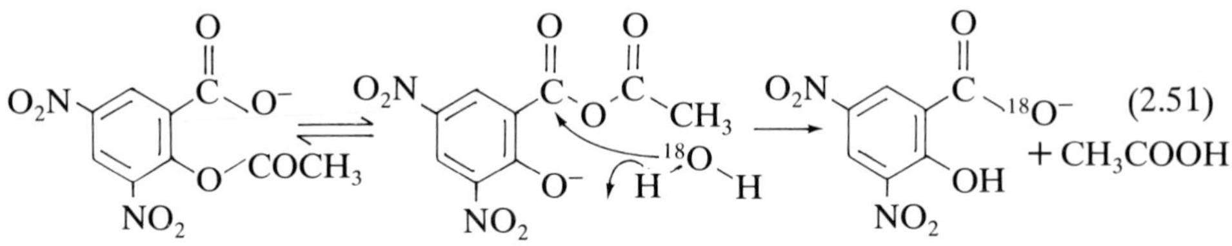

$$(2.51)$$

Other examples of the rapid intramolecular transfer reactions of the aryl groups in anions can be found in the review articles [131, 132].

In the thermal Smiles rearrangement, the stage of an aryl migration in the anion generated by the action of a base also plays a key role [133, 134].

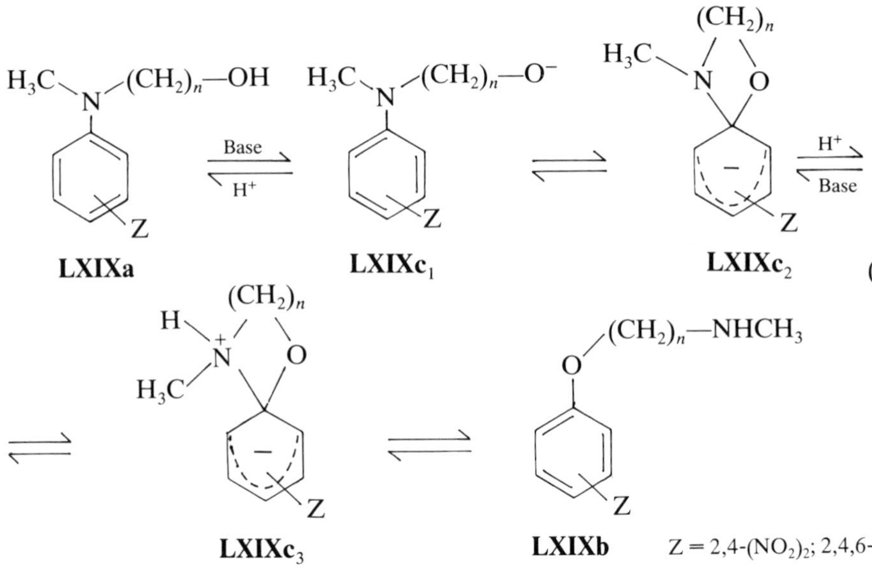

$$(2.52)$$

LXIXa   **LXIXc$_1$**   **LXIXc$_2$**

LXIXc$_3$ **LXIXb**   Z = 2,4-(NO$_2$)$_2$; 2,4,6-(NO$_2$)$_3$

The anionic Meisenheimer spiro σ-complexes of type **LXIXc$_2$** act as intermediates possessing in some cases a quite high stability. Compounds of this type are also characteristic of a number of other reactions of the intramolecular nucleophilic substitution associated with the transfer of aryl groups in anions [135]. An important example thereof are the well-studied 1,2-migrations of the aryl groups in carbanions (Grovenstein–Zimmerman rearrangement) [136] proceeding according to the following overall scheme.

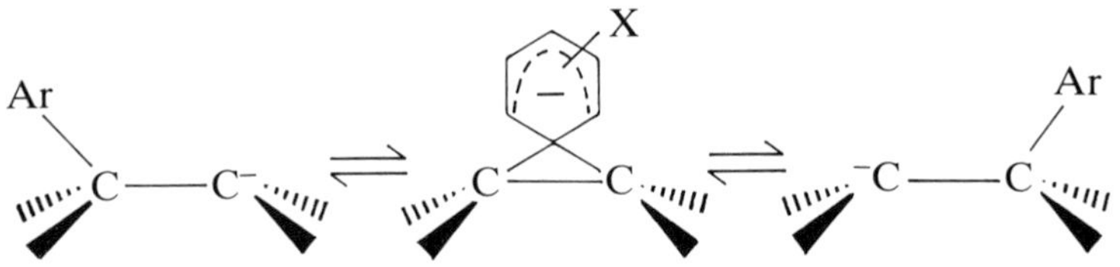

$$(2.53)$$

LXXa   **LXXc**   **LXXb**

Most known reactions of the acyl and aryl transfer in anions belong to irreversible rearrangements. The data on reversible, i.e. truly tautomeric reactions of this type are so far nonexistent; however, such transformations must, apparently, occur, at least, in degenerate molecular systems having an appropriate geometry of the reactive site such as:

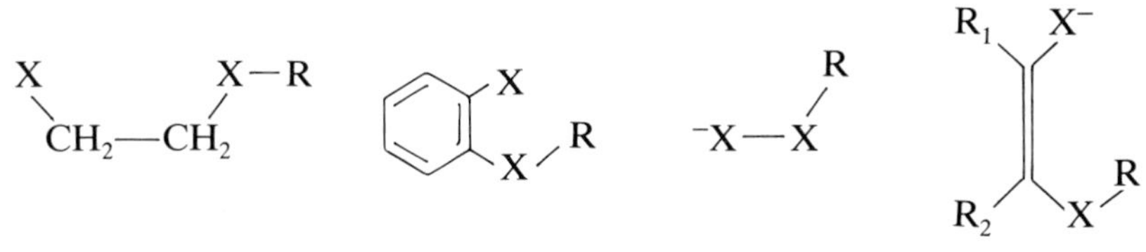

| LXXI | LXXII | LXXIII | LXXIV |

X = O, S, NR′   R = Ac, Ar

5.2. *Acylotropy in Phenoxyl Radicals*

Studying the ESR spectrum of the stable radical 3,6-di(tert.-butyl)-2-methoxy-phenoxyl **LXXV** obtained through oxidation of the corresponding derivative of catechole has shown that up to 200°C no characteristic broadening of the spectral lines occurs and, consequently, no methyl transfer reaction is observed [137].

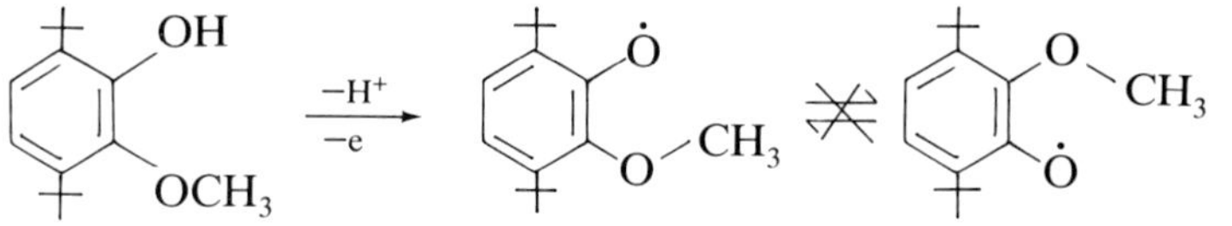

$$\hspace{6cm}(2.54)$$

| LXXVa | LXXVb |

Likewise, in the stable radical **LXXVI**, the exchange processes associated with migration of the C_{sp^3}-centered groups are absent [138].

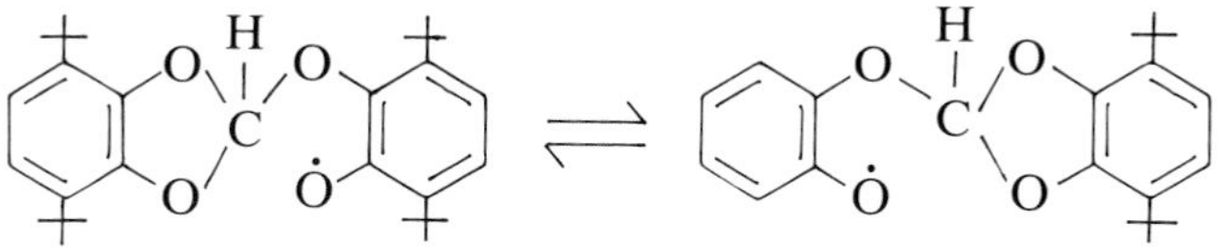

$$\hspace{6cm}(2.55)$$

| LXXVIa | LXXVIb |

By their intrinsic mechanism both above-shown reactions belong to the intramolecular homolytic S_H2 substitution, the MERP of which demands a linear arrangement of the bonds formed by the attacking and leaving groups [82] as is the case with the S_N2 reactions. Since such an arrangement cannot obtain in compounds **LXXV**, **LXXVI**, the activation energies of the alkyl group transfer reactions amount to extremely high values. A rough estimate [139] by the MINDO/3 method yields the value of 61.5 kcal/mol for the energy of the symmetrical structure of a transition state in the methyl group transfer taking place in the model radical **LXXVII**.

(2.56)

LXXVIIa   **LXXVIIb**

In contrast to $S_H 2$ reactions at the C_{sp^3} atom, the MERP of the homolytic substitution reactions at the C_{sp^2} atom is characterized by an angular orientation of the forming and breaking bonds at the C_{sp^2}-center for which favorable conditions are available in 2-acyloxy derivatives of the phenoxyl radicals **LXXVIII**.

(2.57)

LXXVIIIa   **LXXVIIIb**

Indeed, the ESR spectra of compounds **LXXVIII** show very fast carbonotropic rearrangements (2.57). This reaction is subject to a general acid catalysis. Table 2.XV lists the data on the kinetics of this reaction.

Still higher rates have been observed for the metallotropic rearrangements in derivatives of the phenoxyl radicals [141]. Under analogous conditions, the frequency of O,O-migrations of the trimethylsilyl group is 2.5×10^6 s^{-1} whereas that of the trimethylstannyl group exceeds 10^9 s^{-1}.

Table 2.XV. Kinetical Data on Acylotropic Tautomerism in 3,6-di(*tert*-butyl)-2-acyloxyphenoxyl Radical (2.57). (Solvent: Toluene) [140, 141].

No.	Compound **LXXVIII** R	k_{20}, s^{-1}	A, s^{-1}[a]	$E_a (\pm 0.5)$, kcal/mol
1	CH_3	10^4	0.9×10^{13}	11.4
2	CH_3[b]	2.3×10^6	6.7×10^9	4.7
3	C_2H_5	10^4	1.1×10^{13}	13.0
4	CH_2Br	2.6×10^5	2.1×10^{13}	10.0
5	$CH_2C_6H_5$	10^5	0.6×10^{12}	9.15
6	$CH(C_6H_5)_2$	3×10^4	2.3×10^{12}	10.7

[a] Preexponential factor in the Arrhenius equation.
[b] In 1 M solution of CF_3COOH in toluene.

5.3. *Tautomeric Rearrangements of Carbocations.*

Very fast carbonotropic rearrangements accompanied by migrations of the alkyl, aryl and acyl groups take place in carbonium ions. In degenerate or nearly degenerate reaction systems these rearrangements are reversible. Therefore, in accordance with the energy criterion they may be regarded as carbonotropic tautomerism.

In terms of its intrinsic mechanism, the migration of a carbon-centered group to the electron-deficient center, which occurs during the carbocation rearrangement, corresponds to the intramolecular S_E2 reaction. As contrasted with the geometry of the transition state in the reaction of nucleophilic substitution at an sp^3-carbon atom which is rigidly fixed in the form of a trigonal bipyramid, the intermediate structures with the pentacoordinate carbon atom emerging in the reaction of electrophilic substitution at a C_{sp^3} atom differ but slightly from each other in ther relative stability. Thus, for a methonium ion CH_5^+ forming in the simplest reaction of electrophilic substitution ($CH_4 + H^+$), the *ab initio* calculations at the highest level of approximation using the extended basis of orbitals and taking into account the correlation corrections (MP4 SDQ/6-311 G*//MP2/6-31 G*) yield the following order in the stabilities of various CH_5^+ structures [142].

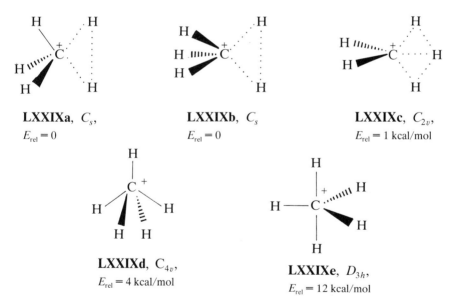

LXXIXa, C_s,	**LXXIXb,** C_s	**LXXIXc,** C_{2v},
$E_{rel} = 0$	$E_{rel} = 0$	$E_{rel} = 1$ kcal/mol

LXXIXd, C_{4v},
$E_{rel} = 4$ kcal/mol

LXXIXe, D_{3h},
$E_{rel} = 12$ kcal/mol

The most favored C_s-structures **LXXIXa, b** correspond to transition structures in the S_E2 reactions expected in these in view of the rule of retention of stereochemical configuration [82]. The orientation of the forming and breaking bonds corresponding to **LXXIXa, b** is well adjusted to intramolecular 1,2-shifts of the alkyl groups typical of carbocation rearrangements. Thus, the 1,1,1,2,2-pentamethylethyl cation **LXXXa** generated in a solution of SO_2ClF—SO_2F_2 suffers a degenerate carbonotropic tautomeric rearrangement with the frequency of over 10^6 s^{-1} even at $-136°C$ ($G^{\neq}_{-136} = 3.8$ kcal/mol) [11].

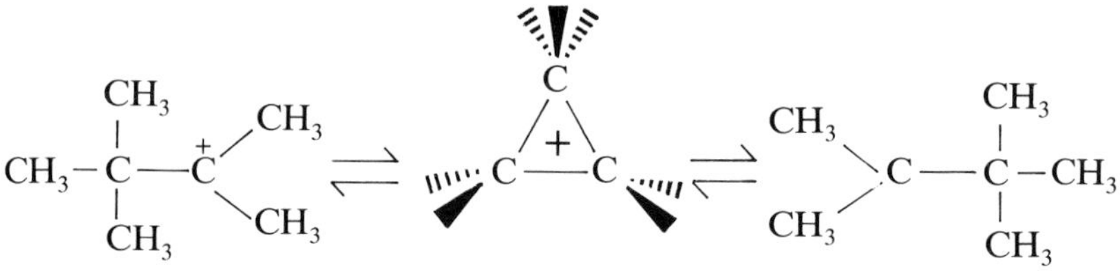

$$\text{(2.58)}$$

LXXXa   **LXXXc**   **LXXXb**

Such a low energy barrier (approximately 3 kcal/mol) for the rearrangement (2.58) can be accounted for by the facility with which the transition (or intermediate) structure **LXXXc** is obtained in the 1,2-shift of the methyl group. Another important factor is here the degree of electron deficiency of the electrophilic center in the carbocation. The inclusion of a cation center in the σ-system lowers the magnitude of the positive charge and reduces the rearrangement rate. This is illustrated by kinetical characteristics of tautomeric 1,2-shifts of the methyl group in the series of arenonium ions presented in Table 2.XVI.

Table 2.XVI. Rate Constants and Free Activation Energy for Degenerate Reactions of the 1,2-Shift of a Methyl Group in Arenonium Ions.

No.	Arenonium Ion	Acid System	k_{25}, s^{-1}	$\Delta G^{\neq}_{25}$, kcal/mol	Reference
1	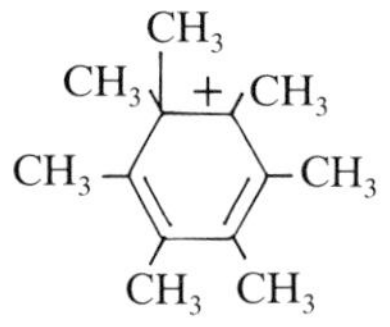	HSO$_3$Cl	6 × 10^{-1}	18.1	[143]
2	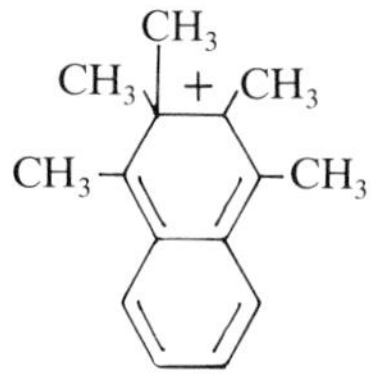	HSO$_3$Cl	—	22.7($-125°C$)	[144]
3	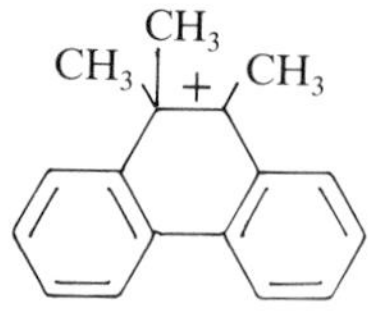	HSO$_3$F + SO$_2$ClF	2.3 × 10^4	11.5	[145]

Table 2.XVI. (continued)

No.	Arenonium Ion	Acid System	k_{25}, s^{-1}	$\Delta G_{25}^{\neq}$, kcal/mol	Reference
4	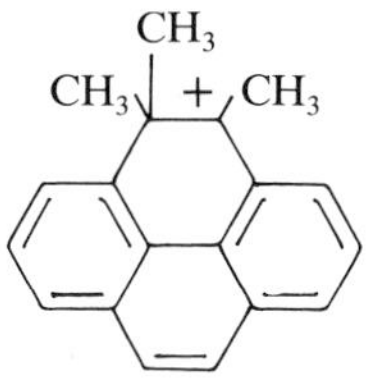	$HSO_3F + SO_2ClF$	1.7×10^4	11.7	[145]
5	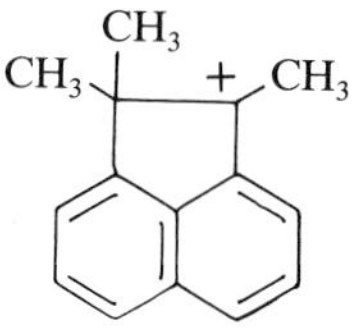	CH_3SO_3H	2×10^{-1}	18.4	[146]

Since the magnitude of the positive charge at a carbon atom correlates with the value of the chemical shift of ^{13}C nuclei in the ^{13}C-NMR spectrum, the linear relationship between the free activation energy of the 1,2-shift of a methyl group in arenonium ions shown in Table 2.XVI and the magnitude of the chemical shift of a carbon atom to which the migrant is moving can be easily understood [10, 146].

$$\Delta G_{25}^{\neq}(CH_3) = 53.7 - 0.181 \delta_c \text{ (kcal/mol)}$$
$$r = 0.995, \quad s = 0.29 \tag{2.59}$$

For the sp^2-hybridized carbon atom carrying a single positive charge, the value of δ_c lies around 300 ppm [147]. According to Equation (2.59), the energy barrier for 1,2-transfers of a methyl group in such cations must be close to zero, as is the case with the rearrangement (2.58), or even have a negative value suggesting that the intermediates of type **LXXXc** are extra stabilized as compared to the classical structures **LXXXa, b**. Indeed, according to *ab initio* calculations at various levels of approximation, the corner-protonated cyclopropane form of the $C_3H_7^+$ cation is 5—15 kcal/mol more stable than its classical 1-propyl isomer [142].

1,2-Migrations of the aryl groups in arenonium ions require a still lower activation energy than the 1,2-migrations of the alkyl groups [9,10]. As can be seen from the correlation equation (2.60), the rate of the 1,2-rearrangement of a phenyl group in the arenonium ion depends to a greater degree on the magnitude of the positive charge at the neighboring C_{sp^2}-atom than in the case of an alkyl rearrangement.

$$\Delta G_{25}^{\neq}(C_6H_5) = 65.0 - 0.244 \delta_c \text{ (kcal/mol)}$$
$$r = 0.995, \quad s = 0.36 \tag{2.60}$$

As befits the electrophilic character of the reaction, donor substituents in the aryl nucleus accelerate the rearrangement. This is, for instance, apparent from the data in Table 2.XVII which contains some results on the kinetics of 1,2-transfer reactions of aryl groups in phenanthrenonium ions **LXXXI**.

Table 2.XVII. Rate Constants and Free Activation Energy for the Degenerate Reactions of the 1,2-Shift of Aryl Groups in Phenanthrenonium Ions **LXXXI** ($HSO_3F + SO_2ClF$ acid system) [9, 10].

No.	R	k_{25}, s^{-1}	$\Delta G_{25}^{\neq}$, kcal/mol
1	p-CF_3	2.6×10^3	12.8
2	m-F	1.5×10^4	11.8
3	p-Cl	1.7×10^5	10.3
4	H	$4 \ \times 10^5$	9.8
5	p-F	4.8×10^5	9.7
6	p-CH_3	1.7×10^6	8.9
7	$pOCH_3$	$1 \ \times 10^8$	6.5

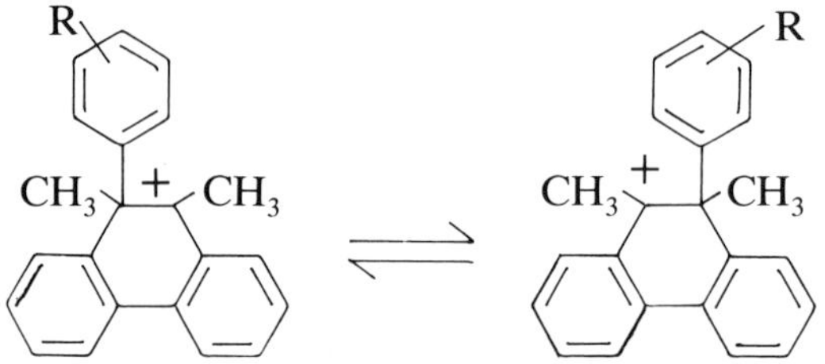

$$(2.61)$$

LXXXIa   **LXXXIb**

Upon going from arenonium to carbonium ions, in which the delocalization of the positive charge in the C_{sp^2}-center does not occur, the rate of the aryl transfer reactions sharply drops and, provided there are donor substituents R in the aryl nucleus, the phenonium structure **LXXXIIc** corresponding to the cation σ-complex is stabilized [4—8].

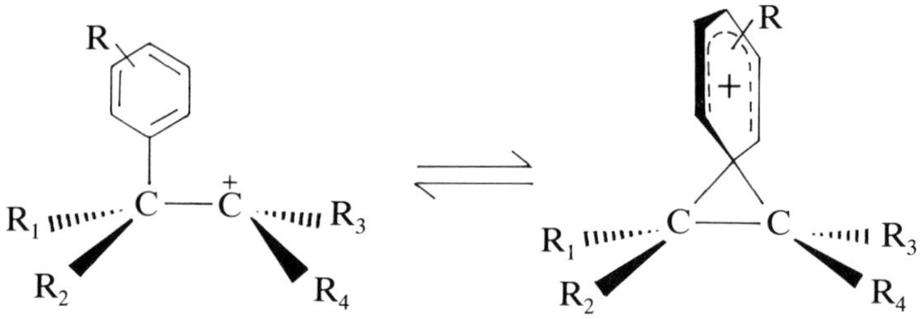

$$(2.62)$$

LXXXIIa   **LXXXIIc**

Thus, carbonotropy of the electrophilic type exhibits the same principal features which characterize nucleophilic carbonotropy, i.e., the need for favorable geometric factors for the realization of an intramolecular rearrangement and a strong dependence of the rate and the character of the energy profile of a reaction on electronic factors of the structure. An interesting feature of carbonotropic tautomeric rearrangements in the carbonium ions is that in this case there is no sharp difference between the migratory aptitudes of the C_{sp^3} and C_{sp^2}-centered groups since the compounds of both classes are characterized by the same type of transition geometry as the S_{E^2} reactions.

As regards the aliphatic carbocations and arenonium ions, they readily undergo reactions of intramolecular electrophilic substitution also at other than carbon centers, *viz.* at halogen atoms, at nitrogen in a nitro group or at sulfur in a sulfo group. This explains the fast 1,2-shifts of these groups along an aromatic ring or aliphatic chain, which corespond to the tautomeric scale of activation energies.

5.4. *Acyl and Aryl Migrations in Cyclopolyenes*

There is one more group of reactions in which reversible carbonotropic migrations occur, whose energy parameters correspond to a tautomeric scale. It is the so-called circumambulatory rearrangements of 1,3-cyclopentadiene derivatives which belong to thermally allowed, 1,5-sigmatropic shifts. An important factor for the occurence of these reaction are secondary orbital interactions between the π-system of a cyclopentadiene moiety and the frontier orbitals of a migrating group.

As could be expected on the basis of the equations (2.59) and (2.60), the migration of methyl and phenyl groups along the cyclopentadiene ring is practically inhibited. The free activation energies of the 1,2-transfers of a methyl group in various cyclopentadiene derivatives are not less than 40 kcal/mol [148—150]. For an analogous degenerate 1,2-rearrangement of a phenyl group in the 1,2,3,4,5-pentamethylcyclopentadiene derivative this value amounts to 37.8 kcal/mol [151]. On the other hand, 5-formylcyclopentadiene **LXXXIII** (R = H) exhibits a ¹H-NMR spectrum which is largely averaged at ambient temperature due to a very rapid circumambulation. Analogous behavior is evident in trifluoroacetyl (R = CF₃) and trichloroacetyl derivatives of **LXXXIII**.

$$ \tag{2.63} $$

LXXXIIIa   **LXXXIIIb**   **LXXXIIIc**

	k_{25} = 90 s⁻¹	$\Delta H^{\neq}$ = 13.2 kcal/mol	$\Delta S^{\neq}$ = −5.1 e.u.	$\Delta G_{25}^{\neq}$ = 13.8 kcal/mol
R = H [152]				
R = CF₃ [153]	4.7 × 10⁻³	18.2	−8.2	20.5
R = CCl₃ [153]	4.5 × 10⁻⁶	21.6	−10.3	24.7

The presence of electron-donating substituents R in the acyl group of the compounds **LXXXIII** and of their analogs practically inhibits the rearrangement (2.63). Even the acetyl group ($R = CH_3$) possesses such a low migratory aptitude that the (2.63) reaction cannot be induced with the aid of the NMR spectra. This rearrangement becomes, however, extremely fast when the acyl oxygen atom is complexed with a Lewis acid such as aluminum trichloride. It has been shown that the reaction proceeds via a bicyclic zwitterionic complex **LXXXV** of the bicyclo[3.1.0]hexenyl type [154].

$$
\begin{array}{ccccc}
\text{LXXXIVa} & & \text{LXXXVa} & & \text{LXXXIVb}
\end{array}
\tag{2.64}
$$

$R = CH_3$	$k_0 = 10^4\ s^{-1}$	$\Delta G^{\neq} = 10.9$ kcal/mol*
$R = C_2H_5$	$k_{-90} = 8$	9.7
$R = OC_2H_5$	$k_{70} \leqslant 17$	$\geqslant 18.2$
$R = C_6H_5$	$k_{-90} = 710$	8.0
$R = C_6H_4OCH_3\text{-}p$	$k_{-90} = 0.15$	11.2
$R = C_6H_4CH_3\text{-}p$	$k^{-90} = 17$	$\Delta G^{\neq} = 9.8$
$R = C_6H_4C(CH_3)_3\text{-}p$	$k_{-90} = 16$	9.6
$R = C_6H_4Cl\text{-}p$	$k_{-90} = 540$	8.3
$R = C_6H_4CF_3\text{-}p$	$k_{-90} = 4.5 \times 10^5$	5.0

Since the zwitterion **LXXXV** is a stabilized tetrahedral intermediate for the acyl transfer serving as an analog of **XLVIIIc, LVc**.

The reaction (2.64) is characterized by the same dependence on parameters of the substituents R which is typical of the acylotropic rearrangements (2.35), (2.41). Plotting $\log k_{-90}$ against σ^+ gives a straight line with a slope ρ of +4.6. The positive value of ρ in this correlation dependence parallels that for the correlation of the acyl group O,O'-transfer in O-acyl derivatives of acetylacetone **XLVIII** shown in Fig. 2.10.

The most important orbital interaction affecting the height of the energy barrier in reactions of the 1,2-shift of a migrant in the cyclopentadiene ring accompanied by the formation of bicyclic η^2-intermediates or type **LXXXV** transition-states is the donation of the electron density by one of the e-orbitals of the cyclopentadiene ring to a vacant p or sp^n-orbital of the migrant [155, 156]. In the case of the acyl migrant, the role of the latter is assumed by the vacant π^*-orbital of the carbonyl

---

* $\Delta G^{\neq}$ of the reaction (2.64) was assumed to be temperature-independent, which is validated by the low $\Delta S^{\neq}$ values obtained [154]. CH_2Cl_2 and $CHClF_2$ were used as solvents.

bond. When a deformation of the structure of acylcyclopentadiene corresponds to the direction of the 1,2-shift of the migrant **LXXXVI**, the factors favoring the rearrangement will be, in accordance with the concepts of the perturbation MO theory, the large value of the overlap integral $S_{\pi\pi^*}$ and the small energy difference between the interacting orbitals.

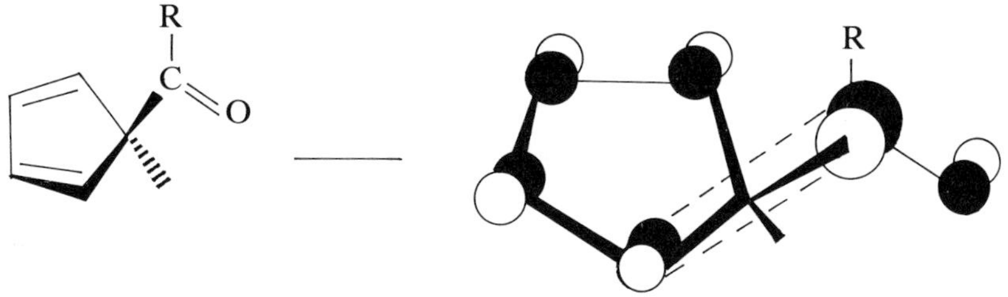

LXXXVI

π (cyclopentadiene) — π^* (carbonyl) interaction

The largest value of the overlap integral is attained upon complete localization of the π^*-orbital, i.e. when R = H, which would correspond to a particularly high mobility of the formyl group. Complexation with $AlCl_3$ involves a considerable drop in the energy level of the π^*-orbital and a reduction in the energy gap between the interacting π and π^*-orbitals, which explains the enormous acceleration of circumambulation of complexed acyl groups in compounds **LXXXIV**. The nature of this effect is the same as that of the abovementioned catalysis of the *O,O'*-transfer reactions of acyl groups in the derivatives of 1,3-diketone *cis*-enols (Fig. 2.11).

Upon annelation of the benzene ring to cyclopentadiene in the derivatives of indene, the order of migratory aptitudes of the C_{sp^2}-centered groups is retained [157, 158].

$$HCO \gg CH_3CO > PhCO > H > COOCH_3 > CH{=}CH_2 \qquad (2.65)$$

Only the formyl group migrations fall within the activation scale of tautomeric reactions. Even slower are the acyl rearrangements in cyclohexadienes [148]. A detailed study has been carried out regarding the effect of the substituent R on the rates of intramolecular 1,2-shifts of the methoxycarbonyl migrant in the derivatives of pentamethoxycarbonyl cyclopentadiene **LXXXVII** [159].

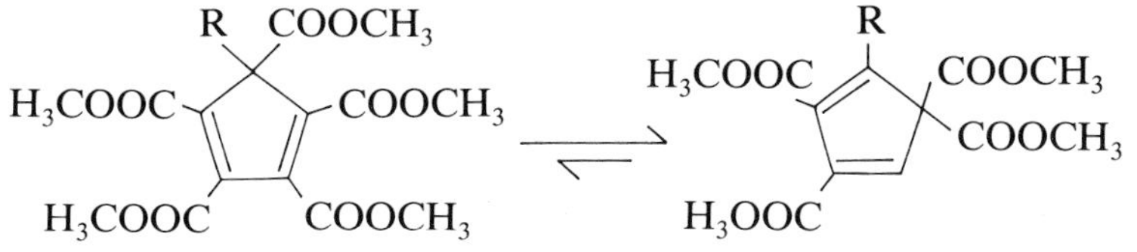

$$(2.66)$$

LXXXVIIa   **LXXXVIIb**

The results of the research into fast and reversible acylotropic and arylotropic rearrangements examined in this Chapter show that the task of creating carbono-tropic tautomeric systems, which were totally unknown before 1970, has now been resolved. The synthesis of these systems was successfully approached on the basis of structural modification of compounds capable of fast rearrangements of the acyl, aryl and similar groups through an adjustment of their molecular structure to energy requirements of the tautomeric reactions (1.5). The requirement for thermodynamic equilibrium is satisfied automatically in the case of degenerate transformations. The achievement of a sufficiently low activation barrier for a tautomeric rearrangement is conditional upon low energy conformational reorgani-zation of the ground state structure to a structure that would resemble the transition state of the reaction. This condition requires both sterical proximity of the interacting centers and a favorable orientation of the orbitals participating in the bond-breaking and bond-making processes.

These factors of proximity and of a suitable mutual orientation are a necessary but not a sufficient condition for the occurence of rapid (on a tautomeric scale) carbonotropic transformations. These can take place only when favorable enough electronic factors of reactivity are operative, e.g. a sufficiently high electron-deficiency of the carbonyl carbon atom in a migrating acyl group of acylotropic systems. Only by varying the number and position of nitro groups in the migrating aryl group when optimum geometry of the reaction site of the tautomeric system is provided, as is the case with tropolone derivatives **LXI** is it possible to embrace the entire energy scale of tautomerism and even mesomerism ($\Delta G^{\neq} = 30-40$ kcal/mol). In case the geometry of a tautomeric system is not optimal (triazenes, amidines) for a reaction of the carbon-centered group transfer to fall within the tautomeric activation scale, there is a need for such a variation of reactivity as would ensure a gain in activation energy of up of $40-50$ kcal/mol.

The carbonotropic, particularly acylotropic, tautomeric systems may be regarded as simple models of group transfer steps (transacylation) of enzymic reactions. Their important advantage is the possibility of isolation of the key step, i.e., of the transfer of a group not complicated by secondary acid-base equilibria.

Acceleration of reactions of intramolecular transfer of the carbon-centered groups induced only by varying the structure of the migrant and of the rest of the molecule can be comparable with or even exceed the acceleration due to enzymatic catalysis (up to 10^{14} times). As is apparent from Table 2.XVIII, where data are summarized on some of the acylotropic systems considered above, the optimization of geometry parameters of a tautomeric system and the modification of electronic characteristics of the nucleophilic oxygen atom toward which the acetyl migrant is shifting allow one to embrace an extremely wide range (10^{20} s^{-1}) of the rates of intramolecular migration. With fixed geometry of a tautomeric system (tropolone derivatives **LV**), through varying the group R in the acyl migrant, the O,O'-transfers can be speeded up 10^{9} times (Table 2.IX). All this

Table 2.XVIII. Rate Constants and Free Activation Energy of Reversible Migrations of an Acetyl Group.

No.	Compound	k_{25}, s^{-1}	$\Delta G_{25}^{\neq}$, kcal/mol	Reference
1	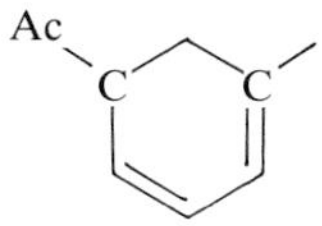	2.4×10^{-14}	35.9	[160]
2	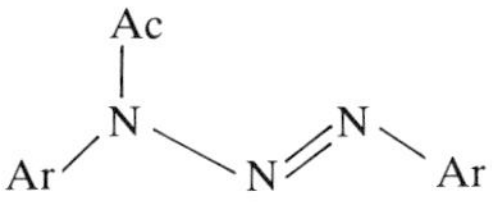	1.2×10^{-13}	35.0	[40]
3	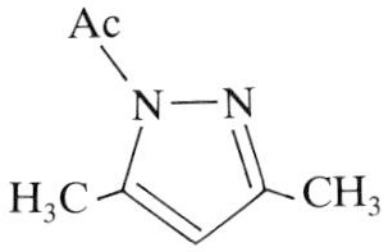	10^{-7}	26.0	[67]
4	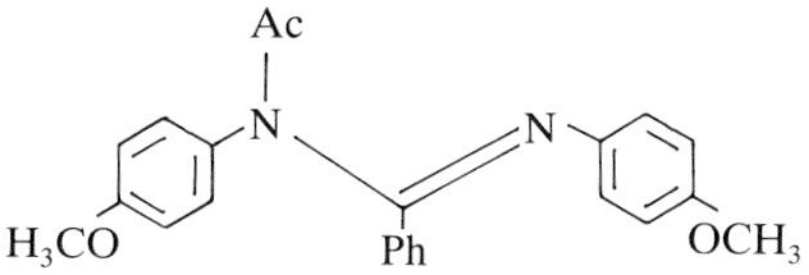	1.2×10^{-6}	25.5	[71]
5	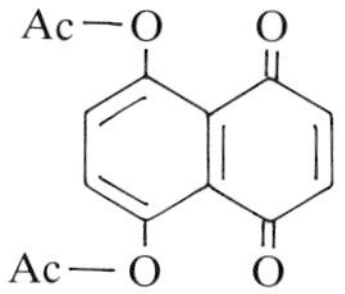	6×10^{-6}	24.5	[66]
6	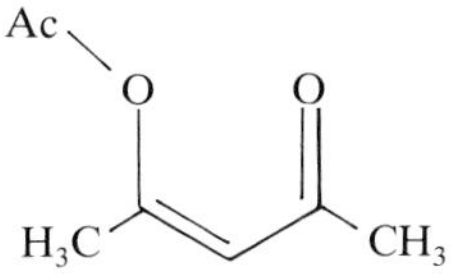	3.5×10^{1}	15.3	[68, 92]
7	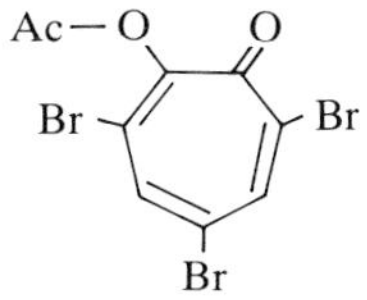	4.5×10^{3}	12.5	[113]
8	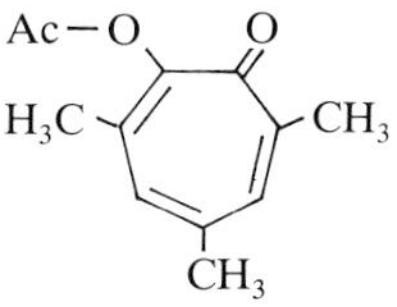	1.2×10^{6}	9.2	[110]

clearly shows that the colossal rates (and accelerations) of chemical reactions, in which there occur processes of breaking and making of the C—X bonds, are by no means an exclusive monopoly possessed by enzymatic systems in living organisms; they are attainable in simple model systems as well. The route leading to construction of such tautomeric systems should be based on an analysis of the intrinsic mechanism of a given reaction — in other words, on the analysis of the minimal energy reaction path.

A formal approach to the construction of tautomeric systems consisting in a simple replacement of a migrant by some other group does not appear to hold much promise. A number of compounds synthesized by us in accordance with this principle where, by analogy with prototropic, acylotropic and metallotropic tautomeric rearrangements, it had been expected to find migrations of the alkyl, phosphoryl and sulfonyl groups, proved to be structurally rigid, with no manifestations of tautomerism.

In Chapter 3, an attempt will be made to account for these results as well as to develop a generalized approach to construction of tautomeric sigmatropic systems with broadly varied migrants. This approach will be rooted in basic principles underlying the theory of reaction mechanisms and of molecular structure. Chapter 4 will deal in detail with mechanisms and stereochemistry of the reactions of nucleophilic substitution at atoms of the main group elements with the view of their possible adaptation for the design of intramolecular tautomeric systems.

REFERENCES

1. Zhdanov Yu. A., Minkin V. I., Olekhnovich L. P., and Malysheva E. N., *Zh. Org. Khim.*, **6**, 554 (1970).
2. Minkin V. I., Olekhnovich L. P., and Zhdanov Yu. A. *Molecular Design of Tautomeric Systems*. Publ. House of Rostov on Don University. Rostov on Don. 1977.
3. Bethell D. and Gold V. *Carbonium Ions. An Introduction*. Academic Press. London and New York. 1967.
4. Olah G. A. *Carbocations and Electrophilic Reactions*. Academic Press. New York. 1974.
5. Brouwer D. M. and Hogeveen H., *Progress Phys. Org. Chem.*, **9**, 189 (1972).
6. Saunders M., *Acc. Chem. Res.*, **6**, 53 (1973).
7. Brown H. C., *Tetrahedron*, **32**, 179 (1976).
8. *Contemporary Problems of Carbonium Ion Chemistry* (Ed. V. A. Koptyug). Izd. Sybir. Otd. Akad Nauk SSSR. Novosibirsk. 1975.
9. Schubin V. G., *Topics Curr. Chem.*, **116/117**, 267 (1984).
10. Koptyug V. A., *Topics Curr. Chem.*, **122**, 1 (1984).
11. Saunders M. and Kates R. M., *J. Amer. Chem. Soc.*, **100**, 7082 (1978).
12. Minkin V. I., Olekhnovich L. P., and Zhdanov Yu. A., *Zh. Vsesoyuzn. Khim. D. I. Mendeleev Obschz.*, **22**, 274 (1977).
13. Minkin V. I., Olekhnovich L. P., and Zhdanov Yu. A., *Acc. Chem. Res.*, **14**, 210 (1981).
14. Pinnavaia T. J., Collins W. T., and Howe J. J., *J. Amer. Chem. Soc.*, **92**, 4544 (1970).
15. Bock B., *Angew. Chem. Intern. Ed. Engl.*, **10**, 225 (1971).
16. Martin J. C. and Basalay R. J., *J. Amer. Chem. Soc.*, **95**, 2572 (1973).
17. Sommer L. H., *Stereochemistry, Mechanism and Silicon*. McGraw-Hill. New York. 1965.

18. Brook A. G., *Acc. Chem. Res.*, **7**, 77 (1974).
19. Tobe M. L., *Inorganic Reaction Mechanisms*. Th. Nelson and Sons Ltd. London. 1972.
20. Corriu R. J. P. and Guerin C., *J. Organometal. Chem.*, **198**, 231 (1980).
21. Baybitt P., *Molecular Phys.*, **29**, 389 (1975).
22. Reich H. J. and Murcia D. A., *J. Amer. Chem. Soc.*, **95**, 3418 (1973).
23. Litchman W. M., *J. Amer. Chem. Soc.*, **101**, 545 (1979).
24. Hegazy M. F., Quinn D. M., and Schowen R. L., in: *Transition States in Biochemical Processes* (Eds. Gandour R. D. and Schowen R. L.). Plenum Press. New York. 1978, p. 355.
25. Sheppard W. A. and Sharts C. M., *Organic Fluorine Chemistry*. Benjamin. New York. 1969.
26. *Biogenesis of Natural Compounds* (Ed. Bernfeld P.). Pergamon Press. Oxford-London-New York-Paris. 1963.
27. Ingram V. M., *Biosynthesis of Macromolecules*. Benjamin. Menlo Park, California. 1972.
28. Wall M. C. and Laidler K. J., *Arch. Biochem. Biophys.*, **43**, 299 (1953).
29. Shaw W. H. R. and Walker D. C., *J. Amer. Chem. Soc.*, **80**, 5337 (1958).
30. Lipscomb W. N., *Acc. Chem. Res.*, **15**, 232 (1982).
31. Bruice T. C. and Benkovic S. J., *Bioorganic Mechanisms*. Vol. 1. Benjamin. New York-Amsterdam. 1966.
32. Jencks W. P., *Catalysis in Chemistry and Enzymology*. McGraw-Hill. New York. 1969.
33. Milstein S. and Cohen L. A., *Proc. Natl. Acad. Sci. USA*, **67**, 1142 (1970).
34. Storm D. R. and Koshland D. E., *J. Amer. Chem. Soc.*, **94**, 5805, 5815 (1972).
35. Volkenstein M. V., *Molecular Biophysics*. Ch. 6. Nauka. Moscow. 1975.
36. Menger F. M., *Tetrahedron*, **39**, 1013 (1983); *Acc. Chem. Res.*, **18**, 128 (1985).
37. Wheeler H. Z., Goffman J. B., and McFarland L. F., *J. Amer. Chem. Soc.*, **25**, 790 (1903).
38. Forster M. O. and Garland C. S., *J. Chem. Soc.*, **95**, 2051 (1909).
39. Dimroth O. and Hartman M., *Ber.*, **40**, 2404, 4460 (1907); **41**, 4012 (1908).
40. Curtin D. Y. and Engelmann J. H., *J. Org. Chem.*, **37**, 3439 (1972).
41. Curtin D. Y. and Routsma D. S., *J. Amer. Chem. Soc.*, **84**, 4887, 4892 (1962).
42. Curtin D. Y. and Dreclener I. D., *J. Org. Chem.*, **33**, 1552 (1967).
43. Hauthal H. G., *Tetrahedron Lett.*, 759 (1966).
44. Curtin D. Y. and Miller L. L., *Tetrahedron Lett.*, 1869 (1965).
45. Curtin D. Y., Gibbs E. L., and McCarty C. G., *J. Amer. Chem. Soc.*, **88**, 2775 (1966).
46. Curtin D. Y. and Engelmann J. H., *Tetrahedron Lett.*, 311 (1968).
47. Schwartz, *J. Org. Chem.*, **37**, 2906 (1972).
48. Chan A. W. K. and Grow W. D., *Austral. J. Chem.*, **21**, 2967 (1968).
49. Bushagen H. and Geiger W., *Chem. Ber.*, **103**, 123 (1970).
50. McKillop A., Lelesko M. J., and Tayceor E. C., *Tetrahedron Lett.*, 4945 (1968).
51. Fleming J. and Philippides D., *J. Chem. Soc. C*, 2426 (1970).
52. Mumm O. and Hesse H., *Ber.*, **43**, 2505 (1910).
53. Mumm O., Hesse H., and Volghartz, *Ber.*, **48**, 379 (1915).
54. Chapman A. W., *J. Chem. Soc.*, 127, 1992 (1925); 2296 (1926); 1743 (1927); 2133 (1929); 2548 (1930).
55. Schulenberg and Archer S., *Org. Reactions* **14**, 1 (1965).
56. Wiberg K. B. and Rowland B. I., *J. Amer. Chem. Soc.*, **77**, 2205 (1955).
57. Sauer J. and Huisgen R., *Angew. Chem.*, **72**, 309 (1960).
58. Chapmann A. W. and Perrott C. H., *J. Chem. Soc.*, 2462 (1930); 1770, 1775 (1932).
59. Stevens T. S., *J. Chem. Soc.*, 1932 (1932); Schönberg A., *Lieb. Ann. Chem.*, **483**, 107 (1930).
60. Newman M. S., *J. Org. Chem.*, **31**, 3980 (1966); Kwart H. and Evans E. R., *J. Org. Chem.*, **31**, 410 (1966).
61. Kemp D. S. and Velassio S., *J. Org. Chem.*, **40**, 3003 (1975).
62. Dewar M. J. S., *The Molecular Orbital Theory in Organic Chemistry*. McGraw-Hill. New York. 1969.
63. Olekhnovich L. P., Nivorozhkin L. E., and Minkin V. I., *Zh. Org. Khim.*, **6**, 1681 (1968).
64. Woodward R. B. and Olofson, *J. Amer. Chem. Soc.*, **83**, 1007, 1010 (1961).

65. Mumm O. and Bergell C., *Ber.*, **45**, 3040 (1912); Mumm O. and Hornhardt H., *Ber.*, 70, 1930 (1937).
66. Calder J. C. and Cameron D. W., *J. C. S. Chem. Commun.*, 360 (1971).
67. Castells J., Merino M. A., and Moreno-Manas M., *J. C. S. Chem. Commun.*, 709 (1972).
68. Minkin V. I., Olekhnovich L. P., Zhdanov Yu. A., Kiselev V. V., Voronov M. A., and Budarina Z. N., *Zh. Org. Khim.*, **9**, 1319 (1973).
69. Kravtsov D. N., Fedorov L. A., Peregudov A. S., and Nesmeyanov A. N., *Dokl. Akad. Nauk USSR*, **196**, 110 (1971).
70. Minkin V. I., Olekhnovich L. P., Zhdanov Yu. A., Mikhailov I. E., and Metlushenko V. P., *Zh. Org. Khim.*, **11**, 448 (1975).
71. Minkin V. I., Oleknovich L. P., Zhdanov Yu. A., Mikhailov I. E., Metlushenko V. P., Ivanchenko N. M., and Borisenko N. I., *Zh. Org. Khim.*, **12**, 1261 (1976).
72. McKennis J. S. and Smith P. A., *J. Org. Chem.*, **37**, 4193 (1972).
73. Rappoport Z. and Ta-Shma R., *Tetrahedron Lett.*, 5281 (1972).
74. Capon B., Ghosh A. K., Grieve D. McL., *Acc. Chem. Res.*, **14**, 306 (1981).
75. Olekhnovich L. P., Mikhailov I. E., Ivanchenko N. M., Metlushenko V. P., Zhdanov Yu. A., and Minkin V. I., *Zh. Org. Khim.*, **14**, 340 (1978).
76. Borisenko N. I., Olekhnovich L. P., and Minkin V. I., *Teor. Experiment. Khim.*, **12**, 825 (1976).
77. Noe E. A. and Raban M., *J. Amer. Chem. Soc.*, **96**, 1598 (1974).
78. Krueger P. J. and Fulea A. O., *Tetrahedron Lett.*, 1563 (1975).
79. Santono A. V. and Mickeviaus G., *J. Org. Chem.*, **44**, 117 (1979).
80. Bürgi H.-B., *Angew. Chem. Intern. Ed. Engl.*, **14**, 460 (1975).
81. Kletskii M. E., Minyaev R. M., and Minkin V. I., *Zh. Org. Khim.*, **16**, 686 (1980).
82. Minkin V. I., Simkin B. Ya., and Minyaev R. M., *Quantum Chemistry of Organic Compounds. Mechanisms of Reactions.* Publ. House Khimiya. Moscow. 1986.
83. Burshtein K. Ya. and Isaev A. N., *J. Mol. Struct. THEOCHEM.*, **133**, 263 (1985).
84. Scheiner S., Lipscomb W. N., and Kleier D. A., *J. Amer. Chem. Soc.*, **98**, 4770 (1976).
85. Bürgi H.-B. and Dunitz J., *Acc. Chem. Res.*, **16**, 153 (1983).
86. Furmanova N. G. and Kompan O, E., *Problems of Crystallochemistry*, p. 122. Nauka. Moscow. 1985.
87. Dunitz J. D., *X-Ray Analysis and the Structure of Organic Molecules.* Cornell University Press. Ithaca and London. 1979.
88. Simkin B. Ya., Kletskii M. E., Minyaev R. M., and Minkin V. I., *Zh. Org. Khim.*, **19**, 3 (1983).
89. Minkin V. I., Olekhnovich L. P., Zhdanov Yu. A., Kiselev V. V., Voronov M. A., Nivorozhkin L. E., and Budarina Z. N., *Dokl. Akad Nauk USSR.*, **204**, 1363 (1972).
90. Minkin V. I., Olekhnovich L. P., Zhdanov Yu. A., Kiselev V. V., Voronov M. A., Nivorozhkin L. E., and Budarina Z. N., *Zh. Org. Khim.*, **8**, 1563 (1972).
91. Minkin V. I., Olekhnovich L. P., Zhdanov Yu. A., Kiselev V. V., Metlushenko V. P., Borisenko N. I., *Zh. Org. Khim.*, **10**, 2248 (1974).
92. Mannschreck A. and Dvorak H., *Tetrahedron Lett.*, 547 (1973).
93. Minkin V. I., Olekhnovich L. P., Zhdanov Yu. A., Kiselev V. V., Borisenko N. I., and Levkovich M. M., *Zh. Org. Khim.*, **11**, 1163 (1975).
94. Hartke K., Krampitz D., and Uhde W., *Chimia*, **27**, 209 (1973).
95. Hartke K., Matusch R., and Krampitz D., *Lieb. Ann. Chem.*, 1237 (1975).
96. Kruszynski M. and Kupryszewski G., *Polish J. Chem.*, **52**, 1183 (1978).
97. Hartke K. and Wachsen E., *Liebigs Ann. Chem.*, 730 (1976).
98. Hartke K., Krampitz D., and Uhde W., *Chem. Ber.*, **108**, 128 (1975).
99. Wachsen E. and Hartke K., *Chem. Ber.*, **108**, 138, 683 (1975); **109**, 1353 (1976).
100. Avang D. V. C., *Can. J. Chem.*, **51**, 3752 (1973).
101. Albright T. A., Burdett J. K., and Whangbo M. H., *Orbital Interactions in Chemistry.* J. Wiley. New York. 1985.
102. Olekhnovich L. P., Voloshin N. A., Tikhonova M. E., Borisenko N. I., Minkin V. I., and Zhdanov Yu. A., *Zh. Org. Khim.*, **21**, 2555 (1985).
103. Svensson C., Abrahams S. K., Bernstein J. L., and Haddon R. C., *J. Am. Chem. Soc.*, **101**, 5759 (1979).

104. Rossetti R., Haddon R. C., and Brus E. L., *J. Amer. Chem. Soc.*, **102**, 6913 (1980).
105. Kunze K. L. and De la Vega J. R., *J. Amer. Chem. Soc.*, **106**, 6528 (1984).
106. Schaefer J. P. and Reed L. L., *J. Amer. Chem. Soc.*, **93**, 3902 (1971).
107. Furmanova N. G., *Zh. Struct. Chem.*, **27**, 182 (1986).
108. Masamune S., Kemp-Jones A. V., Green J., and Rabenstein D. L., *J. C. S. Chem. Commun.* 283 (1973).
109. Minkin V. I., Olekhnovich L. P., Zhdanov Yu. A., and Metlushenkov. P., *Tetrahedron Lett.*, 563 (1974).
110. Minkin V. I., Olekhnovich L. P., Zhdanov Yu. A., Budarina Z. N., Metlushenko V. P., and Orenstein I. B., *Zh. Org. Khim.*, **13**, 777 (1977).
111. Elguero J., Muller R. N., Blade-Font A., Faure R., and Vincent E. J., *Bull. Soc. Chim. Belge*, **89**, 193 (1980).
112. Olekhnovich L. P., Borisenko N. I., Budarina Z. N., Metlushenko V. P., Zhdanov Yu. A., and Minkin V. I., *Zh. Org. Khim.*, **18**, 1785 (1982).
113. Budarina Z. N., Kurbatov S. V., Borisenko N. I., Zhdanov Yu. A., Minkin V. I., and Olekhnovich L. P., *Zh. Org. Khim.*, **21**, 233 (1985).
114. Bagli J. F. and St-Jacques J., *Can. J. Chem.*, **56**, 578 (1978).
115. Strauss M. J., *Chem. Rev.*, **70**, 667 (1970).
116. Artamkina G. A., Egorov M. P., and Beletskaya I. P., *Chem. Rev.*, **82**, 427 (1982).
117. Terrier F., *Chem. Rev.*, **82**, 77 (1982).
118. Minkin V. I., Olekhnovich L. P., Zhdanov Yu. A., Mikhailov I. E., Budarina Z. N., and Ivanchenko N. M., *Dokl. Akad Nauk USSR*, **219**, 357 (1974).
119. Minkin V. I., Olekhnovich L. P., Zhdanov Yu. A., Mikhailov I. E., Metlushenko V. P., and Ivanchenko N. M., *Zh. Org. Khim.*, **12**, 1271 (1976).
120. Mikhailov I. E., Minkin V. I., Olekhnovich L. P., Boiko V. I., Ignatt'ev I. V., and Yagupolskii L. M., *Zh. Org. Khim.*, **20**, 454 (1984).
121. Wolff H.-M. and Hartke K., *Tetrahedron Lett.*, 3453 (1977); *Archiv Pharmazie*, **313**, 266 (1980).
122. Newman M. S., *Acc. Chem. Res.*, **5**, 354 (1972).
123. Furmanova N. G., Struchkov Yu. T., Kompan O. E., Budarina Z. N., Olekhnovich L. P., and Minkin V. I., *Zh. Struct. Khim.*, **21**, 83 (1980).
124. Olekhnovich L. P., Furmanova N. G., Minkin V. I., Struchkov Yu. T., Kompan O. E., Budarina Z. N., Yudilevich I. A., and Eruzheva O. V., *Zh. Org. Khim.*, **18**, 465 (1982).
125. Furmanova N. G., Olekhnovich L. P., Minkin V. I., Struchkov Yu. T., Kompan O. E., Budarina Z. N., Metlushenko V. P., and Eruzheva O. V., *Zh. Org. Khim.*, **18**, 474 (1982).
126. Olekhnovich L. P., Kurbatov S. V., Budarina Z. N., Minkin V. I., and Zhdanov Yu. A., *Zh. Org. Khim.*, **21**, 2550 (1985).
127. Olekhnovich L. P., Mikhailov I. E., Minkin V. I., Furmanova N. G., Kompan O. E., Struchkov Yu. T., and Lukasch A. V., *Zh. Org. Khim.*, **18**, 484 (1982).
128. Mikhailov I. E., Kompan O. E., Makarova N. I., Yanovskii A. I., Struchkov Yu. T., Olekhnovich L. P., Knyazhanskii M. I., and Minkin V. I., *Zh. Org. Khim.*, **21**, 237 (1985).
129. Fehrst A. R. and Kirby A. G., *J. Amer. Chem. Soc.*, **90**, 5118 (1968).
130. Gandour R. D. and Schowen R. L., *J. Amer. Chem. Soc.*, **96**, 2231 (1974).
131. Pavlova L. A. and Rachinskii F. Yu., *Uspekhi Khim.*, **37**, 1369 (1968).
132. Acheson R. M., *Acc. Chem. Res.*, **4**, 177 (1971).
133. Sekiguchi S. and Okada K., *J. Org. Chem.*, **40**, 2782 (1975).
134. Bernasconi C. F., Gehriger C. L., and de Rossi R. H., *J. Amer. Chem. Soc.*, **98**, 8451 (1976).
135. Drozd V. N., *Zh. Vsesoyuzn. D. I. Mendeleev Khim. Obschz.*, **21**, 266 (1976).
136. Grovenstein E., *Adv. Organometal. Chem.*, **16**, 167 (1977).
137. Prokof'ev A. I., Prokof'eva T. I., Bubnov N. N., Solodovnikov S. N., Belostockaya I. S., Ershov V. V., and Kabachnik M. I., *Dokl. Akad Nauk USSR*, **239**, 1367 (1978).
138. Prokof'eva T. I., Prokof'ev A. I., Vol'eva V. B., Ershov V. V., and Kabachnik M. I., *Izvest. Akad Nauk USSR (Ser. Khim.)*, 1324 (1985).
139. Raevskii I. I. and Borisov Yu. A., *Dokl. Akad Nauk USSR*, **288**, 664 (1986).
140. Prokof'ev A. I., Malysheva I. A., Bubnov N. N., Solodovnikov S. N., Belostockaya I. S., Ershov V. V., and Kabachnik M. I., *Dokl. Akad Nauk USSR*, **229**, 1128 (1976).

141. Kabachnik M. I., Bubnov N. N., Solodovnikov S. N., and Prokof'ev A. I., *Tautomerism of Free Radicals*, in: *Itogi Nauki i Techniki*, vol. 5. Ser. Organic Chemistry (Eds. Volpin M. E. and Mironov E. A.).
142. Hehre W. J., Radom L., Schleyer P. v .R., and Pople J. A. *Ab Initio Molecular Orbital Theory*. J. Wiley. New York. 1986. Section 7.3.
143. Koptyug V. A., Shubin V. G., Rezvukhin A. I., Korchagina D. V., Tret'yakov V. P., and Rudakov E. S., *Dokl. Akad Nauk USSR,* **171**, 1109 (1966).
144. Morozov S. V., Shakirov M. M., Shubin V. G., and Koptyug V. A., *Zh. Org. Khim.,* **15**, 770 (1979).
145. Borodkin G. I., Shakirov M. M., Shubin V. G., and Koptyug V. A., *Zh. Org. Khim.,* **12**, 1297 (1976).
146. Bushmelev V. A., Shakirov M. M., and Koptyug V. A., *Zh. Org. Khim.,* **12**, 2480 (1976); **13**, 2161 (1977).
147. Stothers J. B., *Carbon-13 NMR Spectroscopy*. Academic Press. New York. 1972. p. 219.
148. Spangler C. W., *Chem. Rev.,* **76**, 187 (1976).
149. Childs R. F., *Tetrahedron,* **38**, 567 (1982).
150. Borodkin G. I., Suscharin E. R., and Shubin V. G., *Zh. Org. Khim.,* **19**, 1004 (1983).
151. Borodkin G. I., Suscharin E. R., Shakirov M. M., and Shubin V. G., *Zh. Org. Khim.,* **21**, 451 (1985).
152. Bushby R. J. and Jones W. D., *J. C. S. Chem. Commun.,* 688 (1979).
153. Mikhailov I. E., Dushenko G. A., Ustynyuk Yu. A., Olekhnovich L. P., and Minkin V. I., *Zh. Org. Khim.,* **20**, 2626 (1984).
154. Childs R. F. and Zeya M., *J. Amer. Chem. Soc.,* **96**, 6418 (1974).
155. Anh N. T., Elian M., and Hoffmann R., *J. Amer. Chem. Soc.,* **100**, 110 (1978).
156. Schoeller W. W., *J. C. S. Dalton Trans.,* 2233 (1984).
157. Jones D. W. and Kneen G., *J. C. S. Perkin Trans. I,* 1313 (1977).
158. Field D. J. and Jones D. W., *J. C. S. Chem. Commun.,* 688 (1977).
159. Backes J., Hoffmann R. W., and Steuber F. W., *Angew. Chem.,* **87**, 587 (1975).
160. Schiess P. and Funfschilling B., *Tetrahedron Lett.,* 5191, 5195 (1972).

3. General Principles of the Design of Tautomeric Systems

Analysis of the concept of tautomerism, of the domain of chemical transformations regarded as tautomeric and the detailed examination of the carbonotropic tautomerism reactions makes it possible to attempt the formulation of general principles of target-oriented construction of tautomeric systems on the basis of migration processes.

In its essence, this task consists in the prediction and design of a reversibly-equilibrated molecular system whose thermodynamic and activation parameters correspond to the commonly accepted scale of tautomeric processes (Equation (1.5)). A simplified, though costly, approach is, in principle, possible: i.e. straightforward calculation of differences between free energies of tautomers as well as between activation energies for the forward and reverse potentially tautomeric reactions. This approach, while not requiring any efforts of chemical imagination, is fraught with practical difficulties and complications associated with the choice of the appropriate methods of quantum mechanical calculations.

1. Possibilities of Quantum Mechanical Prediction of Tautomeric Reactions

As soon as a reaction which is regarded as a candidate for a tautomeric system is selected, it is necessary to find out if (a) the difference between the free energies of the tautomers lies in the range of 5—6 kcal/mol, and (b) whether the free activation energies of the forward and reverse reactions do not exceed the value of 20—25 kcal/mol

$$A \rightleftharpoons B \tag{3.1}$$

The former task is the easier one for the quantum mechanical calculations since it requires the determination of parameters of only two structures A and B identifying with the minima of the potential energy surface (PES) of the system (3.1). Rigorous calculations, however, may run into certain limitations.

1.1. *Theoretical Evaluation of Free Energies of Tautomeric Equilibria*

The difference between the free energies of the tautomers A and B at the temperature T

$$\Delta G_T = H_T^B - H_T^A + \frac{1}{2} \sum_{i \in B} h\nu_i - \frac{1}{2} \sum_{j \in A} h\nu_j -$$

$$- T(S_{vib}^B - S_{vib}^A + S_{rot}^B - S_{rot}^A + S_{t_r}^B - S_{t_r}^A) \qquad (3.2)$$

where the first and the second differences correspond to the difference between the enthalpies of the tautomers A and B and to that between the energies of their zero point vibrations, respectively, the last term comprising the differences between the vibrational, the rotational and the translational components of entropy.

The dependence of reaction enthalpy on temperature is given by Kirchhoff's law:

$$\Delta H_{B-A}^T = \Delta H^0 + \int_0^T (\Delta C_p) \, dT \qquad (3.3)$$

where C_p is the thermal capacity whose value, similarly to that of entropy, can be broken up into the vibrational, rotational and translational components.

In the isomerization reactions of type (3.1) (in which interconverting compounds have equal masses), contributions of the rotational and translational terms to Eq. (3.2) cancel out. Assessment of the contribution made by vibrational terms is most important requiring calculation of vibration frequencies of both tautomers. The technique of this calculation employed as the usual harmonic approximation is well developed. It is based on the evaluation of second derivatives of the function (1.1), i.e. of the potential energy surface curvature in the region of minima [1, 2]. In tautomeric reactions, in which the structures of interconverting compounds differ, mainly, in the position of only one atomic group, contributions by vibrational terms in (3.2) and (3.3) are usually neglected. In this case, the enthalpy of the reaction (3.3) is evaluated as simply the difference between internal energies at $T = 0$ K and the expression (3.1) is reduced to

$$\Delta G_{B-A} = \Delta E_{B-A} + \frac{1}{2} \sum_{i \in B} h\nu_i - \frac{1}{2} \sum_{j \in A} h\nu_j \qquad (3.4)$$

In Table 3.I results obtained by various semiempirical and *ab initio* methods are compared with the experimental data on tautomeric equilibria of 2- and 4-pyridones.

Table 3.I. Tautomerization Energies (kcal/mol) for 2-pyridone and 4-pyridone [3].

Calculational method	ΔE_{B-A} 2-pyridone	ΔE_{B-A} 4-pyridone
MINDO/3	3.7	4.0
STO-3G	15.4	18.6
3-21 G	−1.7	0.7
6-31 G*	−0.4	2.3
MP2/6-31 G*[a]	−1.2	1.7
Zero point vibration[b]	0.8	0.7
Best theoretical estimate	−0.4	2.4
Best experimental estimate	−0.3 ± 0.3	7 ± 2

[a] Split 6-31 G basis with addition of polarization *d*-functions. Correlation energy is taken into account through second-order Møller — Plessett perturbation theory.
[b] MINDO/3 frequencies calculated at MINDO/3 geometries.

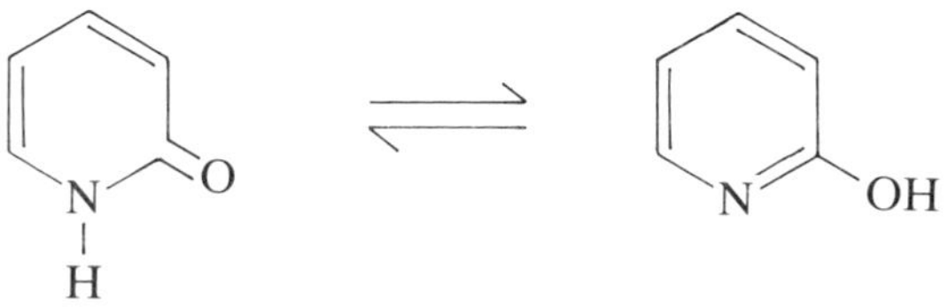

Ia   Ib   (3.5)

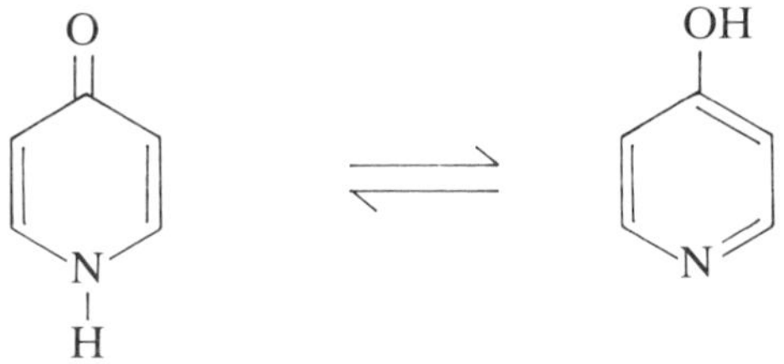

IIa   IIb

Tautomerisation energies being in fact differences between large magnitudes, high accuracy is required when evaluating these magnitudes. As is evident from Table 3.I, satisfactory reproduction of experimental data may be expected only at fairly high levels of approximation. A purely theoretical approach to tautomeric equilibria involves additional difficulties related to the necessity to take account of the role played by association equilibria and by the solvent. In condensed phases, the equilibria (3.5) are considerably shifted towards the pyridone tautomers due to the difference between the solvation energies of tautomers being rather high, amounting, for example, for **IIa** and **IIb** in water to 5—6 kcal/mol [4]. The solvent molecules must be, directly or indirectly, included in theoretical calculations of solvation effects. The most important approaches to their evaluation have to do with the supermolecule approximation, with constructing model Hamiltonians and developing methods of calculating the structure and energetics of a solution [1, 5—7]. One of the most promising is the Monte Carlo method. Monte Carlo calculations [8] on the **IIa** ⇌ **IIb** tautomeric system carried out with 110 molecules of water produced very good agreement with the experimental data on solvation energy.

Thus, it may be concluded that the ways to fairly reliable theoretical evaluations of the positions of tautomeric equilibria, both in gas phase and in solution are, in principle, known. Of course, these evaluations require a great deal of computer time and involve quite a few assumptions which have still to be checked.

1.2. *Theoretical Calculation of Activation Energies of Tautomeric Reactions*

The problem of the theoretical calculation of activation energies of tautomeric interconversions is much more complex than the calculation of the equilibrium free energy. In order to solve it, it is necessary to calculate the energy of the transition state of a reaction which can be done only after identifying its structure.

In the general case, the problem consists in calculating the potential energy surface (PES) of the reaction (3.1) as a function of $(3q-6)$ coordinates where q is the total number of all atoms forming the reacting system. If, as is typical, the system is examined in its lower electron and vibrational states, the next step is the finding of the minimum energy reaction path (MERP) on the PES. The reaction pathway is selected in such a way that it starts from the point corresponding to the geometric configuration of one tautomer and ends up with the point corresponding to that of the other with each advancing step following the line of the minimal energy increase. The saddle point on this path represents the transition state of reaction.*

The relative energy of the transition state structure with respect to reagents determines the rate of their interconversion. The structure of the transition state controls the stereochemical result of the reaction. The position of the transition state structure on the potential energy surface of a reaction affects the energy distribution in the reaction products.

It is impossible, in principle, not only to isolate but even to simply detect the transition states of a reaction by any physical methods.** The only ways to study them are by theoretical calculations. Difficulties with which one may be confronted

---

* Unambiguous definition of the MERP requires, as a preceding step, the localization of the saddle point of the transition state. The paths of the steepest descent from this point to the points of the potential energy surface minima correspond to the MERP [1, 9, 10].

** Statements to the contrary can still be found in some sources, see, for example [11]. Definitive analysis of this problem was given by Wolkenstein [12].

in performing them are evident from the so-called $3q$-6 problem. To calculate, for example, the potential energy surface of the simple prototropic tautomeric system **Ia** ⇌ **Ib** requires calculation of the total energy of **I** as a function of 30 ($q = 12$) coordinates. Of course, such a comprehensive approach is unrealistic. A number of special analytical methods have been developed to search for transition state structures in many important reactions yielding quite a few data on their geometries and energies.

McIver [13] once figuratively compared discussions concerning the structure of transition states to a dispute as to the intimate habits of dinosaurs in which the only available evidence is the fossil fragments of those monsters. Even though this is on the whole true, the transition state structure of many important reactions is not any longer that mysterious.

When the structure of a transition state is identified, it is not difficult to calculate its vibrational spectrum in which there is present one imaginary frequency. At $T = 0$ K:

$$\Delta G^{\neq} = \Delta E + \tfrac{1}{2} \sum_{i \in TS} h\nu_i - \sum_{j \in A} h\nu_j \qquad (3.6)$$

where ΔE is the difference between the internal energies of the transition state and reagents (or products) and ν_i and ν_j are the frequencies of normal vibrations of respective structures. The term with imaginary frequency is omitted for the transition state. Corrections for arbitrary temperature, i.e. calculation of entropy of activation, are made as in Equation (3.2).

According to *ab initio* $(3 - 21G)$ calculations [14], the structure of the transition state for the monomolecular reaction **Ia** ⇌ **Ib**

A large barrier against proton migration is found, being 49 kcal/mol with respect to 2-hydroxypyridine. This means that the actual mechanism of interconversion of tautomers **I** is different from the scheme presented. In agreement with the experimental data [4], the calculations [14] show that in the dimeric associate **III** this barrier drops to 11 kcal/mol thus fitting the scale of tautomeric reactions.

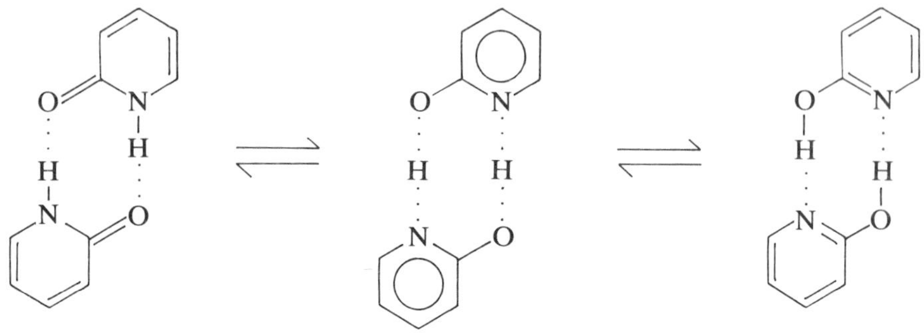

IIIa   IIIc(TS)   IIIb

The reason for such a significant lowering of the activation barrier, i.e., decrease in the relative energy of the transition state structure, is, apparently, related to realization of the optimal linear geometry in **IIIc** for the proton shifting along the N—H—O bridge.

The 2-pyridone—2-hydroxypyridine system may serve as an example of a sufficiently thorough quantum mechanical examination of a tautomeric system. There are other examples of this kind, and they will be analyzed in the coming chapters. In any case, it is obvious that the comprehensive calculational approach cannot replace the qualitative structural concept of tautomeric system construction. It should rather supplement and specify structural ideas.

2. Stereochemical Approach to the Design to Tautomeric Compounds

We shall consider this approach as applied to the most important type of tautomeric reactions, i.e., the reactions of atomic group transfer between two nucleophilic centers of a molecular system. Reactions of this type are almost without exception the *i,j*-sigmatropic rearrangements, and the problem is to advance a well-founded principle of selecting those fitting the thermodynamic and activation scale of tautomeric transformation energies.

As has already been noted in Section 1.1, the problem of evaluating relative stability of two tautomeric forms, i.e., free energy of equilibrium, is at present quite amerable to solution by the methods of contemporary quantum chemistry. We therefore, without losing much generality, shall in the following restrict ourselves to degenerate tautomeric systems, in which case this problem, obviously, does not exist.

The principal points of the generalized approach to the design of the tautomeric systems can be formulated as follows [15, 16].

For the *i,j*-sigmatropic tautomeric transfer of an atomic group between two centers within a molecule or an ion to take place, the structure of the migrating group carrier system has to ensure:

(a) favorable electronic factors and steric conditions for coordination of the central atom of the migrating group by the accepting nucleophilic center in the direction corresponding to the reaction path for intramolecular transfer of this group.

(b) the possibility of the cyclic electron transfer leading to a redistribution of σ, π-bonds in the main molecular system and to rupture of the old and formation of the new σ-bond of the migrating group and to a rearrangement of bonds.

(c) favorable steric conditions for the formation of the transition state whose structure is determined by the type of the substitution reaction at the migrating center and by the nature of the migrating group.

(d) in the case of nonconcerted mechanism, the possibility of a low-barrier polytopal rearrangement of the cyclic intermediate appearing upon coordination of the central atom of the migrating group with the nucleophilic center.

Of the requirements the structure of the molecular system attached to a migrant has to meet, those in paragraphs (a) and (b) deal with the electronic and, partly, stereoelectronic factors whereas (c) and (d) refer to the stereochemistry of the intramolecular substitution reaction at the central atom of the migrating group.

2.1. *Intramolecular Coordination of the Migrating Group*

We have already noted earlier (Section 1.7) that tautomeric transformations can take place only if there is a possibility of intramolecular coordination of the central atom of a migrant with the atomic center towards which the migratory group is shifted.

If conditions for such coordination do not exist, intramolecular tautomeric rearrangements do not occur. Thus, one should not expect that e.g. *N*-chloroamidines would undergo tautomeric rearrangements of the following type

which are analogous to the prototropic and carbonotropic tautomerism of *N*-substituted amidines (Chapter 2). The reason for this lies in the fact that the nucleophilic coordination of imine group by chlorine is impossible due to the absence of low-lying vacant orbitals at the chlorine atom and to the repulsion of the electron lone pairs at these centers.

The possibility of intramolecular coordination must be ensured also by the molecular shape of a tautomeric system. Earlier (Section 2.3), some examples were cited of inhibition of tautomeric transfer processes in acyl and picryl derivatives of pyrazole resulting from unfavorable orientation of the lone pair orbital of the nitrogen atom in the heterocycle. An additional interesting example is associated with the unusual stability of *O*-acyl isoamide derivatives obtained from the nitrilium cations

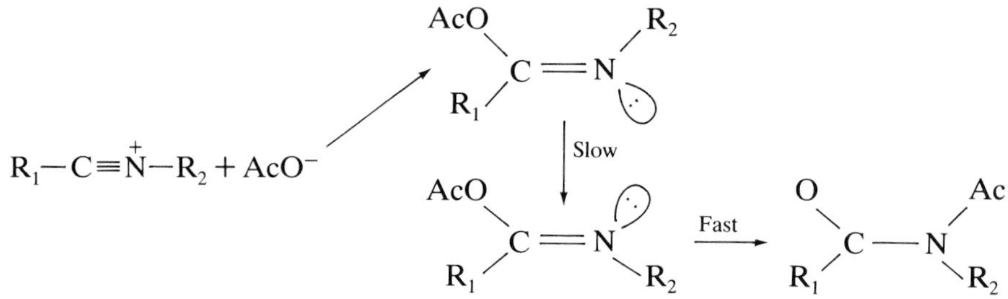

Reaction of acylate ions with the latter was shown [17] to produce exclusively the *Z*-isomers of *O*-acyl isoamides, the reaction rate of the acyl group *O,N*-transfer depending on the rate of the $Z \rightleftharpoons E$ topomerization. The *E*-isomer of *O*-acyl isoamide, in which the nitrogen lone pair orientation is favorable, rapidly rearranges into the diacylamide isomeric form.

2.2. Cyclic Electron Transfer in the Sigmatropic Reaction and Multigraphs of Tautomeric Systems

Zefirov and Trach [18] emphasized that all main types of tautomerism in electrically neutral systems could be described by formal equations with the cyclic electron transfer.

Cyclic electron transfer is a concept of classical structural and electronic theories of organic chemistry. It has to reflect the synchronized process of electron density redistribution within a molecule (or a conglomerate of molecules) caused by rupture of the old and formation of the new bonds in the course of, respectively, intra- and intermolecular reactions. Real and hypothetical structures shown below where arrows denote the directions of the σ and π bond electron shifts give an idea of the cyclic electronic transfer in the tautomeric reactions including 4—6-membered rings

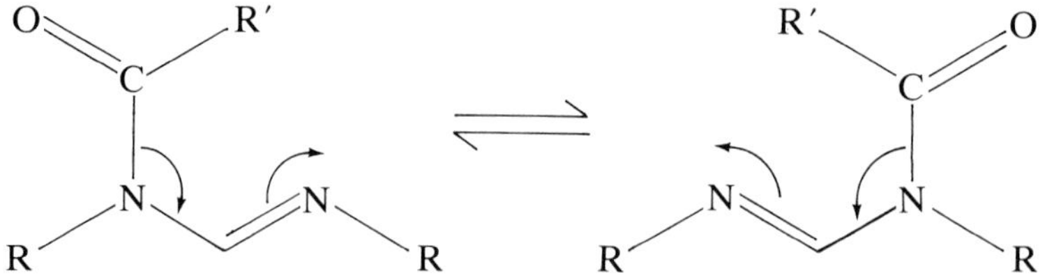

(3.7)

IVa   **IVb**

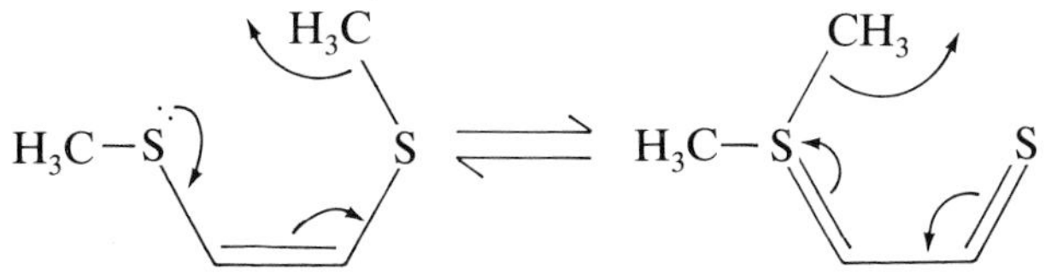

$$ \text{(3.8)} $$

Va   Vb

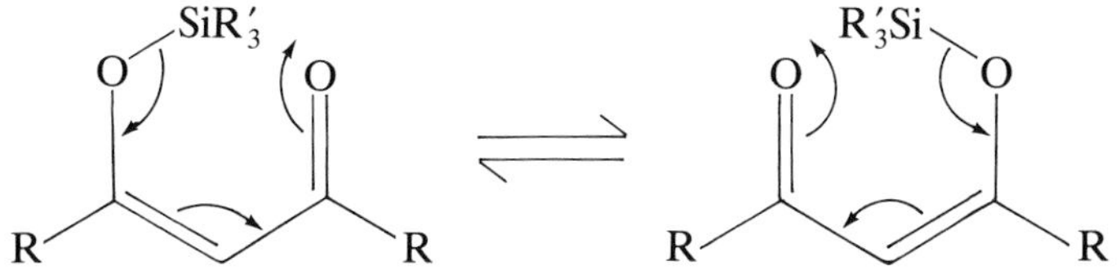

$$ \text{(3.9)} $$

VIa   VIb

Besides σ- and π-bonds, the lone pairs in **V** can serve as donors producing electronic shifts in the fragment with respect to which the σ-bond migrates. In the latter case, the sigmatropic transitions between potentially tautomeric structures entail changes in the formal valency of individual atomic centers. In the formalization [20] these centers are called specific and denoted as X-centers in the symbolic equations of sigmatropic reactions. Those centers whose total valence remains constant are nonspecific, and they are denoted by thick dots. Equations of the reactions (3.7)–(3.9) correspond to generalized symbolic equations (3.7a)–(3.9a), respectively.

$$ \text{(3.7a)} $$

$$ \text{(3.8a)} $$

$$ \text{(3.9a)} $$

The essential feature of this approach is the idea of discriminating between bonds common to both reagent and product and bonds which change their place during the reaction [21, 22]. The terms of the symbolic equations are in fact acyclic multigraphs [23, 24] whose edges are determined by the bond connectivities in conjugated tautomeric forms. Multigraphs are characterized by the presence of multiple edges (multiplicity corresponds in these graphs to the formal bond order not exceeding three). Graphs of the real structures must also contain edges of cyclic compounds, which increases the cyclomatic number of a graph [24]. The degree of the vertex (point) will be limited to four in number, corresponding to the maximum valency usually manifested by atomic groups in organic compounds.

Table 3.II contains structural graphs of all possible tautomeric forms, the interconversion of which is associated with 1,3- and 1,5-sigmatropic reactions in conjugated, electrically neutral systems.

Table 3.II. Graphs of 1,3- and 1,5-sigmatropic rearrangements*

Sigmatropic Shift	Graph		Example of tautomeric compound

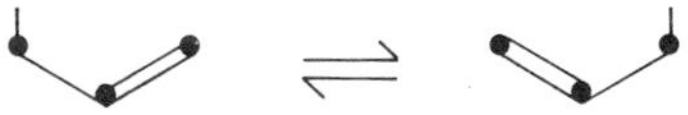

The graphs and structures are shown for 1,3- and 1,5-sigmatropic shifts, with examples (2.32), (2.42), the M—N⟶N structure, (2.35), (2.29), and the M—N—O structure.

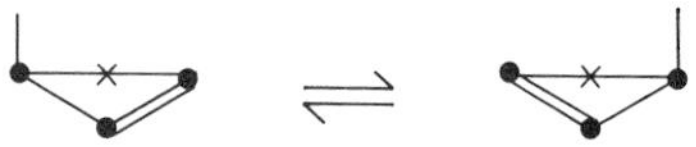

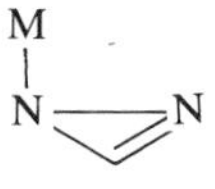

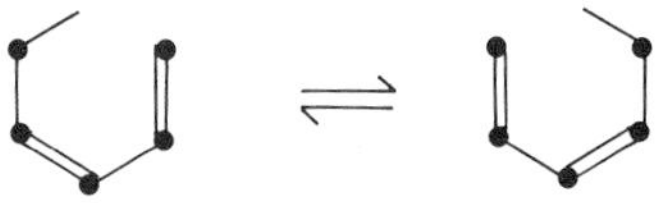

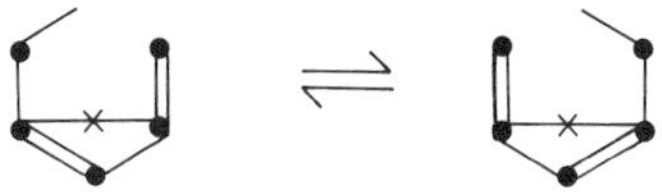

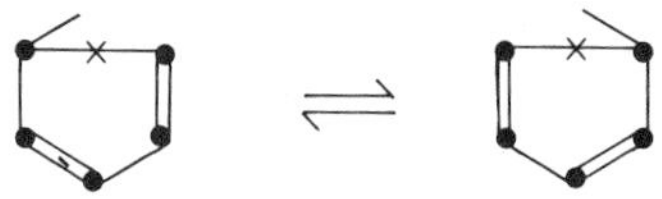

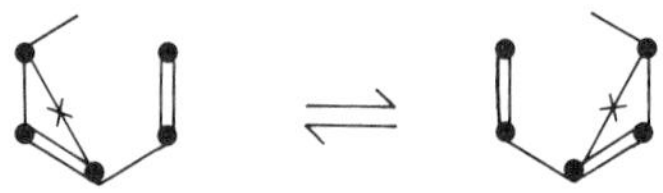

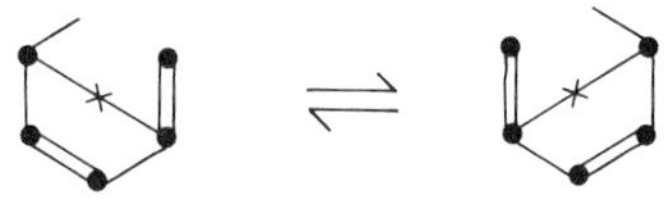

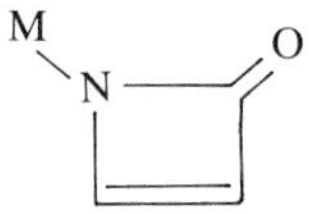

* To form a molecular structure from the graph a vertex in the latter is replaced by the following atomic group:

[a] —• M—O, M—S, M—N(R), M—P(R), M—CH₂···· where M is migrating group;

[b] —•< M—N< M—C(R)< ;

[c] ⟨══ O=, S=, N(R)=, CH₂=····;

[d] —•══ —CH=, —N=, —O⁺=····;

[e] >•══ >C=, >N⁺=····;

[f] —×— stands for any single bond or conjugated chain +•═══•+ₘ.

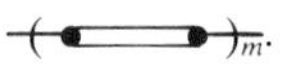

The technique of predicting an expected sigmatropic tautomeric structure consists in the replacement of a vertex in a graph by the symbol of a real atomic group whose valency coincides with the degree of the substituted vertex, and the multiplicity of whose edges coincides with that of the bonds formed by this group (isomorphic substitution). The main groups to be substituted in the graph points are listed in the footnote to Table 3.II.

A similar approach in principle is now used fairly extensively to formalize and systematize a number of structures (in such cases as, for instance, the search for the number of valence isomers of anulenes [25], for resonance forms of polycyclic compounds [26], isomerization reactions [27] and perycyclic reactions [22, 28].

If the *i,j*-sigmatropic migration proceeds in the direction of the atomic center carrying a positive or a negative charge, then this process is not associated with the redistribution of bond orders in the molecular system carrying a migrant, as in the scheme of the cyclic electron transfer. The formal structure of the main molecular system does not suffer any alterations and the value of *j* in conjugated compounds can only be even ($j = 2\,K$)

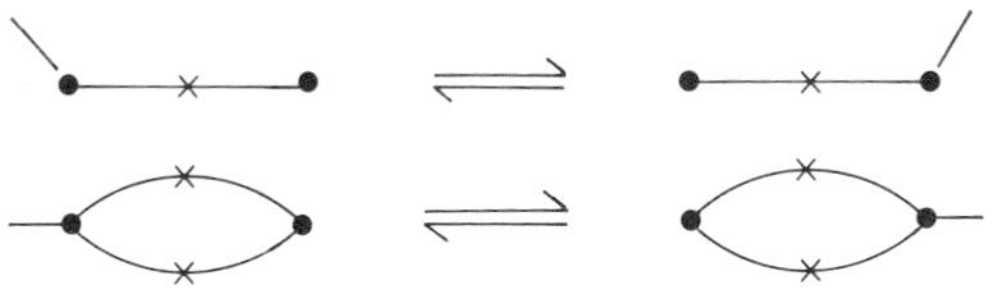

$$\text{(3.10)}$$

$$\text{(3.11)}$$

where   ⟶●   corresponds to $\bar{O}, \bar{S}, \bar{N}R, \bar{C}H_2, \bar{C}(R_1R_2), \ldots$,
$$\overset{+}{S}, \overset{+}{N}R, \overset{+}{C}H_2, \overset{+}{C}(R_1R_2);$$

  ⟩●   corresponds to $-\bar{N}-, -\bar{C}H-, -\bar{C}(R)-, \ldots, -\overset{+}{N}-, -\overset{+}{C}H-,$
$$-\overset{+}{C}(R)-, -\overset{+}{S}(R_1R_2)-$$

and   ⟶✕⟶   is explained in the footnote to Table 3.II.

Equation (3.10) describes, for instance, the 1,2-shifts of the alkyl groups in carbenium ions and the migration of trimethylsilyl groups in the *N*-anions **VII**

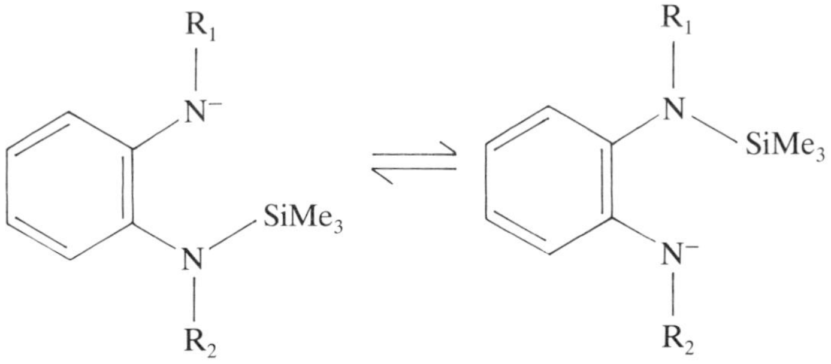

VIIa   **VIIb**

The (3.10) and (3.11) type reactions represent elementary stages in the well-known acyl rearrangements of derivatives of aminoalcohols, carbohydrates and other bifunctional compounds susceptible to nucleophilic catalysis [30]. For example, the S → O-acetyl migration in the thioglycol derivative can be represented as follows:

COCH₃   COCH₃   H₃COC   H₃COC

$\begin{array}{c} \text{COCH}_3 \\ \text{S} \quad \text{OH} \\ \xrightarrow{\text{B:}} \\ \text{CH}_2-\text{CH}_2 \end{array}$ $\begin{array}{c} \text{COCH}_3 \\ \text{S} \quad \text{O}^- \\ \longrightarrow \\ \text{CH}_2-\text{CH}_2 \end{array}$ $\begin{array}{c} \text{H}_3\text{COC} \\ \bar{\text{S}} \quad \text{O} \\ \xrightarrow{\text{H}^+} \\ \text{CH}_2-\text{CH}_2 \end{array}$ $\begin{array}{c} \text{H}_3\text{COC} \\ \text{HS} \quad \text{O} \\ \\ \text{CH}_2-\text{CH}_2 \end{array}$

Membership of the class of one of the graphs listed in Table 3.I is the necessary condition for the occurence of the *i,j*-sigmatropic shift of the migrant in the electrically neutral molecular system. Except for the two last cases, the formal rearrangements listed in the Table are symmetrical and, consequently, the reactions corresponding to them are degenerate. Hence they satisfy the thermodynamic criterion. However, the correspondence to the structural graphs of Table 3.II as well as fitting the thermodynamic criterion are not yet sufficient conditions for the assignment of the molecular system to the tautomeric type. In addition to that, the requirement for a low enough activation barrier of tautomeric interconversion has to be implemented.

2.3. *Orbital Symmetry and Activation Barriers of Sigmatropic Shifts Between Heteroatomic Centers*

Sigmatropic shifts of the atomic groups migrating between the heteroatomic centers possessing electron lone pairs, as considered in the previous chapters, do not, in fact, belong to pericyclic reactions, and the Woodward—Hoffmann rules [31] are, in general, not applicable to them. In these reactions, for which the name pseudopericyclic has been proposed [32], the bonding orbital of a migrant interchanges with the nonbonding orbital of the donor center. The effects of intersection of π-orbital levels in the conjugated system prove to be minor in comparison with the requirement for stereochemical correspondence between the initial structure and the transition state structure inherent in the substitution at the migratory center.

Straightforward proof of this was obtained upon studying the sigmatropic silyl rearrangements. Using the data of the ¹H and ¹³C NMR spectra, it was shown [33, 34] that the 1,5- and 1,9-sigmatropic shifts of the prochiral silyl groups in the *O*-derivatives of acetylacetone and tropolone proceed with retention of configuration at the migrating center. Therefore they belong to suprafacial shifts and should be best viewed as internal nucleophilic displacement processes. The reversible suprafacial 1,5-sigmatropic shift of the trimethylsilyl group in the pyrazole derivative **VIII**, which is allowed by the rules of orbital symmetry retention, requires a greater activation energy than the thermally forbidden 1,2- and 1,3-migrations in the derivatives of hydrazine **IX** and triazene **X**.

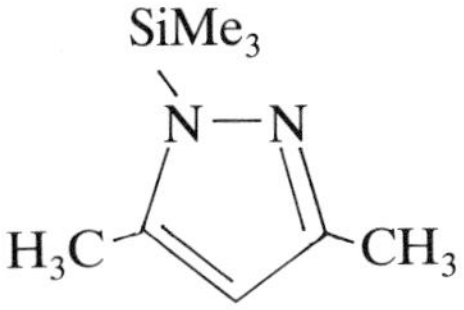

VIII $E_a = 24$ kcal/mol [35] 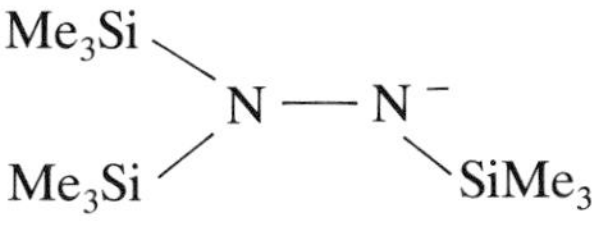   **IX** $\Delta G^{\neq} = 10.9$ kcal/mol [36] 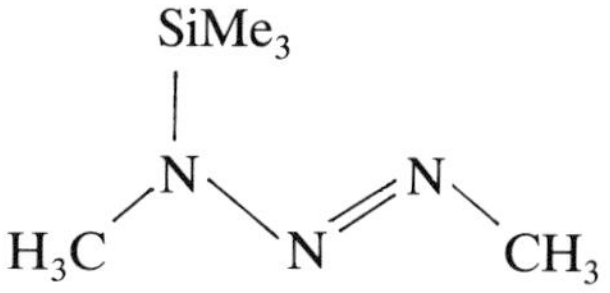   **X** $\Delta G^{\neq} = 16.1$ kcal/mol [37]

The data referred to imply that the orientation effects of the lone pair orbital forming a new bond with the central atom of a migrant play the decisive role in determining the height of the activation barrier. Effects of additional stabilization of the transition state structure resulting from the interaction between the π-orbitals of the carrier system and the orbitals of the migrating bond, and controlled by the orbital symmetry rules, have a subordinate function.

2.4. *Steric Demands of the Transition State Structures*

The overall scheme of tautomeric reactions whose nature is related to intramolecular migrations of complex atomic groups MR_n, in which the central atom M is an element located to the right of carbon in the Periodic Table may be represented by the following equation:

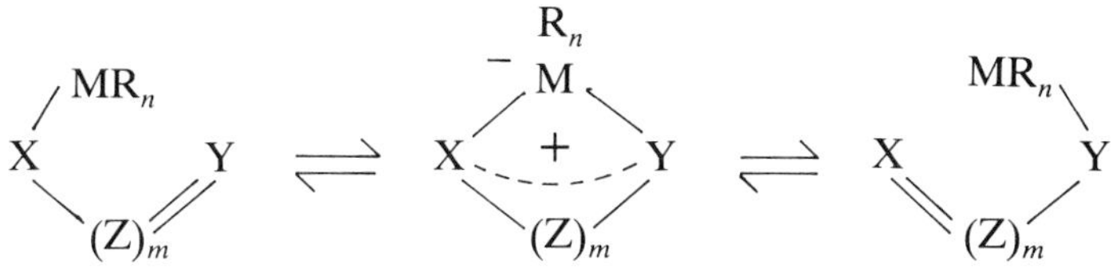

$$\text{XIa} \rightleftharpoons \text{XIc} \rightleftharpoons \text{XIb} \tag{3.12}$$

XIa   **XIc**   **XIb**

Where X, Y = O, S, NR
—Z— = CR, N
MR_n = COR, Ar, NO, NO$_2$, PR$_2$, PR$_3^+$, POR$_2$, AsR$_2$, SR, SOR, SOR$_1$R$_2$.

Investigation of the carbonotropic tautomeric systems (Chapter 2) and the above consideration of the structural factors governing the magnitudes of energy barriers in these enable us to formulate a general idea of the molecular design of the systems susceptible to fast and reversible intramolecular rearrangements [15, 16]. It is necessary for the molecule, either to be rigidly fixed in a structure similar in its geometrical characteristics to the structure of the transition state of the reaction, or to display a high conformational mobility permitting it to take on such a structure without overcoming significant energy barriers.

Since reactions of type (3.12) correspond in their mechanism to an intramolecular nucleophilic substitution at the central atom M of the migrating group, the problem of the synthesis of suitable systems consists of the inclusion of reacting centers (X, Y, M) in the corresponding chain (Z), whose dimensions and conformation will correspond to the requirements of the transition state of S_N2 or *AdE* (**XIc** is the intermediate structure) reactions.

The stereochemistry of nucleophilic addition of the migrating group MR_n to the central atom M will be considered first.

2.4.1. THE BOND CONFIGURATION OF THE CENTRAL ATOM

Detailed characteristics of MERPs and transition state structures of the reaction (3.12) can be obtained only by quantum mechanical calculations. However, examining the reactions (3.12) in an addition—elimination scheme (*AdE*), and regarding the intermediate **XIc** as being topologically similar to the transition state structure, it is easy to derive the principal characteristics of **XIc** proceeding from the highly predictive qualitative model of the VSEPR (valence state electron pair repulsion) theory of Gillespie and Nyholm [38]. According to this theory, which is supported by numerous experimental data, the configuration of bonds in molecules and ions of the MR_n type (where M is the central atom and R are the various groups connected with M by ordinary and multiple bonds) is determined solely by the number of pairs of valence electrons at the σ-bonds and nonbonding orbitals.

Before considering the steric structure of the system carrying a migrant, one's attention should be focused on the important question concerning relative orientation in transition structures of the entering and leaving nucleophilic centers. This question does not arise in the case of two- and three-coordinated structures (with the sole exception of the T-shaped structure No 6 in Table 3.III) as well as tetracoordinated tetrahedral structures. But in regard to tetracoordinated structures of the bisphenoidal and square-planar types, to penta- and hexacoordinated structures, several variants of the relative positions of nucleophilic centers are possible.

It is necessary to know which of these variants is preferable for achieving a transition state since, depending on whether angular or linear configuration of the migrating center bonds with the nucleophilic atoms is established, the conformation of the molecular system carrying the migrant has to satisfy different requirements.

Let us consider the variants of different relative positions of nucleophilic centers leading to nonequivalent configurations:

(1) T-shaped structure

$$N\text{---}M\text{---}N \qquad\qquad N\text{---}M\text{---}$$
$$\vert \qquad\qquad\qquad\qquad\qquad \vert$$
$$\qquad\qquad\qquad\qquad\qquad\qquad\qquad N$$

XIIa   **XIIb**

(2) Bisphenoidal structure

XIIIa   **XIIIb**   **XIIIc**

(3) Square-planar structure

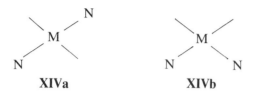

XIVa   **XIVb**

(4) Trigonal bipyramidal structure

XVa   **XVb**   **XVc**

(5) Octahedral structure

XVIa   **XVIb**

Among these structures only those with the suffix **a** possess a linear configuration of bonds with the nucleophilic centers, while in all other cases angular
configurations are established. This follows from the generalized polarity rule
based on the quantum mechanical calculations of relative energies of the topomers
XII—XVI as well as on orbital analysis [39—43]. Among structures of the
$MN_1N_2R_n$ type (n = 1—4) admitting nonequivalent orientations of bonds of the
central atom X with the most electronegative ligands N_1 and N_2, the structures
with the axial (linear) orientation of these bonds possess greater stability.

In Table 3.III, the expected structures of the transition states (intermediates)
are given for intramolecular addition—elimination reactions with regard to the
migrants MR_n of all principal types. Table 3.III also presents some of the available

Table 3.III. Topology and geometry of intermediates or transition state structures **XIc** predicted by the data of quantum-mechanical calculations of the MERPs[a], by the VSEPR theory, and by structural mapping of the MERP.

MR_n	Migrant	VSEPR	Angle X—M—Y (Calculated and Experimental data)	Method	References
M	H, Li	X—H—Y	180° (H_2O····H—OH); 160—180° (X—H····Y)	*Ab initio*; Neutron and X-ray diffraction	[48]; [49]
MR	—N=O		119° (H^- + HNO); 109° (F^- + HNO)	MINDO/3	[50]
			180° (HS—SH + H^-, F^-, SH^-)	*Ab initio* (DZ + *d*-orbitals)	[51]
MR	—SR		180° (CH_3, SH + SH^-)	6-31 G*	[52]
			180° (H_2S + SH^-; ClSF + F^-)	CNDO/2	[53]
			172° (CH_3SF + F^-)	CNDO/2	[54]
			162 ± 10° (RSX + Y nonbonding atomic contacts)	X-ray	[46, 55]
MR_1R_2	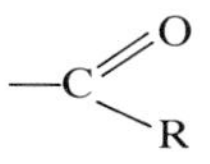	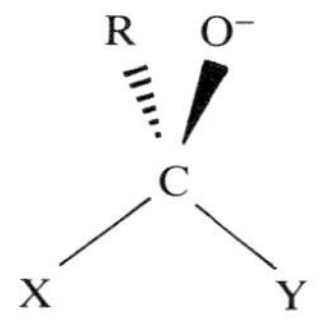	109.5° (H^- + CH_2O)	*Ab initio* (DZ); MINDO/3	[56]; [50]
			99—111° (CH_3O^- + $HCONH_2$)	PRDDO	[57]
			103.5° ($CH_3CHClCHO$ + H^-); 107 ± 5° (aminoketones, intramolecular contacts)	STO-3G; X-ray	[58]; [43, 59]

			Angle (reaction channel)	Method	References
$MR_1R_2R_3$	—Aryl	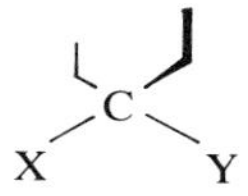 (structure: C with X, Y and ethyl)	107° (Intramolecular contacts)	X-ray	[47, 60]
	—NO_2	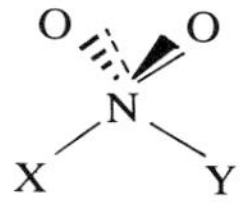 (structure: N with O, O, X, Y)	130° ($HO^- + HNO_3$)	MINDO/3	[50]
	—PR_1R_2	 X—P—Y	180°, 90°ᵇ ($PHClF + F^-$)	CNDO/2	[61]
	—SOR	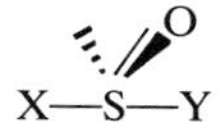 X—S—Y (O)	180°, 90°ᵇ ($SOClF + F^-$)	CNDO/2	[61]
	—$CR_1R_2R_3$	 X—C—Y	180° ($CH_3F + F^-$); 171.5–180° ($CH_2RF + F^-$)	4–31 G; 4–31 G	[62]; [63]
	—POR_1R_2	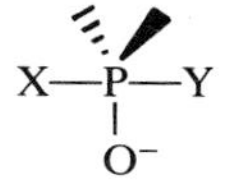 X—P—Y (O^-)	180° (($CH_3O)_2PO_2^- + OH^-$)	STO–3G	[64]
$MR_1R_2R_3R_4$	—SO_2R	X—S—Y (O, O, O^-)	180°	—	—
	—$PR_1R_2R_3R_4$	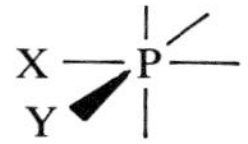 X—P—Y	90°ᶜ ($PH_4CH_3 + H^-$)	MINDO/3	[65]

ᵃ Data are presented for only those calculations for which the reaction paths were properly investigated and the intermediate and transition state structures were identified according to current technique, see [1].

ᵇ Two reaction channels were found in these calculations. The more favorable one corresponds to the formation of a stable intermediate with linear arrangement of F—M—Cl (180°), while the other corresponds to a 90° F—M—Cl configuration.

ᶜ The only principal difference between VSEPR and quantum-mechanical prediction.

data on the quantum-mechanical calculations of the paths of model reactions of the corresponding type. As can be seen, these data are in complete qualitative agreement with the predictions of the VSEPR theory. In individual cases, experimental confirmation of the theoretical predictions has been obtained from an analysis of the structural characteristics of series of compounds containing fragments X, Y, MR_n in intra- or intermolecular contact, arranged in a specific sequence. These sequences very accurately model the reaction channel at the potential energy surfaces [1, 44—47].

The data of Table 3.III show that the intermediate structures of type **Ic** can be divided into two groups, differing in the orientation of the M—X, M—Y bonds, The first of these, i.e. **XII**, includes X—M—Y fragments with an angular configuration, which is present in the bent pyramidal and tetrahedral structures **XVII**. The second group, **XVIII**, composed of the trigonal-bipyramidal bisphenoidal and T-shaped structures **XIc**, contains an X—M—Y fragment with a linear configuration. Both experimental [66, 67] and theoretical [40—43, 68, 69] data indicate that the axial bonds in structures of this type are elongated by 0.2—0.3 Å as compared with the usual bonds.

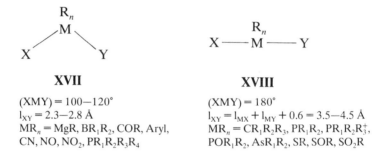

XVII

$(XMY) = 100—120°$
$l_{XY} = 2.3—2.8$ Å
$MR_n = MgR, BR_1R_2, COR,$ Aryl, CN, NO, NO_2, $PR_1R_2R_3R_4$

XVIII

$(XMY) = 180°$
$l_{XY} = l_{MX} + l_{MY} + 0.6 = 3.5—4.5$ Å
$MR_n = CR_1R_2R_3, PR_1R_2, PR_1R_2R_3^+,$
$POR_1R_2, AsR_1R_2, SR, SOR, SO_2R$

Realization of one or the other of the two transition (or intermediate) structures **XVII** and **XVIII** in the reaction (3.12) predetermines the stereochemical course of the reaction, i.e. inversion or retention of configuration at the chiral migrating center. The most important characteristic with regard to the possibilities of the system carrying the migrating group to adapt its configuration to the optimal structure of the transition state is the mutual position of the entering and leaving groups, i.e. of the nucleophilic centers X and Y.

2.4.2. STRUCTURE OF THE MOLECULAR SYSTEM CONNECTED WITH A MIGRANT

To satisfy the steric requirements of the reactive centers **XVII**, **XVIII** dictated by the nature of the migrant MR_n, it is necessary to select corresponding fragments Z in the compounds **XI**. It is convenient to conceive the resulting structures in the form of multigraphs in which each vertex corresponds to a C_{sp^2} center, and derivatives are obtained by morphic substitution with isoelectronic moieties.

For 1,*j*-rearrangements of migrants of type **XVII** in conjugated systems, the Z-fragments **XVIIa—XVIIe** serve as suitable carrier systems:

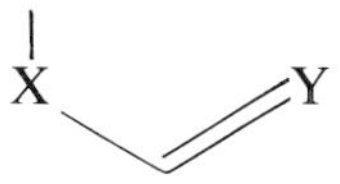

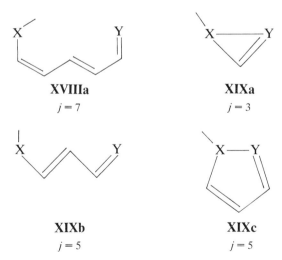

XVIIa *j* = 3 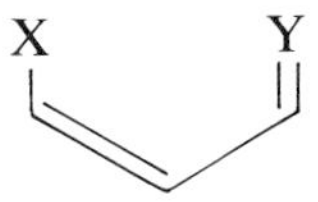   XVIIb *j* = 5 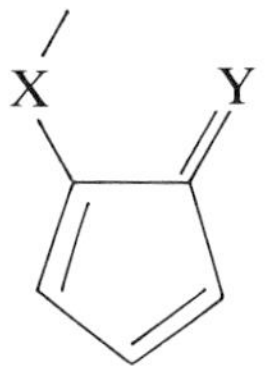   XVIIc *j* = 7

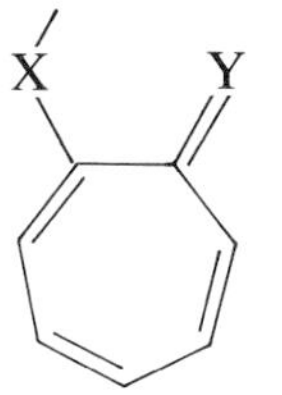

XVIId *j* = 9 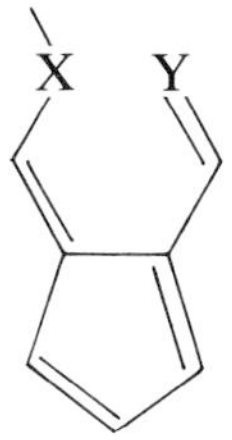   XVIIe *j* = 9

The simplest Z-chain, suitable for carrying out fast rearrangements of migrants of the type **XVIII**, is the structure **XVIIIa**, which ensures the possibility of 1,7-migration.

XVIIIa
j = 7

XIXa
j = 3

XIXb
j = 5

XIXc
j = 5

Structures of the type **XIX** do not satisfy the steric requirements of systems **XI** with fast intramolecular rearrangements of the MR$_n$ groups.

The kinetic parameters of fast acyl and aryl rearrangements of the 1,3, 1,5 and 1,9 types considered above (in Chapter 2) serve as an illustration of the approach described for migrants of type **XVII**. It has also been shown that alkyl, phosphoryl and sulfonyl migrants are inert in the appropriate compounds **XX—XXII** belonging to type **XVII**

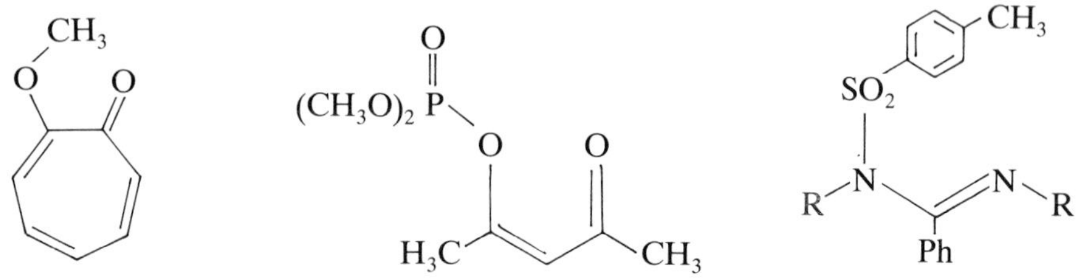

XX [15, 16]   **XXI** [70]   **XXII** [71]

The preparation of compounds possessing the structures with a configuration of the conjugated chain Z of the type **XVIIIa** is accompanied by significant synthetic problems, but it is expected that fast intramolecular migrations of MR_n groups of the type **XVIII** can take place in the cyclic derivatives **XXIII, XXIV** thus ensuring the required orientation of the interacting centers.

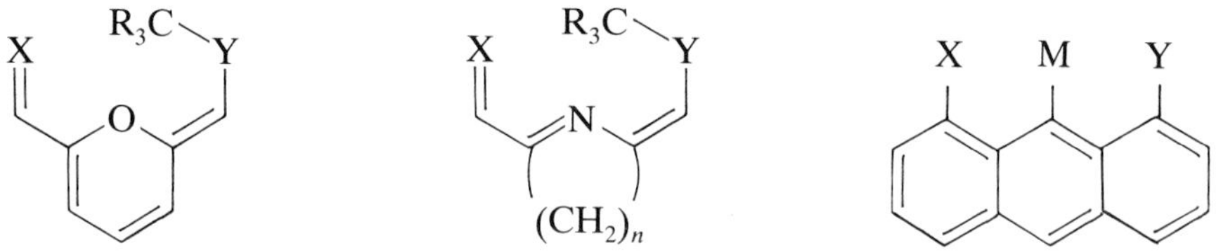

XXIII   **XXIV**   **XXV**

It has been noted [72] that anthracene derivatives of type **XXV** are particularly favorable for the formation of hypervalent structures of type **XVIII**. Exactly this structure was obtained by Martin [73] in the 1,8-di(arylthio)anthracene-9-carbynyl cations, which undergo a fast degenerate rearrangement **XXVIa** ⇌ **XXVIb** ($k_{298} = 10^2$ s^{-1}). Although, strictly speaking, this reaction is not a migration process, it reproduces all specific features of intramolecular substitution at a C_{sp^3} center.

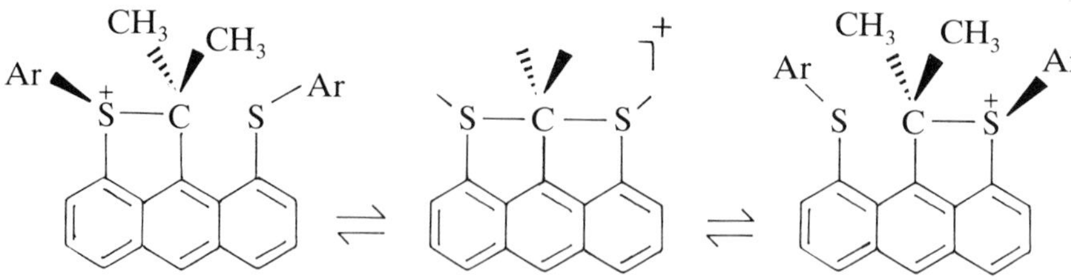

(3.13)

XXVIa   **XXVIc**   **XXVIb**

Similar fast (in the NMR time-scale) rearrangements were recently found [73] for the sulfenyl, sulfinyl and sulfonyl migrants incorporated in structures of the **XVIIIa** type

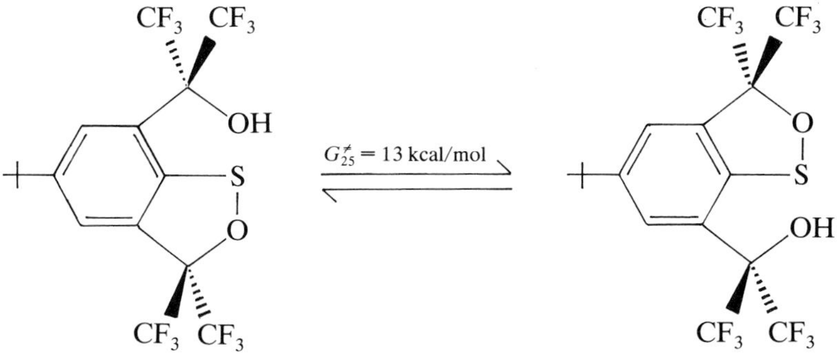

$$G_{25}^{\neq} = 13 \text{ kcal/mol}$$

(3.14)

XXVIIa   XXVIIb

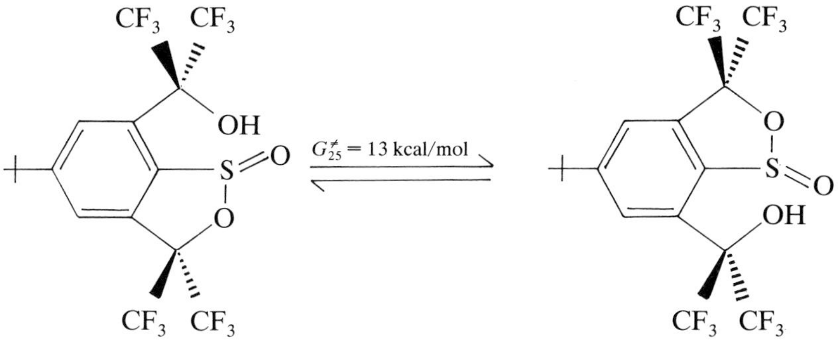

$$G_{25}^{\neq} = 13 \text{ kcal/mol}$$

(3.15)

XXVIIIa   XXVIIIb

$$G_{25}^{\neq} = 14 \text{ kcal/mol}$$

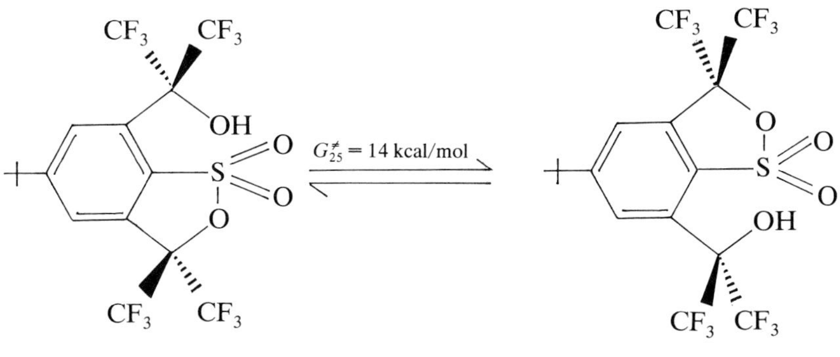

(3.16)

XXIXa   XXIXb

Strong base catalysis of the above interconversions was established, indicating the participation of intermediate sulfuranide, sulfuranide oxide and sulfuranide dioxide anions with the type **XVIII** arrangements of the O—S—O bonds.

In compounds **XXV—XXIX**, the molecules possess a rigidly constrained structure which favors formation of a corresponding transition or intermediate form. The usual structural approach consists in the inclusion of reaction sites **XVII** or **XVIII** in the cyclic system. In the case of noncyclic structures, the situation is somewhat different. To satisfy the steric requirements of the reaction site, the molecule has to be conformationally mobile so that it should be able to pass in its dynamics with sufficiently high frequency (higher than the frequency of tautomeric transitions) through the structure favorable for the occurence of the reaction. *O*-acyl derivatives of *cis*-enols 1,3-diketones **XXXa**, dealt with in detail in Chapter 2 belong, among others, to just such compounds. Four mechanisms of conformational and configurational transformations resulting from the rotation about non-terminal bonds are possible in this case.

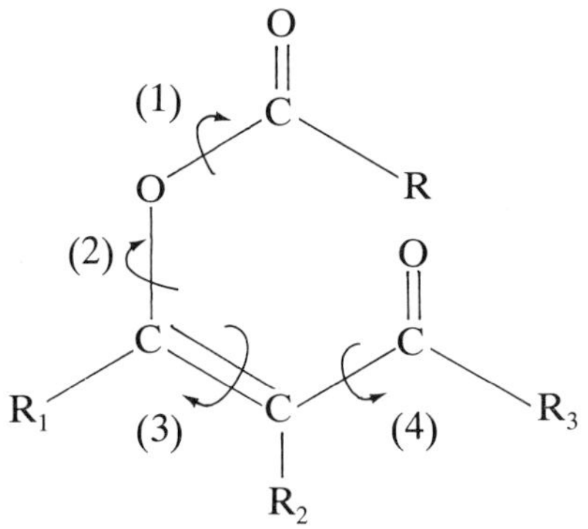

XXXa

The barrier of the rotation (3) about the double bond C=C is large enough, as in the case of all alkenes not belonging to the push—pull substituted type [74], exceeding the limit of activation energies for tautomeric rearrangements (Eq. (1.5)). Therefore, the configuration of the substituents at the double bonds may be regarded as relatively rigid.

The rotations (2) and (4) about ordinary bonds result, even in the case of a *cis*(Z)-diastereomer, in coexistence of four *s-cis—trans*-isomers. Of these, only the conformation **XXXa** meets the condition of sufficient proximity of nucleophilic oxygen atoms both among themselves and to the central atom of the migrating acyl.

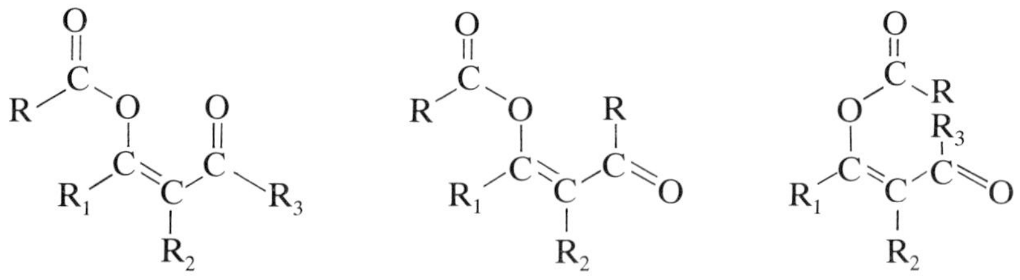

XXXb   **XXXc**   **XXXd**

However, the barriers of rotation about formally single bonds conjugated with the double bonds or heteroatoms separating the conformers **XXXa–d** are not large. In the absence of steric constraints they do not exceed 8—10 kcal/mol [75,

76]. This is also true of the acyl rotation (1) [77]. That the barriers of group migrations are, as a rule, larger than the above values, indicates that an equilibrium population of all conformers is easily established during the life-time of one of the tautomers. If the conformational energy of the structure capable of a tautomeric transformation is not too high — in other words, if the equilibrium concentration of this conformer is not vanishingly small — then the dynamic processes of conformational isomerizations not only do not impede, but, on the contrary, they ensure tautomerization.

It should be noted that in case a conformer capable of tautomerization is not the most stable one, the difference between free conformation energies must be included in the experimentally determined effective value of the activation energy of tautomerization (Fig. 3.1).

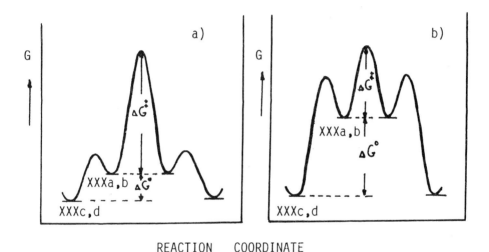

REACTION   COORDINATE

Fig. 3.1. The free energy profile of the reaction of the *O,O*-acylotropic degenerate transformation in the conformer **XXXa** illustrating a possible situation when the stability of *s-trans*-isomers is higher than that of *s-cis*-isomers. For simplicity, the rotations about C—O bonds are not taken into account. (a) $\Delta G° < \Delta G^{\neq}$, $\Delta G^{\neq}_{eff} = \Delta G^{\neq} + \Delta G°$. (b) $\Delta G° > \Delta G^{\neq}$. The active conformer content is small. Tautomerism is not observed.

From the foregoing it follows that, when selecting molecular migrant carrier systems, the following conditions in regard to their conformational characteristics have to be taken into account:

(1) the activation barrier of the conformational transformation into a structure suitable for tautomerization must be lower than the tautomeric reaction barrier;

(2) the difference between free energies of the conformer capable of tautomerization and the stable conformer must be lower than the activation energy of the tautomeric reaction.

2.4.3. ADDITION-REARRANGEMENT-ELIMINATION (*AdRE*) MECHANISM

Structures with the T-shaped **XII**, bisphenoidal **XIII**, trigonal bipyramidal **XV** and octahedral **XVI** configurations of the central atomic bonds are stereochemically nonrigid, they are capable of polytopal rearrangements which change ligand positions in the apexes of the corresponding polyhedron [78, 79]. This fact provides an additional important chance favoring the intramolecular tautomeric reaction of sigmatropic transfer of the type **XVIII** migrants even in those systems where the stable conformation does not ensure realization of the appropriate transition state structure, i.e. in the molecular systems **XVIIa—e**.

In the above structures derived from a trigonal bipyramid, there are two topologically nonequivalent positions, axial and equatorial. Only the first of these is suitable for the approach of a nucleophilic center (X, Y) to the central atom M of a migrating MR_n group, since bond formation and bond rupture processes take place exclusively in axial positions. This was proved by numerous theoretical calculations and by experimental results (see Table 3.III and [1, 62, 80] for reviews) As a consequence of the principle of microscopic reversibility, the cleavage of a leaving group from a trigonal-bipyramidal structure must also take place from an axial position.

Westheimer [81] was the first to indicate that, in nucleophilic substitution reactions, following the *AdE* mechanism, both of the above conditions can be met not only by means of transition states of the type **XVIII** but also in a stepwise reaction. In this case, the attacking nucleophile enters into an axial position with the formation of an intermediate, containing the leaving group in an equatorial position. A polytopal rearrangement of the trigonal-bipyramidal intermediate interchanges the positions of the entering and leaving groups, and the departure of the latter then proceeds in accordance with the principle of microscopic reversibility from the axial position [80—82]. An example of the occurence of this addition-rearrangement-elimination (*AdRE*) mechanism is represented by the tautomeric 1,3-rearrangements of arenesulfenyl groups in amidine derivatives **XXXI** [83, 84].

Because of the steric requirements of the T-shaped structure of the transition state of a nucleophilic substitution reaction at a dicoordinated sulfur atom (Table 3.III), the amidine system **XVIIa** is unfavorable for the occurrence of a concerted intramolecular rearrangement **XXXIa** ⇌ **XXXb**. However, calculations [61] demonstrated another mechanistic possibility, i.e. an unconcerted *AdRE* mechanism, including the formation of an intermediate **XXXIc₁** and its topomerization by rotation of a S—R bond by 90° in the plane, which leads to an interchange of the S—N bond positions.

The presence of the stage of the polytopal intermediate rearrangement in the *AdRE*-mechanism is of crucial importance for determining the stereochemical course of the elementary stage of a tautomeric reaction. Unlike the concerted S_N2

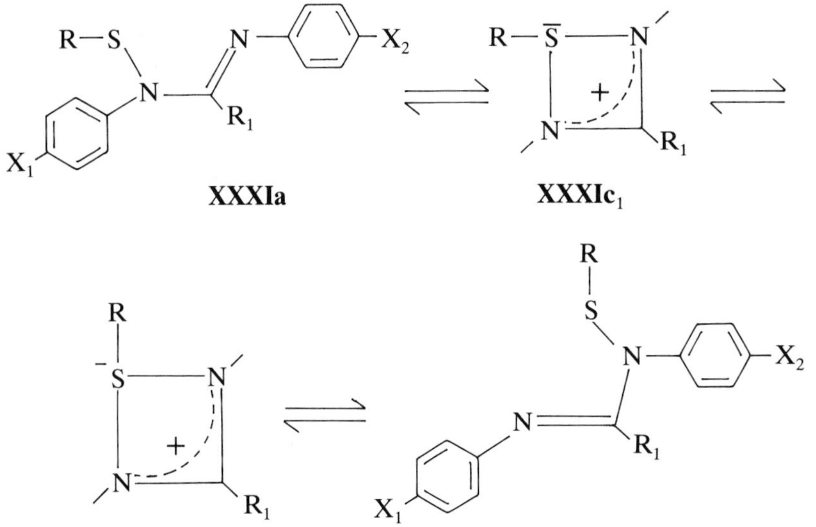

(3.17)

XXXIa **XXXIc₁**

XXXIc₂ **XXXIb**

reaction (**XIc** is a transition state, not an intermediate), the *AdRE* reaction is characterized by retention of the bond configuration of the central atom M in the migrating group. An elegant confirmation of this conclusion was provided by a recent study of the intramolecular acid catalyzed O—O′ migration of the chiral isotopically marked ^{16}O, ^{17}O, ^{18}O phosphoryl group in 2-[(R)-^{16}O, ^{17}O, ^{18}O] phosphopropane-1,2-diol to the 1-position which was found to occur with full retention of configuration [85]

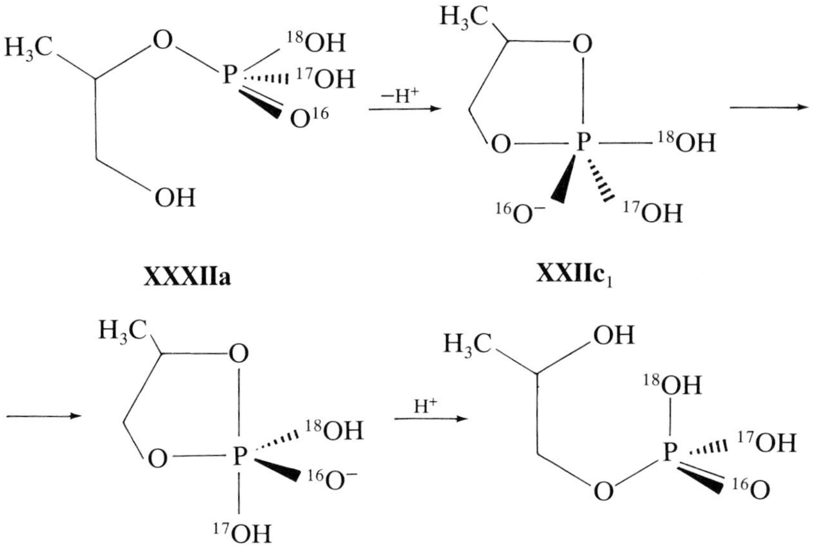

(3.18)

XXXIIa **XXIIc₁**

XXXIIc₂ **XXXIIb**

At the stage **XXXIIc$_1$** ⇌ **XXXIIc$_2$**, there occurs a polytopal rearrangement including the synchronous permutation of two axial and two equatorial ligands. This rearrangement, the so-called Berry pseudorotation [86], is especially typical of trigonal-bipyramidal structures. Some other mechanisms of polytopal rearrangements of intermediates **XIc** and an analysis of the consequences of such rearrangements with regard to the stereochemistry of tautomeric reactions (3.12) as well as to the possibilities of their realization are considered in Chapter 4. Here we only note that if a reaction follows the *AdRE* mechanism, it leads to a great reduction in the structural constraints discussed in the previous Section. The intermediate structures, such as **XXXIc, XXXIIc**, should have a sufficiently long lifetime, i.e. they should correspond to true local minima on the potential energy surface (PES) of the reaction (3.12). This follows from the important statement that the branching of the MERPs can originate only from such regions of the PES [1, 87]. Furthermore, energy barriers at the stage of a polytopal rearrangement must be sufficiently low — at any rate, not exceeding the limits of the activation scale of tautomeric reactions, Eq. (1.5) — which is obvious since the magnitude of the activation barrier at the stage of the **XXXIc$_1$** ⇌ **XXXIc$_2$**, **XXXIIc$_1$** ⇌ **XXXIIc$_2$** rearrangements of intermediates is a part of the total activation energy of the rearrangements (3.17) and (3.18) (Fig. 3.2).

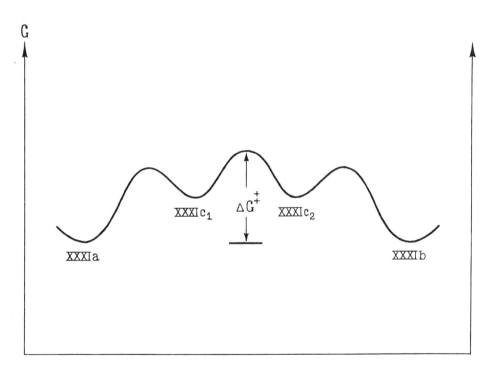

Reaction Coordinate

Fig. 3.2. The free energy profile of the (3.17) type *AdRE* reaction.

In the intermediates **XXXIc** and **XXXIIc**, the size of the cycle rules out a polytopal rearrangement with the formation of a topomer containing both nucleophilic centers in the diaxial position. The minimal size of such a cycle is eight, as in the compounds **XXVI–XXIX**. Emergence of an intermediate of this type is possible, for example, in the 1,7-sigmatropic degenerate rearrangement of the phosphoryl group in the *cis-cis*-enol derivatives of 1,5-diketones.

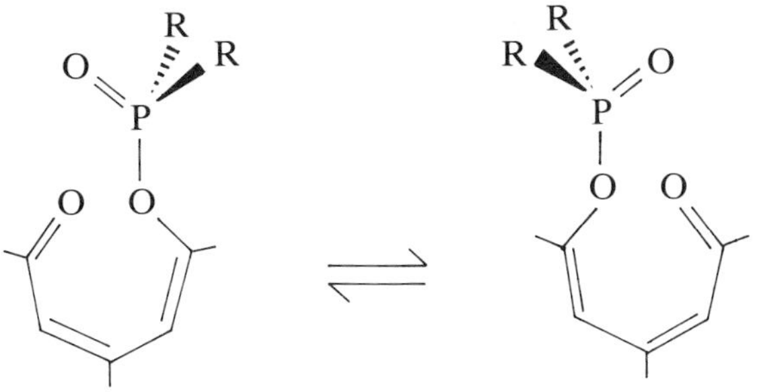

$$(3.19)$$

XXXIIIa   **XXXIIIb**

In phosphoranes with one cyclic bidentate ligand, the axial–equatorial position of electronegative atoms is, as a rule, the most favorable one [80] as is also the case with the primary intermediate **XXXIIIc$_1$**. No fewer than two Berry pseudorotations are needed to provide formation of a less favorable structure **XXXIIIc$_3$** in which rupture of the P—O$_2$ bond leads to the rearrangement product **XXXIIIb***

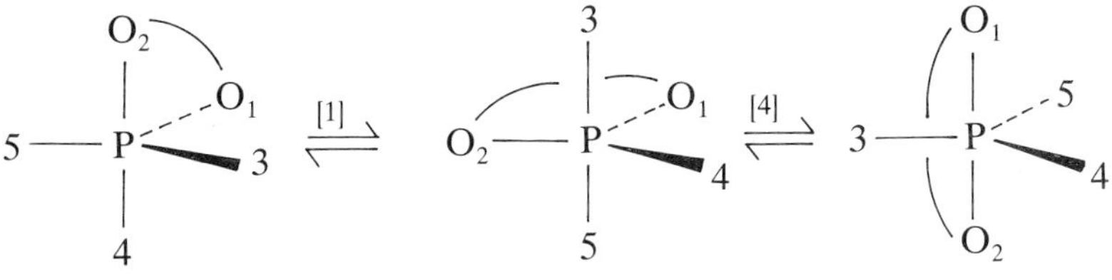

$$(3.20)$$

XXXIIIc$_1$ (24)   **XXXIIIc$_2$** ($\overline{35}$)   **XXXIIIc$_3$** ($\overline{12}$)

Due to the intermediate **XXXIIIc$_3$** emerging in the pathway of the reaction (3.19), its stereochemical result consists in the inversion of the configuration at the phosphorus atom. An essential feature of this process is that, in the reverse direction, the reaction has to follow an enantiomeric pathway via intermediate structures enantiomeric to those which are formed in the forward reaction (Fig. 3.3). This is accounted for by the peculiarities in the shape of the PESs relating to interconversions of enantiomers via chiral intermediates [88, 89]. It has been

---

* In square brackets, the pivot ligand which does not change its position in the course of the Berry pseudorotation is given. A pair of axial ligands is indicated to denote the permutational isomer [80].

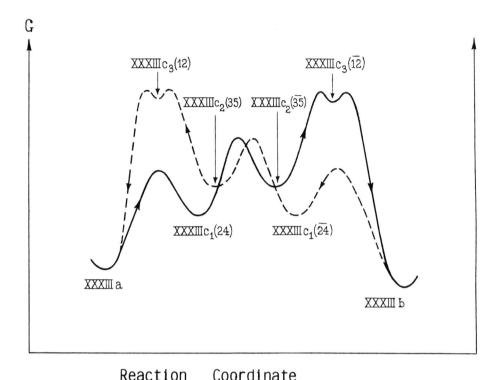

Fig. 3.3. The free energy profile of the reaction (3.19) proceeding by the narcissistic *AdRE* mechanism with inclusion of polytopal rearrangements of the intermediates (3.20)

suggested that the processes with the MERPs such as are shown in fig. 3.3 should be termed narcisstic reactions [90].

2.5. *Analogy in Steric Requirements Between Fast Intramolecular, Enzymic and Topochemical Reactions*

The main idea behind the above described approach to the molecular design of a tautomeric system is to search for compounds whose molecular structure would be sterically best prepared for the occurrence of a transition state of the intramolecular reaction. In order to fulfil such a condition the atomic groups interacting in the course of this transformation must be either rigidly fixed or, owing to low-barrier conformational isomerizations, directed according to the requirements of the MERP of the migrant transfer between the entering and leaving groups [15, 16].

The directionality argument was, apparently, first invoked by Eschenmoser *et al.* [91] to explain their kinetic and crossover experimental data, which exclusively proved the intermolecular nature of the following 1,5-methyl transfers.

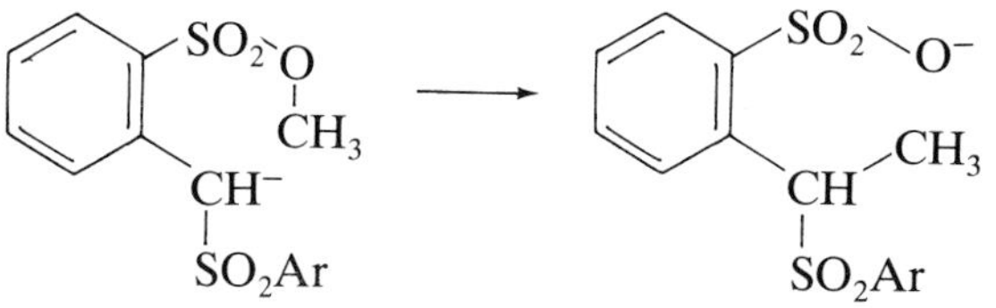

$$(3.21)$$

XXXIVa   **XXXIVb**

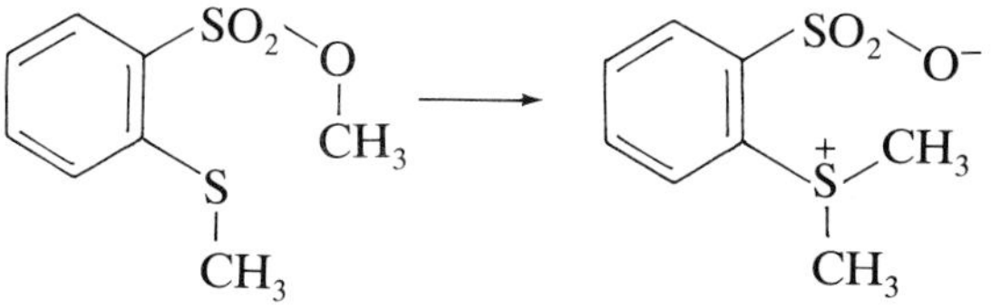

$$(3.22)$$

XXXVa   **XXXVb**

Despite the close proximity of all interacting centers in the six-membered transition state, it is not possible to obtain trigonal-bipyramidal configuration for the transition state of the S_N2 reaction at the tetrahedral carbon atom with the diaxial alignment of the entering and leaving groups (Table 3.I). The proximity of interacting reaction centers coupled with the favorable angular dependence (orbital sterring) formed the basis for Koshland's concepts on the origin of greater accelerations of intramolecular as compared to intermolecular reactions [92]. Table 3.IV contains some data on kinetics of lactonization reactions of a set of hydroxyacids with steadily varied structure which illustrate the concept of orbital steering.

For compounds **XXXVII–XL** angles θ for intramolecular attack on the carboxyl group lie in the region of the 10–15° misalignment of reactant groups, i.e. in an attractive potential zone (*cf.* Fig. 2.5). Therefore, the enormous acceleration of the reaction in the case of compound **XL** can be related to the minimization of disadvantageous entropic factors in this reaction due to the maintenance of the two reactive groups at nearly bonding distances as well as to the freezing of almost all rotations disordering the favorable orientation of these groups.

All these effects ensuring such high intermolecular reactivity also determine the nature of the huge rate enhancements in important biochemical reactions catalyzed by enzymes. According to Pauling [94] "... enzymes are molecules that are complementary in structure to the activated complexes of the reactions that they catalyze." Their primary function is to be operative in bringing together the reactive centers and in providing the steric conditions of the transition-state structure in the strained substrate conformation obtained in the enzyme–substrate complex. The binding free energy of this complex compensates for the losses in the translational and rotational degrees of freedom as well as in the displacement of

Table 3.IV. Relative velocities (related to intermolecular reaction CH$_3$COOH + C$_2$H$_5$OH), activation energy parameters and angular dependence of lactonization reactions [92, 93].

Compound No.	Hydroxyacid	(CO····C=O)	k_{rel}	$\Delta H^{\neq}$, kcal/mol	$\Delta S^{\neq}$, e.u.
XXXVI	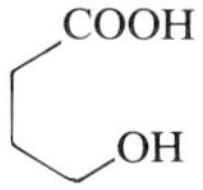	—	80	12.1	−31.5
XXXVII	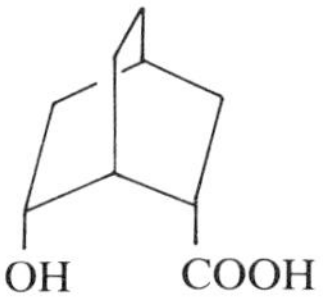	86—91°	870	11.9	−26.5
XXXVIII	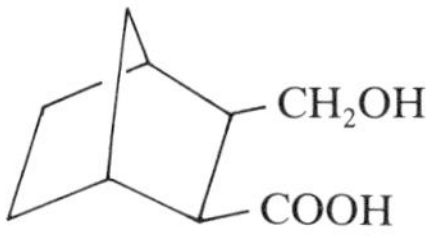	105°	6620	9.7	−30.1
XXXIX	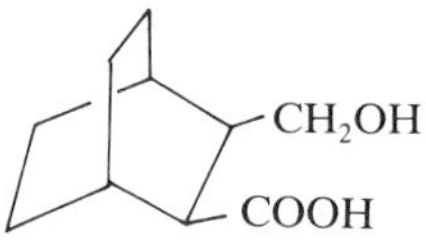	98°	13 940	—	—
XL	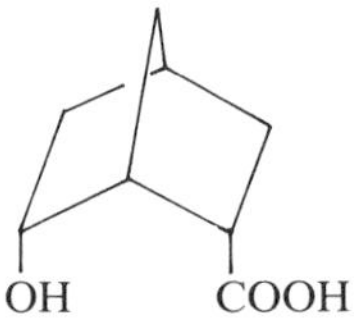	96°	1 030 000	14.9	−2.1

the solvent from the active site [12, 92, 95]. A 10^{10}—$10^{12.5}$-fold enzyme-provoked increase in the reaction rates of typical substrates due to the combined transition-state fitting and entropy effect is observed.* This is about the same increase as in the fastest sterically accelerated intramolecular reactions.

In intramolecular reactions, the drawing together and orientation of reactive centers depends on the conformational possibilities of the molecule, and in the enzyme-catalyzed reactions, adjusting of the reaction site is achieved through specific interactions in the enzyme—substrate complex. There is a third possible mechanism of fitting the reaction site to the steric demands of the MERP and the

* The increase factors depend on the choice of the standard state, see [95].

transition state structure, namely, the ordering of interacting molecules by placing them into a crystalline matrix. According to the so-called topochemical principle [96], reactions in the crystalline phase are accompanied by only minimal atomic and molecular displacements. If, however, the crystalline packing in solids is such that the reacting molecules are held in a fixed orientation favorable to their interaction, the reaction between them may occur at rates which not infrequently exceed those in solution [97]. A typical example is the great acceleration of the methyl transfer reaction in a crystal phase of methyl-*p*-dimethylaminobenzenesulfonate **XLIa** [98, 99].

$$(CH_3)_2N \underset{\textbf{XLIa}}{\underline{}}SO_2OCH_3 \longrightarrow (CH_3)_3\overset{+}{N}\underset{\textbf{XLIb}}{\underline{}}SO_3^- \qquad (3.23)$$

This reaction in the crystal phase proceeds at a considerable rate at room temperature (with the starting **XLI** stored at 20°C, 3.3% of **XLIb** is formed in 24 hours [99]), but in the melt phase or upon heating in a concentrated solution, it progresses rather slowly. The reason for this is revealed upon examination of the crystalline packing of compounds **XLIa** (Figure 3.4).

The molecules **XLIa** are packed in parallel sheets in which the methylsulfonyl and dimethylamino groups are placed in pairs, one over the other, in conformations exactly corresponding to the steric demands of the MERP for the methyl *O,N*-transfer (Table 3.I). These steric demands cannot be satisfied in an analogous intramolecular reaction (3.21).

Thus, the steric adjustment of the reaction site to the directionality of the MERP and to topological and structural peculiarities of the transition state is the crucial requirement for achieving an acceleration of both the intramolecular and intermolecular reactions. For the former, this requirement can be satisfied either in sterically strained or conformationally adjusted structures; for the latter, rigid fixation of reaction centers is achieved in topochemical processes in solids, whereas conformational adjustment is the route for enzyme-catalyzed biochemical reactions.

It should be noted that other explanations of the huge accelerations of some intramolecular and enzyme-catalyzed reactions have been advocated. They include long-range electrostatic and hydrogen bonding effects [100], stereopopulation control [101] and the so-called 'spatiotemporal hypothesis' [102]. However, these explanations either are not general enough or are, in their essence, similar to the view presented above, though expressed in physically rather vague terms.

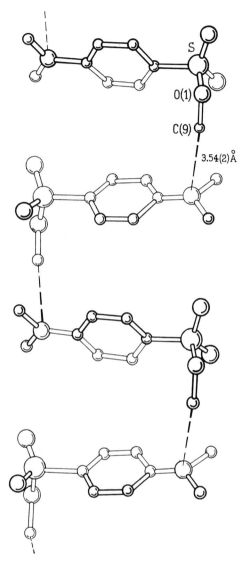

Fig. 3.4. A view of the stacking along one chain of the molecules of methyl-*p*-dimethylaminoben-
zenesulfonate **XLAa** according to an X-ray crystal structure determination [98]. The distance indicated is that between the carbon atom of the methyl group undergoing transfer in
the solid state reaction (3.23) and the nitrogen atom toward which it is moving.

REFERENCES

1. Minkin V. I., Simkin B. Ya., and Minyaev R. M., *Quantum Chemistry of Organic Compounds;
    Mechanisms of Reactions.* Khimiya. Moscow. 1986.
2. Flanigan M. C. Komornicki A., and McIver J. W., In: *Semiempirical Methods of Electronic
    Structure Calculation.* Part B. (Ed. G. A. Segal). Plenum Press. New York and London. 1977.

3. Schlegel H. B., Gund P., and Fluder E. M., *J. Amer. Chem. Soc.,* **104**, 5347 (1982).
4. Beak P., Fry F. S., Steele J., and Lee F., *J. Amer. Chem. Soc.,* **98**, 171 (1976); Beak P., *Acc. Chem. Res.,* **10**, 186 (1977).
5. Simkin B. Ya. and Sheikhet I. I., *J. Mol. Liquids,* **27**, 79 (1983).
6. Jorgensen W. L., *J. Phys. Chem.,* **87**, 5304 (1983).
7. Abronin I. A., Burstein K. Ya., and Zhidomirov G. M., *Zh. Struct. Khim.,* **21**, 145 (1980).
8. Simkin B. Ya., Sheikhet I. I., *Zh. Struct. Khim.,* **24**, 80 (1983).
9. Fukui K., *J. Phys. Chem.,* **74**, 4161 (1970); *Acc. Chem. Res.,* **14**, 363 (1981).
10. Müller K., *Angew. Chem. Intern. Ed. Engl.,* **19**, 1 (1980).
11. Volkenstein M. V., *Molecular Biophysics.* Ch. 6. Nauka. Moscow. 1975.
12. Kurz J. L., *Acc. Chem. Res.,* **5**, 1 (1972).
13. McIver J. W., *Acc. Chem. Res.,* **7**, 72 (1974).
14. Scanlan M. J. and Hillier I. H., *Chem. Phys. Lett.,* **107**, 330 (1984).
15. Minkin V. I., Olekhnovicn L. P., and Zhdanov Yu. A., *Zh. Vsesoyzn. Khim. Obschz. im. D. I. Mendeleev,* **22**, 273 (1977); *Acc. Chem. Res.,* **14**, 210 (1981).
16. Minkin V. I., *Soviet Sci. Rev., B. Chem.* **7**, 51 (1985).
17. Hegarty A. F. and McCormack M. T., *J. C. S. Chem. Commun.,* **168** (1975).
18. Zefirov N. S. and Trach S. S., *Zh. Org. Khim.,* **12**, 697 (1976).
19. Mathieu J., Valls J., *Bull. Soc. Chim. France,* 1509 (1957).
20. Zefirov N. S., Trach S. S., *Zh. Org. Khim.,* **11**, 225, 1785 (1975).
21. Balaban A. T., *Rev. Roumaine Chim.,* **12**, 875 (1967).
22. Hendrickson J. B., *Angew. Chem. Intern. Ed. Engl.,* **13**, 47 (1974).
23. *Chemical Applications of Graph Theory* (Ed. A. T. Balaban), p. 5. Academic Press. New York. 1976.
24. Rouvray D. H., *Chem. Brit.,* **10**, 11 (1974); *J. Chem. Educ.,* **52**, 768 (1975).
25. Balaban A. T., *Rev. Roumain. Chim.,* **17**, 865 (1972).
26. Yen T. F., *Theor. Chim. Acta,* **20**, 399 (1971).
27. Barabas A. and Balaban A. T., *Rev. Roumain. Chim.,* **19**, 1927 (1974).
28. Bauer J., Herges R., Fontain E., and Ugi I., *Chimia,* **39**, 43 (1985).
29. Stewart H. F. and West R., *J. Amer. Chem. Soc.,* **92**, 846 (1970).
30. Acheson R. M., *Acc. Chem. Res.,* **4**, 177 (1971).
31. Woodward R. B. and Hoffmann R., *The Conservation of Orbital Symmetry.* Verlag Chemie. GmbH, Weinheim. 1971.
32. Ross J. A., Seiders R. P., and Lemal D. M., *J. Amer. Chem. Soc.,* **98**, 4325 (1976).
33. Reich H. J. and Murcia D. A., *J. Amer. Chem. Soc.,* **95**, 3418 (1973).
34. McClarin J. A., Schwartz A., and Pinnavaia T. J., *J. Organometal. Chem.,* **188**, 129 (1980).
35. O'Brien D. A. and Hrung C. P., *J. Organometal. Chem.,* **27**, 185 (1971).
36. West R. and Bechlmeir B., *J. Amer. Chem. Soc.,* **94**, 1649 (1972).
37. Wiberg N. und Pracht H. J., *Chem. Ber.,* **105**, 1388 (1972).
38. Gillespie R. J., *Molecular Geometry.* Van Nostrand Reinhold Co. London. 1972.
39. Gillespie P., Hofmann P., Klusacek H., Marguarding D., Pfohl S., Ramirez F., and Tsolis E. A., *Angew. Chem. Intern. Ed.,* **10**, 687 (1971).
40. Ugi I. and Ramirez F., *Chem. Brit.,* **10**, 101 (1972).
41. Rauk A., Allen L. C., and Mislow K., *J. Amer. Chem. Soc.,* **94**, 3035 (1972).
42. Hoffmann R., Howell J. M., and Muetterties E. L., *J. Amer. Chem. Soc.,* **94**, 3047 (1972).
43. Minyaev R. M., Sadekov I. D., and Minkin V. I., *Zh. Obschz. Khim.,* **47**, 2019 (1977).
44. Bürgi H.-B., *Angew. Chem. Intern. Ed. Engl.,* **14**, 460 (1975); Bürgi H.-B. and Dunitz J., *Acc. Chem. Res.,* **16**, 153 (1983).
45. Shefter E., In: *Transition States of Biochemical Processes.* (Ed. Gandour R. D. and Schowen R. L.), p. 341. Plenum Press. New York. 1978.
46. Dunitz J. D., *X-ray Analysis and the Structure of Organic Molecules.* Cornell University Press. Ithaca and London. 1979.
47. Furmanova N. G. and Kompan O. E., in: *Problems in Crystallochemistry* (Ed. Poray-Koshitz M. A.) p. 122. Nauka. Moscow. 1985.

48. Collman P. A. and Allen L. C., *Chem. Rev.,* **72**, 283 (1972).
49. Olovsson J. and Johnson P. G., in: *The Hydrogen Bond.* (Ed. Schuster P., Zundel G., and Sandorfy C.) vol. 2, p. 403. North-Holland Publ. Amsterdam. 1974. Taylor R. and Kennard O., *Acc. Chem. Res.,* **17**, 320 (1984).
50. Kletskii M. E., Minyaev R. M., and Minkin V. I., *Zh. Org. Khim.,* **16**, 686 (1980).
51. Pappas J. A., *J. Amer. Chem. Soc.,* **99**, 2926 (1977).
52. Aida A. and Nagata C., *Chem. Phys. Lett.,* **112**, 129 (1984).
53. Minkin V. I. and Minyaev R. M., *Zh. Org. Khim.,* **13**, 1129 (1977).
54. Schmid G. H. and Hallman G. M., *Intern. J. Sulfur Chem.,* **8**, 607 (1976).
55. Rosenfield R. E., Pathasarathy R., and Dunitz J., *J. Amer. Chem. Soc.,* **96**, 1956 (1974).
56. Bürgi H.-B., Lehn J. M., and Wipff G., *J. Amer. Chem. Soc.,* **96**, 1956 (1974).
57. Sheiner S., Lipscomb W. N., and Kleier D. A., *J. Amer. Chem. Soc.,* **98**, 4770 (1976).
58. Nguyen Trong Anh and Eisenstein O., *Nouv. J. Chim.,* **1**, 161 (1977).
59. Leonard N. J., *Acc. Chem. Res.,* **12**, 423 (1979).
60. Furmanova N. G., Olekhnovich L. P., Minkin V. I., Struchkov Yu. T., Kompan O. E., Budarina Z. N., Yudilevich I. A., and Eruzheva O. V., *Zh. Org. Khim.,* **18**, 465 (1982).
61. Minyaev R. M., Kletskii M. E., and Minkin V. I., *Zh. Org. Khim.,* **14**; 449 (1978).
62. Schlegel H. B., Mislow K., Bernardi F., and Bottoni A., *Theor. Chim. Acta,* **44**, 245 (1977).
63. Wolfe S., Mitchell D. J., and Schlegel H. B., *J. Amer. Chem. Soc.,* **103**, 7692 (1981).
64. Gorenstein D. G., Luxon B. A., and Findley J. B., *J. Amer. Chem. Soc.,* **101**, 5869 (1979).
65. Minyaev R. M. and Minkin V. I., *Zh. Sruct. Khim.,* **20**, 842 (1979).
66. Holmes R. R., *Acc. Chem. Res.,* **12**, 257 (1979).
67. Perozzi F. and Martin J. C., *Science,* **191**, 154 (1976).
68. Gleiter R. and Veillard A., *Chem. Phys. Lett.,* **37**, 33 (1976).
69. Schwenzer G. M. and Schaefer H. F., *J. Amer. Chem. Soc.,* **97**, 1393 (1975).
70. Kolind-Andersen H. and Lawesson O., *Acta Chem. Scand.,* **B29**, 430 (1975).
71. Mikhailov I. E., Ivanchenko N. M., Olekhnovich L. P., Minkin V. I., and Zhdanov Yu. A., *Dokl. Akad. Nauk USSR,* **263**, 366 (1982).
72. Gleiter R. and Hoffmann R., *Tetrahedron,* **24**, 5899 (1968).
73. Martin J. C. and Basalay R. J., *J. Amer. Chem. Soc.,* **95**, 2572 (1973); Perkins C. W., Wilson S. R., and Martin J. C., *J. Amer. Chem. Soc.,* **107**, 3209 (1985).
74. Kalinowski H.-O. and Kessler H., *Topics Stereochem.,* **7**, 295 (1973).
75. Kessler H., *Angew. Chem.,* **82** 237 (1970).
76. Jackman L. M., in: *Dynamic Nuclear Magnetic Resonance Spectroscopy* (ed. Jackman L. M. and Cotton F. A.) p. 203. Academic Press. New York. 1975.
77. Noe E. A., *J. Amer. Chem. Soc.,* **99**, 2803 (1977).
78. Muetterties E. L., *Acc. Chem. Res.,* **3**, 266 (1970).
79. Boldyrev A. I. and Charkin O. P., *Zh. Struct. Khim.,* **25**, 102 (1984).
80. Luckenbach R., *Dynamic Stereochemistry of Pentacoordinated Phosphorus and Related Elements.* G. Thieme Publ. Stuttgart. 1973.
81. Westheimer F. H., *Acc. Chem. Res.,* **1**, 168 (1968).
82. Mislow K., *Acc. Chem. Res.,* **3**, 321 (1970).
83. Olekhnovich L. P., Minkin V. I., Mikhailov I. E., Ivanchenko N. M., and Zhdanov Yu. A., *Dokl. Akad. Nauk USSR,* **233**, 874 (1977).
84. Olekhnovich L. P., Minkin V. I., Mikhailov I. E., Ivanchenko N. M., Zhdanov Yu. A., *Zh. Org. Khim.,* **15**, 1355 (1979).
85. Buchwald S. L., Pliura D. H., and Knowles J. R., *J. Amer. Chem. Soc.,* **104**, 845 (1982).
86. Berry R. S., *J. Chem. Phys.,* **32**, 933 (1960).
87. Pearson R. G., *Acc. Chem. Res.,* **4**, 152 (1971).
88. Wolfe S., Schlegel H. B., Csizmadia I. G., and Bernardi F., *J. Amer. Chem. Soc.,* **97**, 2020 (1975).
89. Brocas J. and Willem R., *J. Amer. Chem. Soc.,* **105**, 2217 (1983).
90. Salem L., *Acc. Chem. Res.,* **4**, 322 (1971).
91. Tenud L., Farooq S., Seibl J., and Eschenmoser A., *Helv. Chim. Acta,* **53**, 2059 (1970).

92. Storm D. R. and Koshland D. E., *J. Amer. Chem. Soc.,* **94**, 5805, 5815 (1972).
93. Moriarty R. M. and Adams T., *J. Amer. Chem. Soc.,* **95**, 4070 (1973).
94. Pauling L. *Nature* (London), **161**, 707 (1948).
95. Lipscomb W. N., *Acc. Chem. Res.,* **15**, 232 (1982).
96. Cohen M. D., *Angew. Chem. Intern. Ed. Engl.,* **14**, 386 (1975).
97. Gavezzotti A. and Simonetta M., *Chem. Rev.,* **82**, 1 (1982).
98. Sukenik C. N., Bonapace J. A. R., Mandel N. S., Lau P.-Y., Wood G., and Bergmann R. G., *J. Amer., Chem. Soc.,* **99**, 851 (1977).
99. Kokura P., *Chem. A. Ind. Chem.* (Japan), **30**, 439 (1977).
100. Pincus M. R. and Sheraga H. A., *Acc. Chem. Res.,* **14**, 299 (1981).
101. Hillery P. S. and Cohen L. A., *J. Org. Chem.,* **48**, 3465 (1983).
102. Menger F. M., *Acc. Chem. Res.,* **18**, 128 (1985).

4. The Mechanisms of Nucleophilic Substitution at the Main Group Elements and Design of Intramolecular Tautomeric Systems

The general principles of the construction of 1,*j*-tautomeric sigmatropic systems developed in the previous chapters are here applied to predict novel compounds of this type. Furthermore, they are used for the analysis of the intramolecular dynamics of recently-obtained tautomeric systems based on transfer reactions of heavy atomic groups. Mechanisms of the elementary stages at which the transition states are formed, which determine the stereochemical outcome and the magnitude of the activation barrier of 1,*j*-sigmatropic reactions, are examined in greater detail. The data obtained are then used for the structural modelling of the tautomeric compounds that are sterically adjusted to mimic the configuration of the reaction centers in the reactive zone of the potential energy surface.

1. Tautomeric Rearrangements of Carbon-containing Groups

Intramolecular tautomeric rearrangements associated with the transfer of other carbon-centered groups than the acyl, aryl and heteroaryl groups considered in Chapter 2 are so far unknown. One can, however, predict the structural conditions under which they may occur, as well as certain interesting stereochemical consequences of these hypothetical tautomerization processes.

1.1. *Alkyl Transfers*

1.1.1. ALKYL REARRANGEMENTS

For a number of known compounds of type **I** that are derived from triad prototropic tautomers, some equilibrium isomerizations associated with intermolecular methyl group 1,3-transfers catalyzed by quaternary salts and proceeding in accordance with the nucleophilic substitution mechanism have been found [1—3].

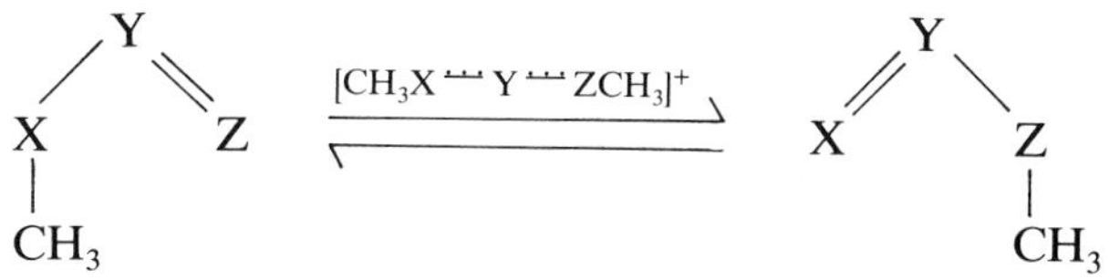

$$X = O, S; Y = C; Z = NR$$

The Y—Z bond in compounds **I** is usually included in the heterocyclic framework (of pyridine or azole types). Calorimetric measurements permitting an adequately accurate determination of the enthalpy term of the free equilibrium energy have shown that in certain cases such as [3]

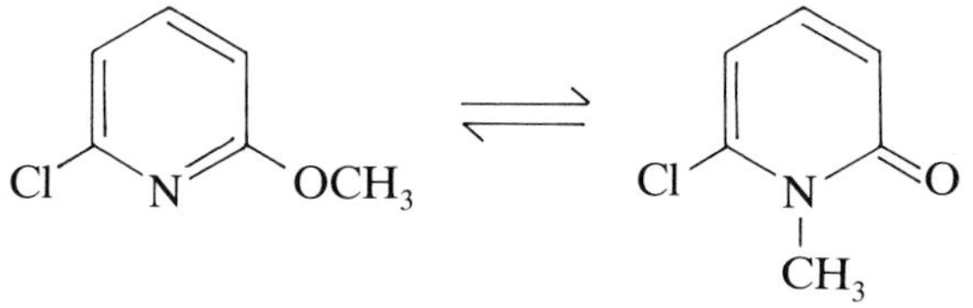

$$H^0_{liquid} = -4.1 \pm 2.2 \text{ kcal/mol}$$
$$H^0_{gas} = +0.3 \pm 3.7 \text{ kcal/mol}$$

the thermodynamic equilibrium parameters correspond to the range of tautomeric values. Activation energies, however, in all cases greatly exceed the necessary limit: equilibrium is established very slowly even at elevated temperatures of 120—150°C.

Some anion 1,2-alkyl group rearrangements apparently proceed intramolecularly. But in this case the character of the reaction mechanism drastically changes. Similarly to the well-studied Stevens' rearrangement, e.g.

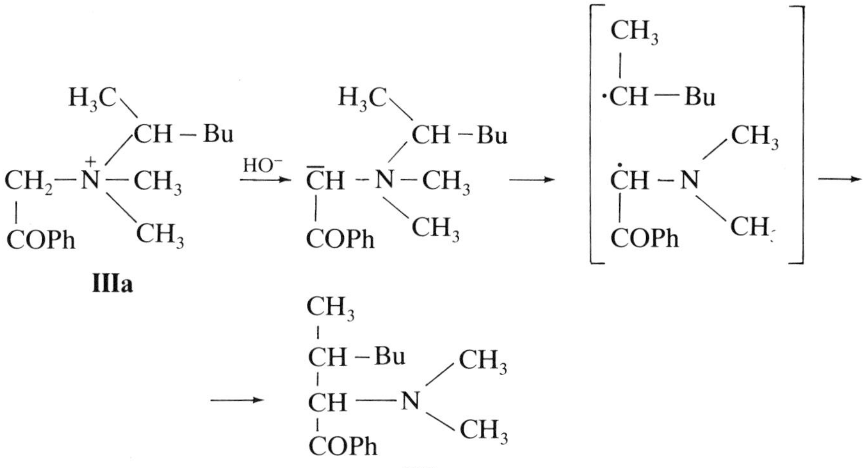

the other processes of this type, such as Wittig or Wawzonek rearrangements, probably include intermediate formation of radical pairs [4—8]. A similar mechanism is operative in the case of 1,4- and 1,9-benzyl transfer reactions [7, 9].

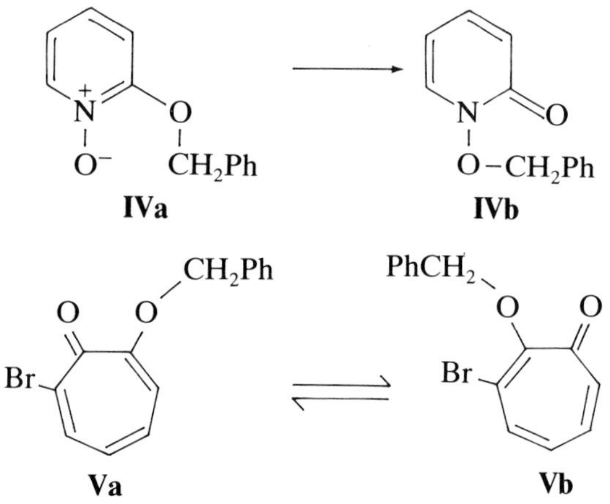

Other intramolecular alkyl rearrangements belonging to the large group of thermal $1,j$-sigmatropic reactions are characterized by extremely high activation barriers (25—50 kcal/mol) even in the cases which are allowed by orbital symmetry rules [10—11]. Moreover, among numerous investigated instances of nondegenerate reactions of this type, equilibrium reactions are very rare (smallness of ΔG^0 requirement for tautomeric systems).

1.1.2. STEREOCHEMISTRY OF NUCLEOPHILIC SUBSTITUTION REACTIONS AT THE sp^3-HYBRIDIZED CARBON ATOM

The reasons for high energy barriers or the intermolecular character of the above considered reactions become clear when examining the ways of formation of a transition state of the nucleophilic substitution reaction at the sp^3-carbon center.

The first organic stereodifferentiated reaction was the bimolecular nucleophilic substitution (S_N2) at the tetrahedral (sp^3) carbon atom studied by P. Walden in 1895 [12]. The well-established inversion of the carbon atom configuration in this reaction was originally associated with the following route:

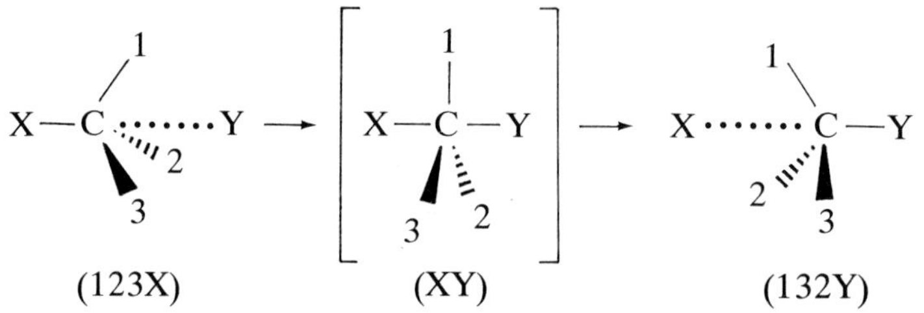

$$(4.2)$$

in which the nucleophilic electronegative group Y attacks the molecule with the initial configuration (123 X) from the rear, displacing the other nucleophilic group X via the transition state (XY)*. This transition state has the structure of a trigonal bipyramid where the electronegative nucleophilic groups occupy axial positions.

Westheimer [15] was the first to point out that in the S_N2-type reactions one more route leading to an inversion of configuration was possible, namely an attack by the nucleophile Y, not upon the rear side of the departing group X, but upon an edge of the initial tetrahedral molecule

$$(4.3)$$

(123X)   (12)   (132Y)

In the transition structure (12), the electronegative nucleophilic groups X, Y occupy diequatorial positions. In addition to the structure shown in (4.3), two more trigonal bipyramids ensuring an inversion of configuration are possible, *viz.*, (13) and (23). The energetic preferability of the (4.2) reaction path and of the corresponding transition state structure (XY) over other mechanistic variants has been demonstrated by means of numerous quantum mechanical calculations of the S_N2 reaction at the sp^3-carbon atom (see Table 3.III). These calculations are considered in detail in the monograph [16] as well as in the reviews [17, 18].

In the initial stage of the MERP, sufficiently stable ion-molecular clusters (Y is an anion) are formed both in the gas phase and in solution (see [19]) so that the potential energy curve takes on a double-well shape and the intrinsic barrier grows.** In the region of the transition state formation, the forming bond is already at a considerable C····Y distance, oriented strictly axially with respect to the breaking C—X bond. Structures **VIa, VIIa** with axial position of the entering nucleophile which emerge in this path are, according to nonempirical calculations [21, 22], 40—60 kcal/mol more favored than the topomers **VIb, VIIb** containing the entering nucleophile in an equatorial position

---

* In a tetrahedral structure, the substituents 1,2,3 are placed clockwise according to the Cahn—Ingold—Prelog sequence rule. For a trigonal bipyramid only two axial ligands are indicated [13, 14].

** MNDO calculations [20] show that the formation of a reactant complex occurs also in the case of electroneutral participants of the reaction.

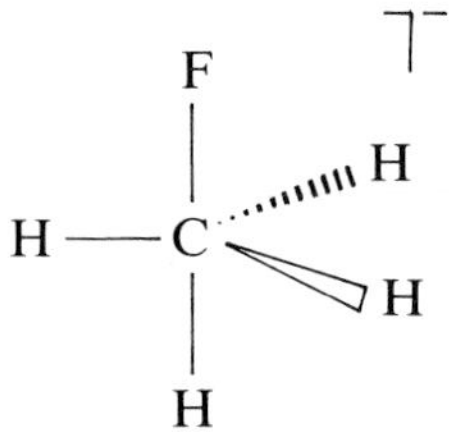

$$\left[\begin{array}{c} F \\ | \\ H-C{\overset{\displaystyle \text{\tiny\bf\textbackslash\textbackslash\textbackslash\textbackslash}H}{\diagdown H}} \\ | \\ H \end{array}\right]^{-}$$

VIa

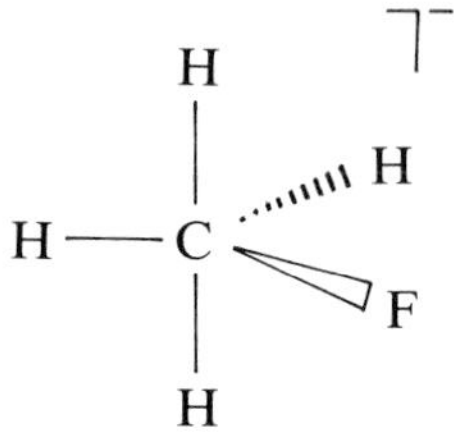

$$\left[\begin{array}{c} H \\ | \\ H-C{\overset{\displaystyle H}{\diagdown F}} \\ | \\ H \end{array}\right]^{-}$$

VIb

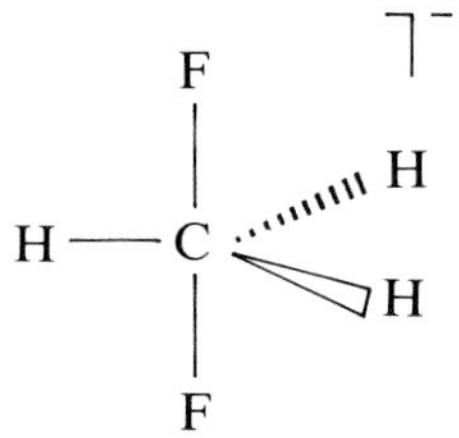

$$\left[\begin{array}{c} F \\ | \\ H-C{\overset{\displaystyle H}{\diagdown H}} \\ | \\ F \end{array}\right]^{-}$$

VIIa

$$\left[\begin{array}{c} H \\ | \\ H-C{\overset{\displaystyle H}{\diagdown F}} \\ | \\ F \end{array}\right]^{-}$$

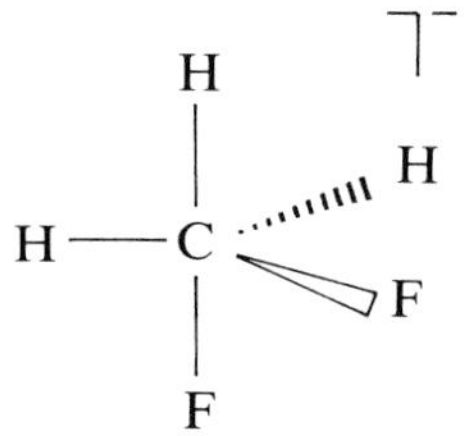

VIIb

This result is easy to understand by considering the shape of MOs of the model trigonal-bipyramidal structure CH_5^- (Fig. 4.1). The MOs a_2'', a_1' occupied by electrons are localized at axial bonds. Hence, the groups with greater electronegativity (richer in electrons) will tend to occupy precisely these bonds.

One more important result from the quantum mechanical calculations is the clear indication of a considerable stretching of the axial bonds. Thus, the length of the axial CH bond in the trigonal-bipyramidal anion CH_5^- is nearly 0.7 Å larger than the equatorial bond length. In the anion **VIIa**, the length of the axial C—F bonds is 0.4 Å larger than the 1.420 Å length of the equilibrium CF bond in fluoromethane. Table 4.I contains data from the most rigorous *ab initio* calculations of the geometric parameters of the trigonal-bipyramidal transition state structures.

Table 4.I Lengths of Equatorial and Axial C—H and C—F Bonds in CH_5^- and $CH_3F_2^-$ Anions considering as Transition State Structures in S_N2 Reactions $CH_4 + H^-$ and $CH_3F + F^-$ according to calculations [23—25].

Orbital Basis Set	H⋯⋯CH_3⋯⋯H		F⋯⋯CH_3⋯⋯F	
	CH_{ax}	CH_{eq}	CF_{ax}	CF_{eq}
3 − 21 G	1.702	1.061	1.776	1.062
4 − 31 G	1.730	1.059	1.830	1.060
6 − 31 G*	1.715	1.062	1.820	1.061
(8,4/4) + p(H) + d(F,C)	1.740	1.060	1.810	1.060

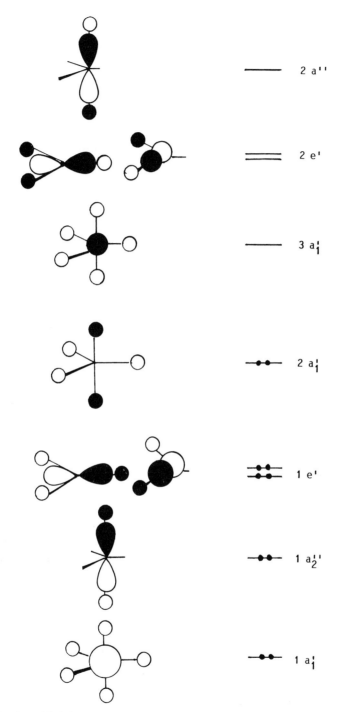

Fig. .4.1. Molecular orbital shapes and levels of the trigonal bipyramidal anion CH_5^- and the related isoelectronic compounds.

Thus, fast reactions of alkyl group transfer between intramolecular nucleophilic centers accompanied by inversion of configuration can, in agreement with the principle of correspondence between the structure of a tautomeric molecule and the transition state of a substitution at the migrating center reaction, take place only when there is a sufficiently large separation (3.5—4.0 Å) between nucleophilic centers and a linear arrangement of the reactive site X····$CR_1R_2R_3$····Y is accomplished. Possible topological graphs of such structures were considered earlier (Chapter 3, Section 2.4), but experimental data on alkylotropic tautomeric systems are so far unknown.

1.1.3. ALKYL MIGRATIONS WITH RETENTION OF CONFIGURATION AT THE sp^3-CARBON ATOM

Synthesis of alkylotropic systems would be made a great deal easier if the reaction could be directed towards a channel which ensured retention of configuration of the migrating sp^3-carbon center. Such a stereochemical effect would result from the front-side approach of the nucleophile Y with formation of the type **VIII** transition state structures possessing angular orientation of the entering and leaving groups [16, 21].

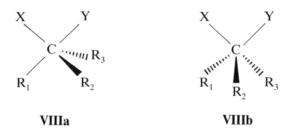

| VIIIa | VIIIb |

Factors facilitating stabilization of structures of the pentacoordinated carbon atom have been enough well analyzed [26, 27]. The following prerequisites are favorable for the frontal approach of a nucleophile: highly electronegative (hard) groups X and Y, enhancement of softness of the central atom due to introduction of electron donating substituents R, and inclusion of the migrating center in the small cycle framework. Indeed, in the case of cyclopropane derivatives, the semiempirical calculations [28] indicate the preferability of a reaction path with retention of configuration. However, the S_N2 reactions of cyclopropane derivatives are characterized by very low rates, and no reliable confirmation of configuration retention at the elementary stage of the reaction could be obtained [29—31].

Another possibility of alkyl migrations with retention of configuration of the migrating center relates to the two-stage *AdRE*-mechanism. Degenerate 1,3 and 1,5-alkyl group transfers served as examples in the theoretical modelling of such a possibility [32]. As was shown earlier (Chapter 3, Section 2.4.3), in this case the intermediate **Xa** with the axial-equatorial position of nucleophilic centers X and Y, which is formed in the first stage of the frontal nucleophilic attack, would have to correspond to the local minimum on the reaction PES.

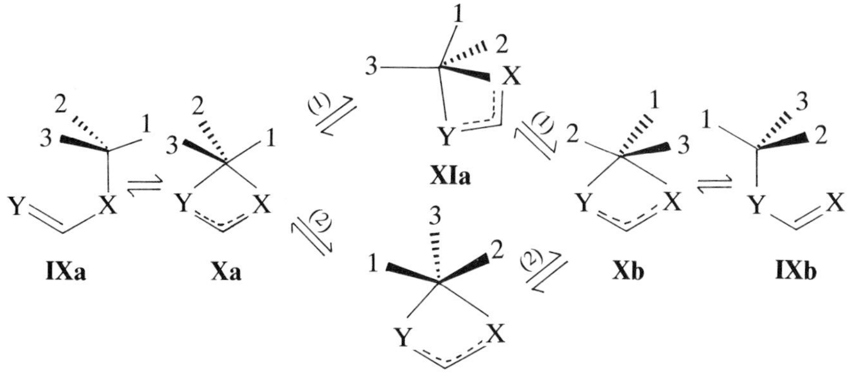

(4.4)

Most probable for this intermediate structure are two types of polytopal rearrangements (see [13]), i.e. the Berry pseudorotation with the transition state **XIa** or the turnstile rotation with the transition state **XIb**. Either path leads to the **Xa** ⇌ **Xb** rearrangement in which the positions of the entering and leaving groups are interchanged. MINDO/3 results for energy barriers of reactions (4.4) are listed in Table 4.II.

Table 4.II. MINDO/3 Calculated [32] Activation Enthalpies for Degenerate Reactions of 1,3- and 1,5-Migration of Alkyl and Fluoroalkyl Groups (kcal/mol).

Migratory Group $CR_1R_2R_3$	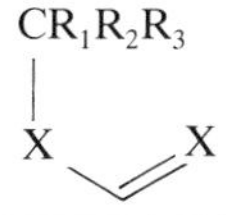 $CR_1R_2R_3$ along X—X		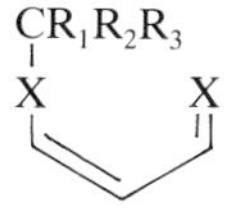 $CR_1R_2R_3$ on X...X	
	X = O	X = NH	X = O	X = NH
CH_3	80.6	60.0	62.0	50.9
CH_2F	80.0	61.3	51.7	44.6
CH_3	60.1	20.7	—	—

In all cases, the calculated transition states of the reaction corresponded to type **XIb** structures. As to the trigonal-bipyramidal **X** forms, they do appear in the MERPs but not as local PES minima. This may, apparently, be explained by excessive angular strain in these structures. The calculated values for energy barriers are too great for tautomeric reactions, albeit a marked trend towards their lowering is observed with the accumulation at the migrating carbon atom of substituents (such as fluorine) possessing a positive electromeric effect. In order to form a more general notion about the relationship between the character of a transition state of intramolecular S_N2 reactions and the steric potentialities of the reacting system, critical portions of the PES of intraanionic alkylotropic migrations have been calculated.

$$(4.5)$$

Values of R and the angle α have been varied. In view of a correlation between the angular parameter α and the possibilities for closure of the cyclic structure, the values of $\alpha < 90°$ correspond to 1,3-shifts and of $\alpha > 90°$ to 1,5- and 1,7-shifts.

Calculation of the reaction path **XIIa** → **XIIb** using two typical values of α, the results of which are given in Fig. 4.2, shows that in the case of 1,3-alkylotropic shifts ($\alpha < 90°$) the transition state with the angular C_s-configuration of the rupturing and forming bonds (**XIIIb**) is clearly favored. Formation of the transition state **XIIIa** with the diaxial position of oxygen atoms would require considerable structural deformation driving the nucleophilic centers apart to the distance $R \geqslant 3.5$ nm. Such a situation cannot be realized in the 1,3-sigmatropic system under any structural variations whatever.

When $\alpha > 90°$, in which case the orientation of bonds and orbitals of nucleophilic centers meets the steric conditions of the 1,5- and 1,7-sigmatropic systems, for all real values of R starting from 2.6 Å, is the structure of the transition state of the trigonal-bipyramidal type **XIIIa** is much more favorable. One should, however, keep in mind that the MINDO/3 method (similarly to other methods based on the zero differential overlap approximation) considerably underestimates the length of axial bonds in trigonal-bipyramidal structures of transition states of the S_N2-type reactions [16] so that the values of R 0.6—0.8 Å greater than those given in Fig. 4.2 are more realistic.

From the foregoing analysis it apparently follows that, regarding the design of the alkylotropic system, the strategy which aims at circumvention of the steric demands inherent in the mechanism of the concerted S_N2-reaction is of little avail.

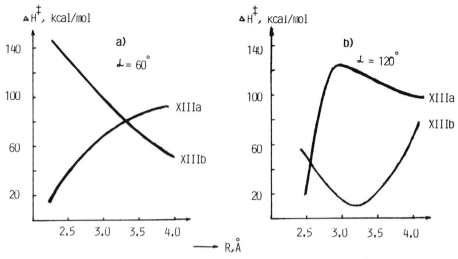

Fig. 4.2. The dependence of energies of the transition states **XIIIa, XIIIb** on the distance R between nucleophilic centers in the reaction (4.5) according to MINDO/3 calculations [32]: (a) $\alpha = 60°$ (the region of $\alpha < 90°$); (b) $\alpha = 120°$ (the region of $\alpha \geqslant 120°$). The angle α ($\alpha = \beta$) is shown on the structure **XIIa**. Similar results have been obtained also for other values of the angle α lying in the above-indicated regions.

1.2. *Vinylotropic Tautomerism*

Chapter 2 has dealt in detail with the experimental data on tautomeric migrations of acyl and aryl groups which contain the sp^2-carbon center. The same hybridization type also characterizes the methine carbon in vinyl derivatives. Following the concepts presented in Chapter 3, this should lead to a tetrahedral transition structure with angular orientation of the forming and rupturing bonds of the reactive site. Quantum mechanical calculations of trajectories of the nucleophilic attack on the sp^2-carbon atom at the double carbon—carbon bond bear out this qualitative conclusion. For the reaction of addition of a hydride-ion to propylene, the optimum angle of approach to the double-bond plane is, according to *ab initio* (3—21 *G*) calculations [33], 123° (**XIV**). In the case of the fluoride ion **XV** attacking ethylene, the angle $\theta = 119°$ [34].

XIV

XV

Similar values have been obtained in all other calculations, regardless of the substrate structure and the type of the attacking nucleophilic particle [16, 35]. Charge transfer from the lone electron pair of a nucleophile to the π^*-orbital of the substrate is the driving force of the reaction, as in the case of addition to the carbonyl group. This π^*-orbital, unlike the π^*_{CO}-orbital (Fig. 2.11) is not concentrated at the carbon atom, and this diminishes the overlap and the attractive interaction with the nucleophile. For activating a nucleophilic attack, it is necessary to polarize the π^*-orbital of alkene thus enhancing the $n-\pi^*_{C=C}$ overlap. This activation is achieved in alkene derivatives containing in the β-position a strong electron-accepting, e.g. aldehyde, group. When the stereochemically dictated geometry of the reaction is taken into account, the vinylotropic tautomerism can be represented as follows:

$$
\underset{\textbf{XVIa}}{\begin{array}{c} CR_1{=}CR_2{-}CH{=}O \\ | \\ X \qquad Y \end{array}} \;\rightleftharpoons\; \underset{\textbf{XVIc}}{\begin{array}{c} R_1 \quad CR_2{=}CH{-}O^- \\ C \\ X \qquad Y \\ + \end{array}} \;\rightleftharpoons\; \underset{\textbf{XVIb}}{\begin{array}{c} O{=}CH{-}CR_2{=}CR_1 \\ | \\ X \qquad Y \end{array}} \tag{4.6}
$$

The presence of a double bond at the migrating center poses an interesting stereochemical question as to whether there occurs a conversion or a retention of configuration (Z, E) at the migrating group carbon linked with the leaving group. An answer to this can be obtained on the basis of a detailed analysis of the mechanism of the nucleophilic substitution reaction at the vinyl sp^2-carbon atom. If the (4.6) reaction proceeds as a concerted one-step shift of the keto- or aldovinyl group with the **XVIc** structure corresponding to a transition state but not to an intermediate, then configuration of the substituents at the double C=C bond is retained and, upon substitution, the *cis*- and *trans*-derivatives are transformed into *cis*- and *trans*-diastereomers, respectively [36—38].

Different stereochemical relations are expected when the stepwise *AdE* mechanism is operative with the bipolar **XVIc** ion as an intermediate product. The stereochemical result of this mechanism may be quite flexible, although in most reactions studied retention of configuration of the nonsaturated carbon atom was noted [36, 39]. That this observation cannot be of general significance is shown by theoretical analysis as well by some experimental evidence, e.g. from the studies of autosubstitution reactions (4.7)

Under conditions of thermodynamic control, irrespective of the choice of the initial *Z*- or *E*-diastereomer, and equilibrium mixture is formed containing 60% *Z*- and 40% *E*-diastereomers **XVIIc** [40]. The (4.7) reaction is intermolecular analog of the degenerate system (4.6), it is a clear example of conversion by means of the *AdE*-mechanism associated with the formation of an intermediate.

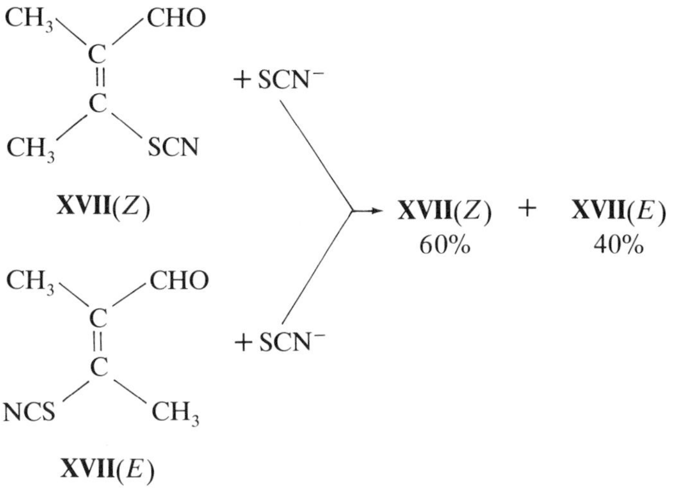

CH₃ — C(=C) — CHO / CH₃, SCN

$$\begin{array}{c} CH_3 \diagdown \quad \diagup CHO \\ C \\ \| \\ C \\ CH_3 \diagup \quad \diagdown SCN \end{array} \qquad + SCN^-$$

XVII(Z)

$$\rightarrow \textbf{XVII}(Z) \; + \; \textbf{XVII}(E) \qquad (4.7)$$
$$60\% \qquad\qquad 40\%$$

$$\begin{array}{c} CH_3 \diagdown \quad \diagup CHO \\ C \\ \| \\ C \\ NCS \diagup \quad \diagdown CH_3 \end{array} \qquad + SCN^-$$

XVII(E)

Indeed, assuming, in accordance with calculation data and the least motion principle, that the approach of the attacking nucleophile Y and the withdrawal of the group X occur as shown in the Scheme (4.8), one may foresee, for instance, in the case of the initial Z-diastereomer **XVII**, the following stereochemical results of (4.7) reactions

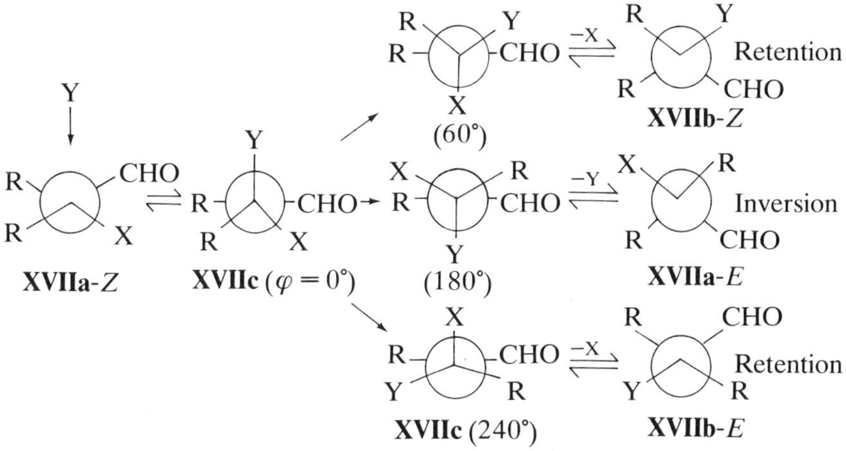

$$ \textbf{(4.8)} $$

Scheme showing Newman projections:

XVIIa-Z ⇌ **XVIIc (φ = 0°)** → branches to:

- (60°) → $\xrightarrow{-X}$ **XVIIb-Z** Retention
- (180°) $\xrightarrow{-Y}$ **XVIIa-E** Inversion
- **XVIIc (240°)** $\xrightarrow{-X}$ **XVIIb-E** Retention

Owing to the symmetry properties of the reaction coordinate, only those rotamers of the intermediate **XVIIc** which correspond to the minima of the potential function of rotation about the C—C bond are capable of splitting off a nucleophilic group, and they, consequently, determine the stereochemical course of the reaction. But, as implied above, not all of them correspond to the potential function minima. If this function has a single minimum, the result of each elementary stage of the reaction can be predicted by means of Scheme (4.8). The presence of several minima complicates the kinetic scheme of the reaction whose result cannot be in this case reduced to the pure inversion or retention of configuration.

It will be noted that scheme (4.8) possesses wide generality. Upon approach of the nucleophile Y, carbanions enantiomeric to **XVIIc** are formed on the second enantiotopic side of the double bond. The stereochemical results of the reaction are thereby not changed. On the other hand, Scheme (4.8) links, via intermediate carbanions, the substitution reactions in the Z- and E-diastereomers. Thus, when equilibrium with the attacking nucleophile is established, then, regardless of whether the Z- or E-diastereomer is chosen as the initial compound, the reaction result (relationship between stereoisomeric products and the initial compounds) must be identical provided the mechanism (4.8) is operative. The results of the reaction (4.7) corroborate these conclusions.

The *AdE*-mechanism of nucleophilic substitution at the sp^2-carbon atom admits of an interesting possibility of stereoselection of diastereomers in the tautomeric vinylotropic system of type (4.6), in which the nucleophilic centers X and Y are included in one molecule. Most interesting is the variant of the degenerate reaction X = Y (O, S, NR), in which of all possible intermediate **XVIIc** structures only the form **XVIIc** (240°) is stable. This means that there exists a reaction path linking the Z-diastereomer **XVIIa** containing a vinylaldehyde group at the center X with the E-diastereomer **XVIIb** containing this group at Y.

Since the free energies of the Z- and E-diastereomers are not equal, their equilibration alters the initial ratio between them in favor of the more stable form.

Thus, the following transformations based on the (4.6) and (4.8) reactions become possible

$$\text{(4.9)}$$

XVIIIa-Z   **XVIIIb**-E

Storing the diastereomer **XVIIIa**-Z solution at an appropriate temperature may lead to conversion of the C=C bond configuration with a quite low activation barrier equal to the barrier of a tautomeric migration. Conversion of the type (4.9) represents a novel intramolecular mechanism for *cis—trans*-isomerisation of the double C=C bond.

2. Silylotropic Tautomerism

Unlike alkylotropic rearrangements, instances of nucleophilic intramolecular migrations of the silyl groups are diverse and numerous [41, 42]. The optimal pathway for nucleophilic attack upon the tetracoordinate silicon atom is identical to the (4.2) path for the tetracoordinate sp^3-carbon atom. The attacking nucleophile Y tends to occupy an axial position in the trigonal-bipyramidal structure **XIXa** forming in the course of the reaction.

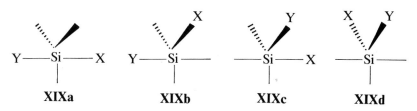

| **XIXa** | **XIXb** | **XIXc** | **XIXd** |

Data of *ab initio* calculations of the reaction of the SiH_5^- formation, when silane is attacked by a hydride ion [43—45], and of the semiempirical MNDO calculation of the S_N2 reaction pathway ($R_3SiX + X^-$; X = halide, R = H, Alkyl) [46] predict a backside approach for the electronegative nucleophilic groups Y with respect to the group X being substituted. The preferability of the trigonal-bipyramidal bond configuration of the pentacoordinate silicon atom **XIXa** is also predicted. These predictions are consistent with the conclusions obtained from the crystallographical mapping of the valence shell expansion at silicon from tetra- to penta-coordination for a large series of compounds taken mostly from the Cambridge structural data base [47, 48]. Of importance, however, is that, unlike trigonal-bipyramidal structures with the central carbon atom emerging in the S_N2-reaction (4.2), their silicon analogs **XIXa** are not transition states but intermediates. Although the **XIXa** intermediates with the diaxial position of the most electronegative groups X and Y are, indeed, the most stable, as has been shown not only by theoretical but also by experimental NMR spectral studies of the pentacoordinated silicon anions $R_3SiF_4^-$, $R_2SiF_3^-$ [49], the type **XIXb—d** structures topomeric to them are energetically sufficiently close to **XIXa**. This judgement is, in particular, supported by stability of a number of 10—Si—5 siliconates containing electronegative groups in equatorial positions [50].

| **XX** | **XXI** |

Equatorial entry of the nucleophile Y is thought [52] to be associated with the formation of a transition state possessing bond configuration of a tetragonal pyramid or of type **VIIIa, b** structures with the central silicon atom. Factors which stabilize such structures are particularly characteristic of silicon compounds (see Section 1.1.3)

$$R_3Si—X \xrightarrow{Y^-} \underset{\underset{Y\quad X}{\diagup\hspace{-0.3em}\diagup}}{Si^-} \longrightarrow \overset{X}{\underset{}{Si^-}}—Y \xrightarrow{-X^-} R_3Si—Y \tag{4.10}$$

XXII (TS)   **XIXc**

In degenerate reactions (X = Y), when equatorial nucleophilic attack is taking place, the microscopic reversibility principle forbids axial departure of the nucleofugal group X. For this reason, the stage of the Si—X bond rupture must be preceded by the polytopal rearrangement of the intermediate **XIXc**. Such rearrangements follow the course of the Berry pseudorotation as was recently demonstrated in the case of inversion at the tetracoordinate silicon which include 10—Si—5 intermediates and are catalyzed by weak nucleophiles [53].

Equatorial attack of the nucleophile in the reactions of substitution at the tetracoordinate carbon atom as well as the *AdRE* substitution mechanism, examined in the preceding Chapter, lead to retention of configuration at the silicon atom. An identical stereochemical result is also achieved with a number of other mechanisms (see [52]) which are not as important as those considered above for the intramolecular tautomeric systems with migrations of the silicon-containing groups. In view of its extraordinary stereochemical flexibility, the silicon atom readily adapts itself to the steric conditions of different nucleophilic substitution mechanisms. The mechanism of the equatorial attack with the **VIII** and **XXII** transition state stuctures does not require that the nucleophilic centers X and Y be driven apart to a considerable distance, it may be realized in small enough cycles which include the X—Si—Y group. A number of electronic factors, *viz.*, great diffuseness and, therefore, high polarizability of valence orbitals of silicon, low electronegativity leading to a substantial positive charge and the presence of low-lying vacant orbitals, ensure great facility of reactions of the tetracoordinate silicon atom with nucleophiles as well as a high migratory aptitude of the groups formed by this atom.

Let us now consider the most important rearrangements of the silyl groups whose intramolecular character has been reliably proved.

The 1,2-anion rearrangements of the type

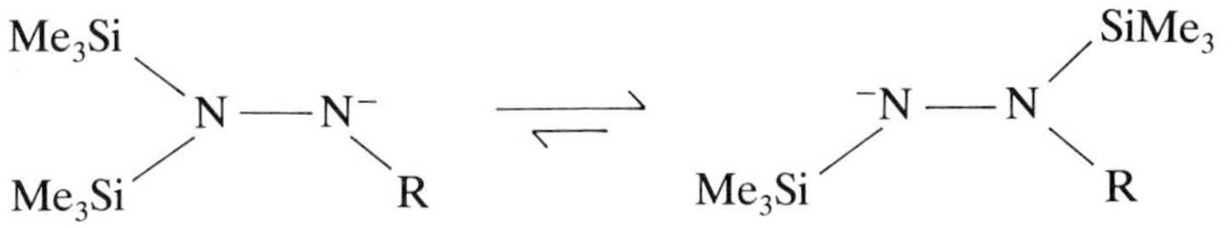

XXIIIa   **XXIIIb**

R = SiMe$_3$, CH$_3$

readily proceed in ether solution at high rates, as determined by the NMR method ($\Delta G^{\neq} = 11.5-15$ kcal/mol) [41]. The rate of intramolecular 1,2-migrations of the trimethylsilyl group in the hydrazine anion has been shown to be 10^{12} higher than that of the phenyl group.

The degenerate 1,3-silyl migrations of silyltriazenes **XXIV** (Z = N) [54] and amidines (Z = CPh) [55] have been studied in detail.

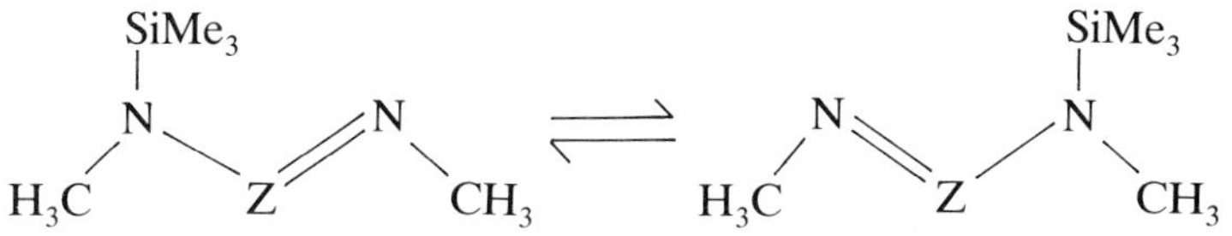

XXIVa   **XXIVb**   (4.11)

Z = N $\Delta G^{\neq}_{25} = 16.1$ kcal/mol
Z = PhC $\Delta G^{\neq}_{43} = 11.7$ kcal/mol

The 1,3-rearrangements **XXIVa** and **XXIVb** are characterized by high rates, but the same migration in imidazoles **XXV** does not occur, even at elevated temperature [56]. This is obviously due to unfavorable steric conditions for an intramolecular rearrangement in **XXV**.

XXVa   **XXVb**

On the other hand, in isoindolenine **XXVI** derivatives, two successive 1,3-shifts of the trimethylsilyl group coupled with the planar inversion of the noncyclic imine nitrogen atom lead to fast intramolecular 1,5-rearrangements [57].

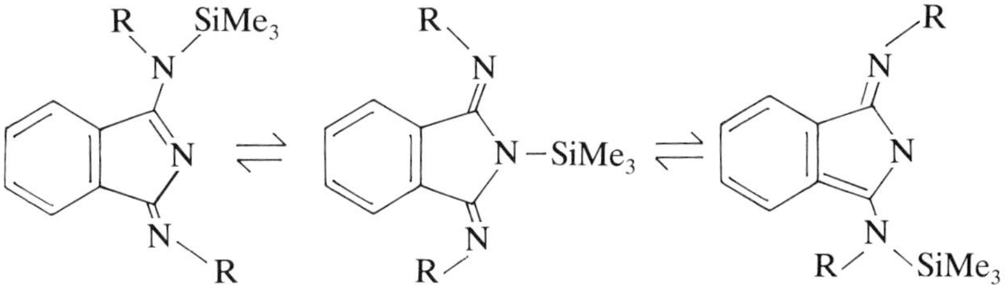

XXVIa   **XXVIc**   **XXVIb**   (4.12)

The free activation energy of the (4.12) reactions amounts at 130°C to 21—23 kcal/mol, and in compounds in which R = Aryl, the 1,3-diaryliminoisoindolenine tautomer **XXVIc** prevails in the equilibrium with **XXVIa, b**.

In the pure form, the tautomeric 1,5-silyl migrations were first detected in the series of the acetylacetone *cis*-enols **XXVII** (R = CH$_3$, R$_1$ = R$_2$ = R$_3$ = CH$_3$) [58].

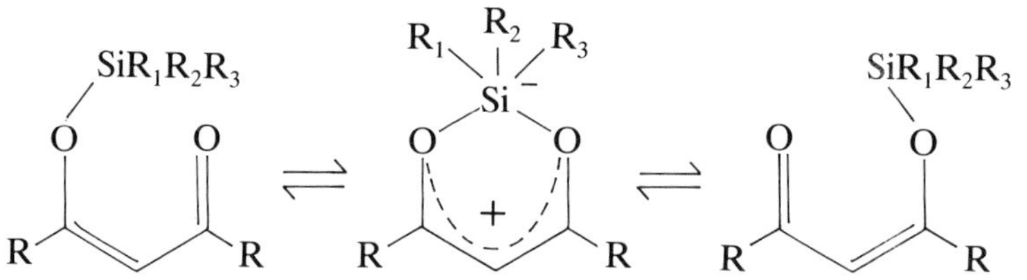

$$\text{(4.13)}$$

XXVIIa   **XXVIIc**   **XXVIIb**

Compounds **XXVII** are obtained by treatment of diketones with silyl chloride in the presence of pyridine. Under these conditions, along with *cis*-diastereomers, there are formed *trans*-diastereomers **XXVII** [58, 59]. Owing to remoteness of the reactive centers the *O,O'*-silyl migrations in **XXVIII** do not occur.

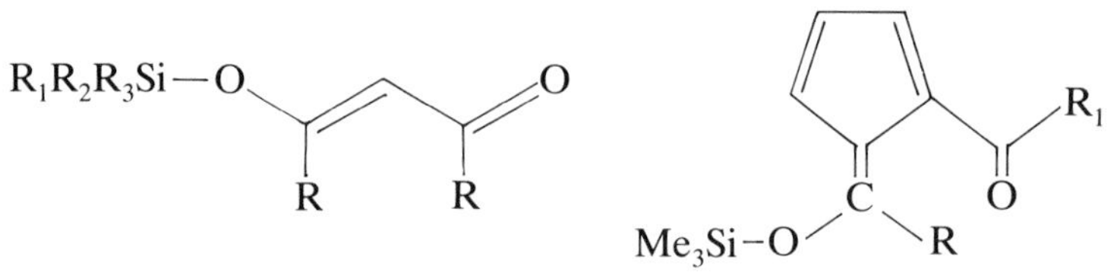

XXVIId   **XXVIII**   R$_3$, R$_1$ = H$_1$, CH$_3$, Aryl

For the same reason, no 1,9-silyl migrations could be observed in the *E*-diastereomers of 1,5-diacylcyclopentadiene derivatives **XXVIII** which are the prevailing isomeric forms. At elevated temperatures, the *E* ⇌ *Z* isomerization takes place, accompanied by the *O,O'*-migration of the trimethylsilyl group observed in the ¹H-NMR spectra and the rapid thermal decomposition of the compounds [51].

Table 4.III provides some NMR kinetical data concerning the (4.13) degenerate rearrangements. They show that the rate of the silyl group migration depends on the angular strain at silicon and on the nature of the diketone substituent R. The difference in free activation energy (at coalescence temperatures) between the silacyclobutane and silacyclohexane derivatives (No. 2 and 4, respectively) is more than 6 kcal/mol.

For the chiral derivative **XXVIIa** (R$_1$ = Me, R$_2$ = Ph, R$_3$ = PhCH$_2$, No. 6 in Table 4.III), the AB pattern for the diastereotopic benzyl protons on silicon in the ¹H-NMR spectrum is retained at temperatures where the 1,5-silyl migrations are fast. This means that the migration occurs with retention of configuration at silicon. It has been shown that the AB pattern can be viewed even at 211°C in chloron-

Table 4.III. Kinetic Data for 1,5-Silyl Migration of type (4.13) in $CHClF_2$ Solution according to [59].

No.	Compound **XXVIIa, b**		$\Delta G_{T_c}^{\neq}$, kcal/mol	k_{25}, s^{-1a}
	$R_1R_2R_3Si$	R		
1	Me_3Si-	CH_3	13.9 ± 0.4^b	4×10^2
2	$CH_3P-CH_2-)_3Si-$	CH_3	< 6.5	$> 1 \times 10^8$
3	$CH_3(-CH_2-)_4Si-$	CH_3	9.3 ± 0.3	1×10^5
4	$CH_3(-CH_2-)_5Si-$	CH_3	12.9 ± 0.3	2×10^3
5	Me_3Si	$C(CH_3)_3$	< 7.2	$> 3 \times 10^7$
6	$(Me)(Ph)(PhCH_2)Si-$	$C(CH_3)_3$	9.4 ± 0.3	8×10^5
7	Me_3Si	Ph	11.5^c	2×10^4

[a] Extrapolated values obtained under assumption of $\Delta S^{\neq} = 0$
[b] In chlorobenzene solution
[c] In CH_2Cl_2 solution

aphthalene solution. By simple extrapolation it has been estimated that fewer than one out of every 10^{13} 1,5-migrations at 25°C results in inversion at silicon. This result points quite definitely to the preferability of the front-side approach of the attacking carbonyl oxygen to the tetracoordinate silicon atom. Regarding the intermediate **XXVIIc**, the authors [59] give preference to the type **XIXb, c** structures.

Retention of configuration at silicon was also observed in the case of the 1,9-silyl migrations in the tropolone derivatives **XXIX** [60]. Separate NMR resonances were found for diastereotopic methyl groups at temperatures, at which very fast silyl migrations occur. A rough estimate from the line broadening of the energy barrier for this migration gave $\Delta G^{\neq}$ as 8.2 kcal/mol.

XXIXa   **XXIXb**

The nucleophilic silylotropic systems considered represent, in a purified form, which is rather convenient for detailed kinetical and mechanistic studies, the most important features of the plethora of molecular rearrangements of organosilicon compounds. With very rare exceptions, they have been shown to proceed as intramolecular reactions with virtually complete retention of configuration at a chiral silicon center [42]. This explains the important role which silatropic shifts have begun to play in regiospecific synthesis.

Tetracoordinate silicon is much more susceptible to nucleophilic attack than carbon, but the next Group IV element, germanium, shows a still greater migratory aptitude. A characteristic example of this is given by the fast 1,5-rearrangements oi the germyl group in the *cis*-enol derivatives of acetylacetone [61].

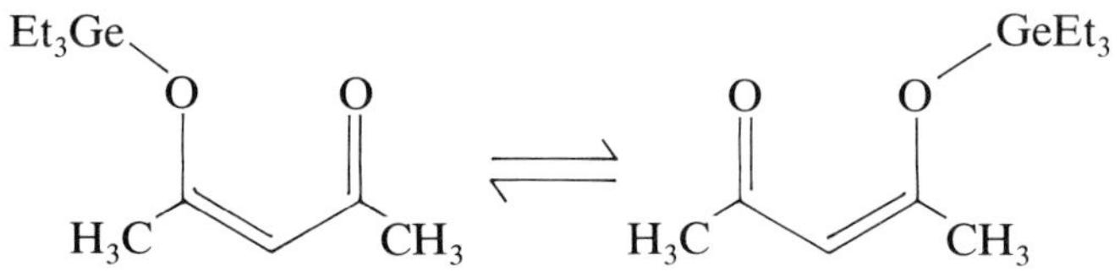

XXXa	**XXXb**

While the type **XXVIII** *trans*-isomer is not subject to an *O,O'*-transfer of the germyl group, in compounds **XXX** the **XXXa** ⇌ **XXXb** transfers do occur at a very great frequency (the values of $\Delta G^{\neq}$ in the −80° to 0°C range attain 11.1— 12.3 kcal/mol) exceeding the corresponding values for analogous migrations of the trimethylsilyl group **XXVIIa** ⇌ **XXVIIb** (Table 4.III).

3. Tautomeric Migrations of Phosphorus-Containing Groups

Fast and reversible 1,3-migrations of the triphenylphosphonio group between nitrogen atoms in the *N,N'*-dimethylbenzamidine derivative **XXXI** (R = Ph, X = R_2) detected in 1970 were one of the first examples of the tautomeric reactions based on shifts of the non-transition element groups [62].

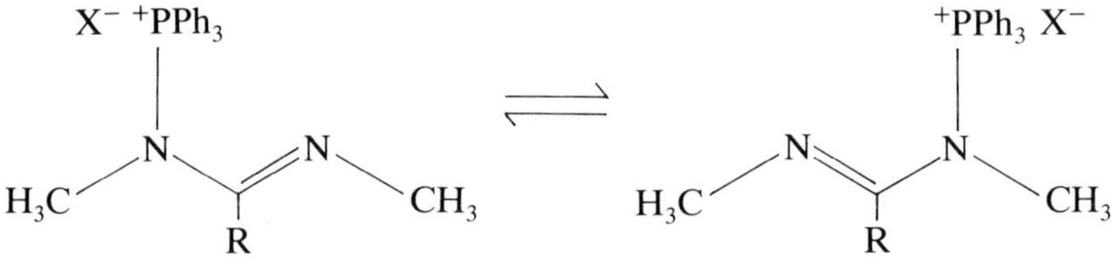

(4.14)

XXXIa	**XXXIb**

During the following decade, the rearrangement (4.14) remained the only instance of intramolecular tautomeric migrations of the P-containing group. Analysis of the possibilities of adaptation of various mechanisms governing the reactions of nucleophilic substitution at the phosphorus atom at various coordination states to the steric conditions of the low-barrier intramolecular rearrangements [63—65] encouraged the chemists to search for the appropriate compounds. Since 1980 rapid accumulation of data on the tautomeric rearrangements of the P^{III}, P^{IV}, and P^{V} — containing migrants has been going on. These are considered in the present Section, starting with tautomeric rearrangements of the migrants with the tetracoordinate phosphorus as the central atom.

3.1. P^{IV}-Migrants

Phosphorylation and the phosphoryl group transfer reactions play a special role in biochemical processeses [66—68]. The construction of a tautomeric system characterized by the rearrangement of these groups between nucleophilic centers is, therefore, of great interest.

In Section 3.10 it has already been noted that the phosphoryl derivatives of *cis*-enols of 1,3-diketones cannot undergo rapid tautomeric O,O-rearrangements of the phosphoryl group, although prolonged heating of their solution brings about a shift of the equilibrium **XXXIIa** ⇌ **XXXIIb** [69].

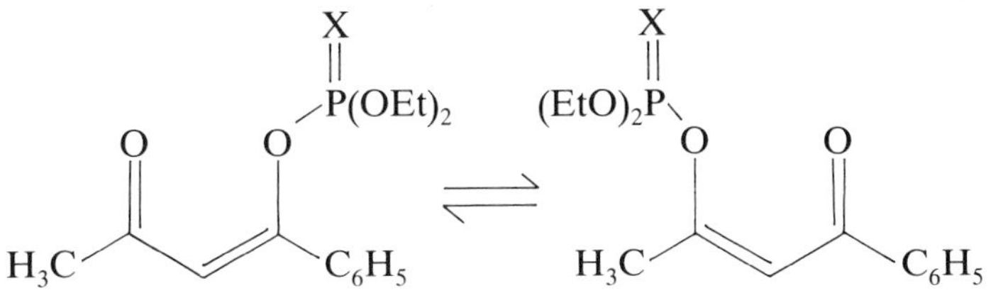

| **XXXIIa** | **XXXIIb** | X = O, S |

The reasons for the low migratory aptitude of the phosphoryl groups in such compounds lie in the structure of the transition state in the reaction of the associated nucleophilic substitution at the tetrahedral phosphorus atom. This mechanism bears a great resemblance to the mechanism of substitution at a tetrahedral carbon (4.2). The attacking nucleophile Y approaches the phosphorus tetrahedron from the back side with respect to the leaving group X.

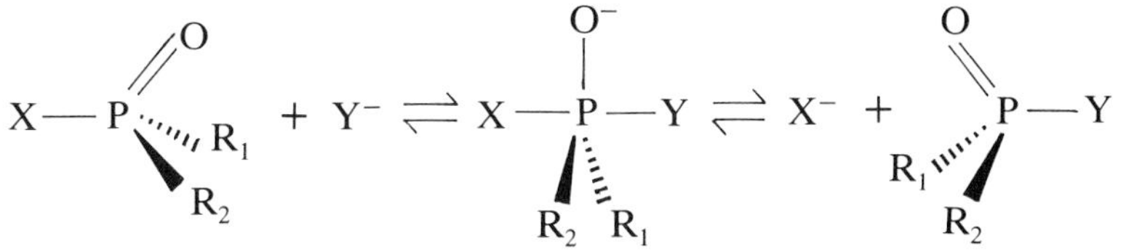

$$(4.15)$$

| **XXXIIIa** | **XXXIIIc** (XY) | **XXXIIIb** |

Being a higher-period element, phosphorus is apt to form hypervalent structures. This was one of the reasons which prompted Westheimer [70] to suggest that the trigonal-bipyramidal structure **XXXIIIc**, unlike the S_N2 reactions of the tetracoordinate carbon atom, is not a transition state but an intermediate in reaction (4.15). The preferability of a backside nucleophile attack (4.15) leading to the stereochemical inversion at phosphorus and accompanied by formation of the intermediate **XXXIIIc** is also predicted by quantum mechanical calculations of model reactions using the CNDO/2 and *ab initio* STO—3G methods [71, 72]. According to the calculated data, the transition states in the stages of the inter-mediate **XXXIIIc** formation and decomposition are, in their structure, very close to it. Their chief distinction is an additional lengthening, by approximately 0.1 Å, of the axial bonds P—X and P—Y. In reactions of nucleophilic substitution at the phosphorus atom in compounds **XXXIIIa**, the **XXXIIIc** intermediates with diapical orientation of the entering nucleophile and the departing nucleofuge could be observed by means of the NMR method, and could even be preparatively isolated [73—75].

The second important type of associative nucleophilic substitution at the tetracoordinate phosphorus atom was outlined in Section 2.4.3 of Chapter 3. If two phosphorus coordination sites are linked in a small four- to seven-membered ring, the forming intermediate phosphorane contains the entering nucleophile Y in the apical position and the leaving group X in the equatorial. In order to provide for the later apical departure condition, such an intermediate should necessarily suffer at least one polytopal rearrangement to allow expulsion of the leaving group. Here, the stereochemical outcome depends on whether the number of polytopal rearrangements is even, which leads to an inversion, or odd. The latter case results in retention of configuration at the phosphorus atom, as exemplified by reaction (3.18).

Certain important generalizations have been made regarding the structure and modes of rearrangements of the pentacoordinate phosphorus compounds through studies of both their crystal structures and NMR behavior. Though virtually all acyclic phosphoranes exist as trigonal bipyramids, various degree, of distortion to another square pyramid topology may be traced in strained mono- and bicyclic compounds. Analysis of this trend allows a crystallographical mapping of the reaction path connecting the two limiting structures **XXXIVa, b** and **XXXIVc** to be carried out [76, 77]. It corresponds to synchronized permutation of two pairs of apical and equatorial ligands in accordance with a pseudorotation scheme first proposed by Berry [78].

$$\text{XXXIVa (12)} \qquad \text{XXXIVc} \qquad \text{XXXIVb (34)}$$

(4.16)

Another topological equivalent possibility for polytopal rearrangement is the so-called turnstile mechanism proposed by Ugi [79].

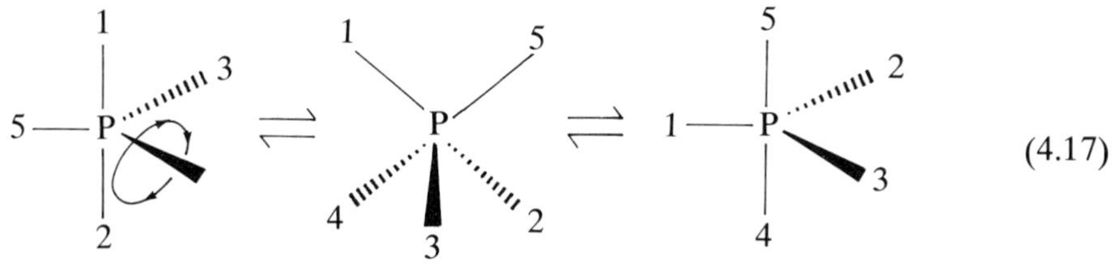

XXXIVa   XXXIVc   XXXVa

$$(4.17)$$

Both experimental [76] and theoretical [80—82] results favor the Berry pseudorotation scheme as being a lower energy process for polytopal rearrangement at pentacoordinate phosphorus. Interpretation of experimental data is greatly facilitated by two important empirical rules also confirmed by molecular mechanics and quantum mechanical calculations. The ring-strain rule [73, 76] states that four- and five-members rings preferentially span one (but not two) apical and equatorial positions in the trigonal bipyramid structures **XXXIVa, b**. The second so-called polarity rule, first formulated by Muetterties [83], implies the tendency of electronegative substituents to occupy apical sites in a trigonal bipyramid. This rule stems largely from the nature of the highest occupied MOs of phosphoranes, which is similar to that of CH_5^- (Fig. 4.1). It involves a nonbonding interaction between central atom and the apical ligands, which causes the electron density to reside on the ligands. Another effect influencing the distribution of ligands between apical and equatorial positions in **XXXIVa, b** is due to the π-donor ability of the ligand. It was predicted [80, 84], and experimentally confirmed, that good π-donors like NH_2, or O^- preferentially take equatorial positions while good π-acceptors, like CN, adopt apical positions.

The parameters of apicophilicity, i.e., the stabilization achieved through the ligand's changing its position from the equatorial to the apical one, have been established on the basis of dynamic NMR studies of cyclic oxyphosphoranes [85] and *ab initio* calculations of monosubstituted phosphoranes PH_4X **XXXVI** [86].

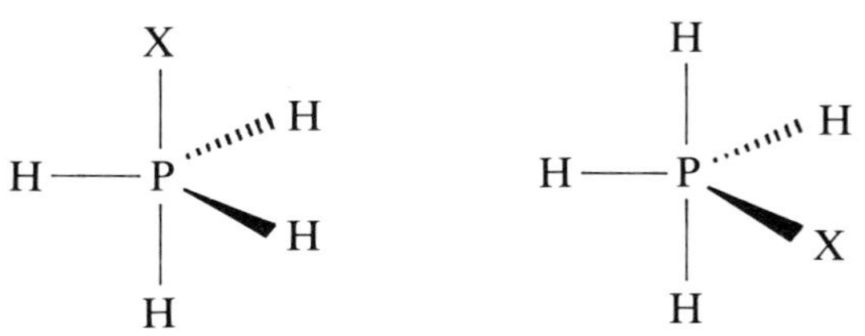

XXXVIa   XXXVIb

The latter method yields the most accurate quantitative expression of apicophilicity free from any contaminating contributions of ring-strain effects. In Table 4.IV, the apicophilicity of various substituents in phosphoranes **XXXVI** is given in descending order.

Table 4.IV. Apicophilicity of Certain Ligands according to *ab initio* (3—21 G*) Calculations [86] for Geometrically Optimized Structures of **XXXVIa** and **XXXVIb**.

No.	Ligand X	ΔE(apical—equatorial), kcal/mol
1	Cl	−15.92
2	CN	−11.24
3	F	−7.55
4	C≡CH	−4.61
5	H	0
6	CH_3	2.32
7	OH	2.70
8	S^-	9.36
9	O^-	9.32
10	NH_2	10.31
11	BH_2	16.56

Polarity and ring-strain rules are quite useful in the analysis of the *AdRE*-mechanism of the tautomeric reactions associated with the formation of cyclic intermediates. Of three possible types of intermediate phosphoranes forming upon the initial nucleophilic attack:

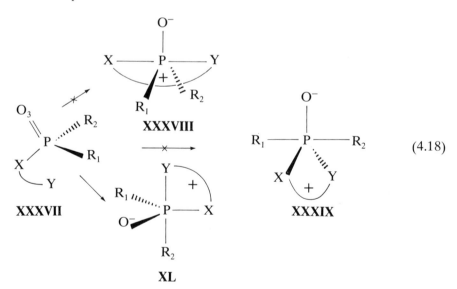

(4.18)

only **XL**, with the equatorial-apical position of reaction sites satisfies the ring-strain rule. **XL** is not, however, the only phosphorane of this type formed from **XXXVIII**. Three out of the four sides of the phosphorus tetrahedron are accessible to nucleophilic attack which is accompanied by the formation of pentacoordinate intermediates **XLa—c** with the apical entering nucleophilic group Y. The overall scheme of their further transformations requiring at least one pseudorotation in order to transfer the departing group into the apical position* with the formation of phosphoranes **XLIa−c** may be represented as follows:

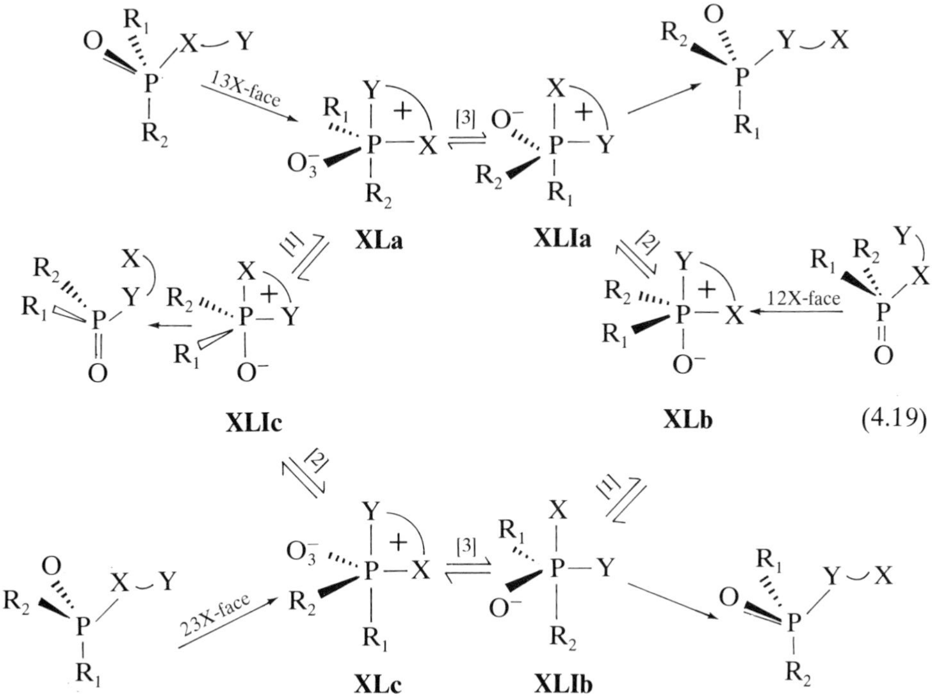

(4.19)

The most probable channel for a degenerate reaction (R_1, R_2 = Alkyl, Aryl) appears to be the formation, in the first stage, of phosphoranes **XLa**, **XLc** while phosphorane **XLb**, with an oxyanion group in the apical position possessing minimal apicophilicity, is energetically unfavored. Apicophilicity values from Table 4.IV visualize energetics and the preferable course of intervening rearrangements of emerging intermediates.

Nucleophilic reactions of substitution at the phosphorus atom of the phosphoryl group involving the intervening polytopal rearrangements of intermediates have been known for quite a time (see [67, 70, 73, 87]), but only recently have reports on tautomerism based on this mechanism begun to appear.

A temperature-variable equilibrium ($\Delta G° < 2$ kcal/mol), quite sensitive to the nature of a solvent, has been found in the case of imidoyl derivatives of thiophosphoric acid [88, 89].

---

* This conclusion, following from the microscopic reversibility principle, is imperative only for degenerate reactions (X = Y), although deviations from it in the case of nondegenerate reactions are, so far, unknown.

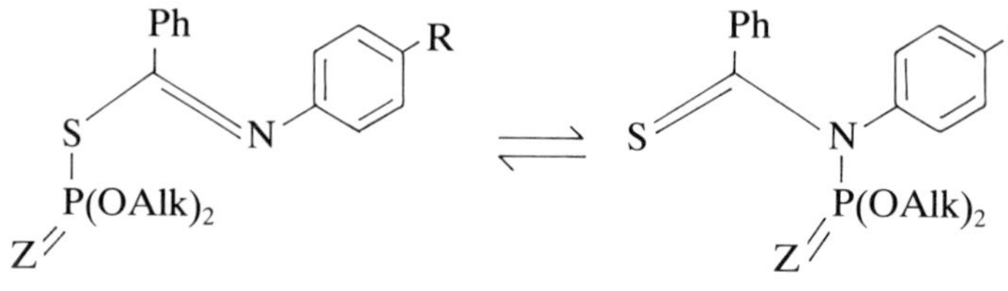

(4.20)

XLIIa   XLIIb

Z = O, S; R = H, Br, CH₃

Passing to amidine derivatives has allowed a more detailed study of the kinetics of phosphoryl rearrangements and the stereochemical course of the reaction. Table 4.V contains the main results of studies [90] of the degenerate systems of phosphorylotropic tautomeric rearrangements.

Table 4.V. Kinetical Data on Degenerate Phosphoryl and Iminophosphoryl Migration (4.21) in C_6D_6 Solution according to NMR Spectral Study [90].

No.	Compound **XLIII**			$\Delta H^{\neq}$, kcal/mol	$\Delta S^{\neq}$, e.u.	$\Delta G_{T_c}^{\neq}$, kcal/mol $(T, {}^{\circ}C)$
	R	Z	R₁, R₂			
1	Ph	O	(OEt)₂	—	—	24.9 (174°)
2	Ph	NPh	(OEt)₂	18.1	−12.4	23.6 (174°)
3	Ph	O	—OCH₂—CH₂—CH₂—	9.5	−26.1	18.6 (77°)
4	Ph	O	—O—C₆H₄—O (*ortho*)	—	—	<8 (−90°)ᵃ
5	Ph	O	F₂	—	—	<8 (−90°)ᵃ
6	CF₃	NPh	—OCH₂—CH₂O—	—	—	21.7 (174°)
7	CF₃	O	Cl₂	21.9	2.0	21.0 (124°)
8	CCl₃	O	(OEt)₂	—	—	> 25 (200°)
9	CCl₃	O	F₂	19.8	0.3	19.7 (124°)
10	SCH₃	O	(OEt)₂	—	—	26.0 (200°)

ᵃ In deuterochloroform solution.

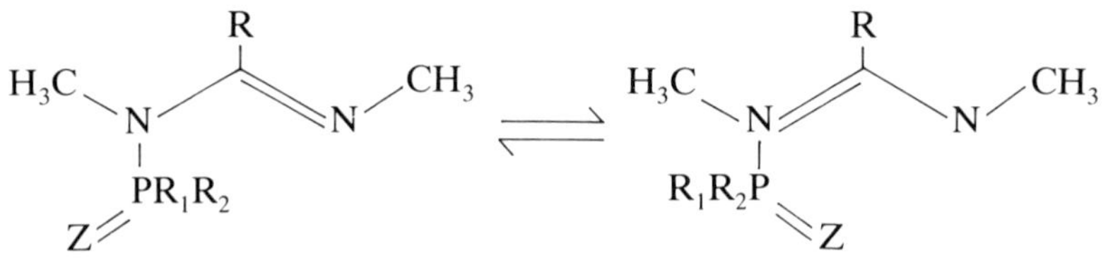

(4.21)

XLIIIa   XLIIIb

Similarly to reactions of the *N,N′*-acyl transfer in amidine systems, the C-phenyl group appreciably lowers free activation energy of the *N,N′*-migration (by 2—7 kcal/mol), which is accounted for by delocalization of the positive charge in the type **XL** intermediate cyclic phosphorane. The most substantial effect upon the rate of reactions (4.21) is exerted by the nature of the ligands R_1, R_2, X at the tetracoordinate phosphorus atom. Migrations are accelerated more than 10^{12} times upon the attachment to phosphorus of apicophilic electron-accepting substituents which stabilize the intermediate phosphoranes **XL** and lower the activation barriers of polytopal rearrangements of the **XLA** ⇌ **XLIa** type. Great acceleration of phosphoryl migrations is produced by incorporating a phosphorus atom into phospholane rings (Nos. 3, 4, 6 in Table 4.V). Acceleration of nucleophilic reactions at the phosphorus atom in cyclic systems is a well-documented fact which is also associated with the additional stabilization of intermediate phosphorane structures [67, 73, 87].

Retention of the phosphorus configuration in intramolecular (4.21) migrations is borne out by the character of evolution experienced by the portion of the NMR spectrum of **XLIII** related to the methylene protons of the dioxaphospholane ring (compound No. 6 in Table 4.V). Proceeding by elevating the temperature of solution from slow to fairly high rates of intramolecular migrations, the multiplet AA′BB′X (X = ^{31}P) of these protons stays unchanged, not transforming into an A_4X pattern, which would occur if each rearrangement act followed the S_N2 mechanism (4.15) leading to inversion of configuration at the phosphorus atom. In addition to this, large negative values of activation entropy of reaction (4.21) are evidence in favor of a two-step mechanism (4.19)

Tautomeric migrations of phosphoryl and thiophosphoryl groups have also been found and investigated in the series of derivatives of cyclic amidines **XLIV** [57] and **XLVI** [91].

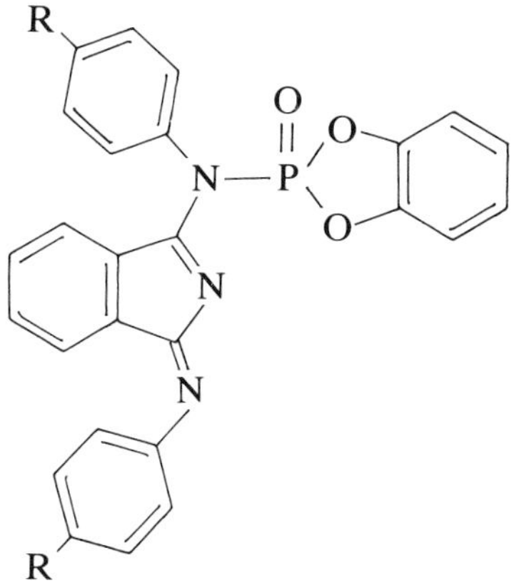

XLIV

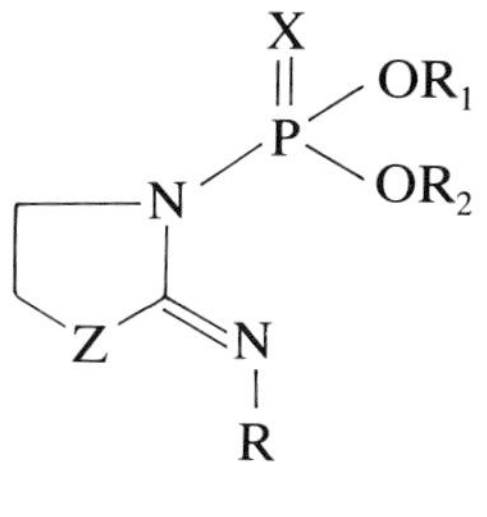

R = Me, Ph; Z = S, CH_2, X = O, S

XLV

The mechanism of a substitution reaction at the tetracoordinate phosphorus atom in the phosphonium salts **XXXI** is stereochemically equivalent to the mechanism (4.19). The kinetics of these reactions and the effect of the counterion

X have been studied by means of the dynamic NMR method [90, 91]. Results of these studies are given in Table 4.VI.

Table 4.VI. Kinetical Parameters of Degenerate Tautomeric Rearrangements (4.14) of Compounds **XXXI** [90].

No.	Compound **XXXI**		Solvent	$\Delta H^{\neq}$, kcal/mol	$\Delta S^{\neq}$, e.u.	$\Delta G_{25}^{\neq}$, kcal/mol
	R	X				
1	Ph	Cl	CDCl$_3$	9.4	−15.2	13.9
2	Ph	Br	CDCl$_3$	14.2	−8.2	17.2[a]
3	Ph	ClO$_4$	CDCl$_3$	9.9	−13.3	13.8
4	Ph	BPh$_4$	(CD$_3$)$_2$CO	10.3	−12.0	14.0
5	CHCl$_2$	ClO$_4$	CDCl$_3$	14.8	−3.6	15.9

[a] At 77°C [62].

3.2. *P^{III}-migrants*

The pyramid structure of compounds of tricoordinate phosphorus **XLVI** being, as a rule, characterized by high configurational stability, the reactions of these compounds may be considered as being similar to reactions of the tetracoordinate phosphorus compounds with the electron lone pair in **XLVI** as a phantom-ligand.

   The CNDO/2 calculation [92] of trajectories of the fluoride-ion nucleophilic attack on phosphine **LVI**, where electronegativity of possible departing groups decreases in the order F > Cl > H, has shown that two pathways of nucleophilic substitution reactions (4.22) are possible.

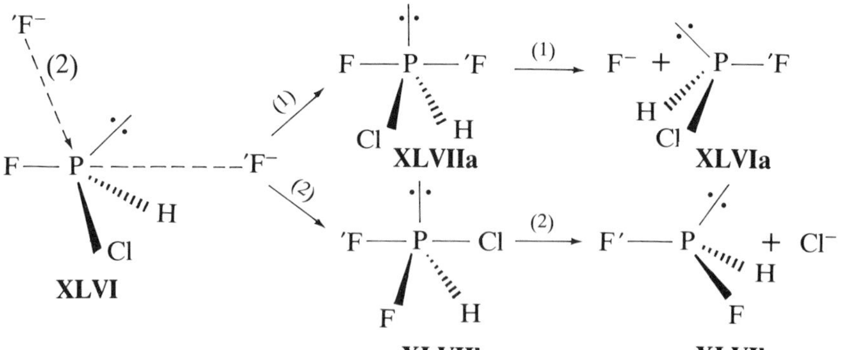

$$(4.22)$$

Either path is associated with the backside approach to the departing group of the attacking nucleophile, the formation of the intermediate tetracoordinate anion **XLVII** and inversion of configuration at the P^{III}-center. The preferable path, however, is a rear-side approach to the more electronegative departing group (2), whereas the rear-side approaches to the P—H bond or along the axis of the electron long pair are energetically unfavorable.

Intermediate anions **XLVII** have a slightly distorted bisphenoidal structure which may be regarded as a truncated trigonal bipyramid **XXXIV** with the phantom-ligand, i.e. the long pair, in the equatorial position. Of all conceivable ligands, the electron lone pair is most apicophobic, which accords with the expectations of VSERP theory. Consequently, the lone pair always serves as a pivot ligand in the Berry pseudorotations of such anions. Semiempirical [92] and *ab initio* (SCF + CI) [93] calculations of the structure and electron distribution in type **XXXIV** anions point to a considerable elongation of their apical bonds and accumulation of the negative charge at apical centers.

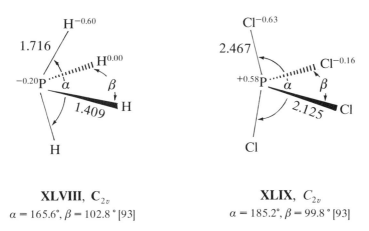

XLVIII, C$_{2v}$

$\alpha = 165.6°, \beta = 102.8°$ [93]

XLIX, C$_{2v}$

$\alpha = 185.2°, \beta = 99.8°$ [93]

The results of various experimental studies on the stereochemistry of substitution at tricoordinated phosphorus indicate that nucleophilic substitution reactions proceed in most cases with complete inversion. When the reactants and products are configurationally stable under the reaction conditions, the overall result is an inverted product, as was found in reactions of substitution of aryl groups in phosphines [94] and alkoxy groups in phosphinites [95] by alkyl lithium as well as in some other reactions [96].

In accordance with these data and with the predictions of Table 3.III, it may be expected that concerted intramolecular rearrangements of phosphinyl groups between nucleophilic centers should be feasible only in the systems which allow axial orientation of the attacking and leaving groups of type **XLVII**. Another possibility is the *AdRE* mechanism of type (4.19) in which the phantom ligand, i.e. the electron lone pair, retains the equatorial position in all stages of polytopal rearrangement of the intermediate. Precisely this mechanistic possibility was successfully realized by Negrebetski *et al.* [90] in the reactions of 1,3-phosphinyl rearrangements of amidine derivatives.

$$R_1 \underset{H_3C}{\overset{R_2}{\diagdown}} P \underset{N_1}{\overset{|}{\diagdown}} \underset{R}{\diagdown} \underset{CH_3}{\overset{N_2}{\diagdown}} \rightleftharpoons \quad La \quad \rightleftharpoons \quad LIa \quad \rightleftharpoons \quad LIb \quad \rightleftharpoons \quad Lb \qquad (4.23)$$

| La | LIa | LIb | Lb |

The reaction is accelerated by electron-accepting substituents at the phosphorus atom creating positive charge at the phosphorus center (compare electron distributions in **XLVIII** and **XLIX**) and by incorporating the migrating phosphorus atom into the cycle. Results from studies on kinetics of tautomeric rearrangements (4.23) are listed in Table 4.VII.

Table 4.VII. Kinetical Parameters of Degenerate Tautomeric Migration of Phosphinyl Groups (4.23) according to [90]. (In C_6D_{12} Solution).

No.	Compound L		$\Delta H^{\neq}$, kcal/mol	$\Delta S^{\neq}$, e.u.	$\Delta G_T^{\neq}$, kcal/mol (T, °C)
	R	R_1, R_2			
1	Ph	—OCH$_2$CH$_2$O—	13.3	−12.6	17.0 (23°)
2	Ph	—OCH(CH$_3$)CH(CH$_3$)O—	13.5	−5.9	15.4 (30°)
3	Ph	—O—C$_6$H$_4$—O—(*ortho*)	10.1	−10.8	13.2 (8°)ᵃ
4	Ph	(OEt)$_2$	14.6	−22.7	24.6 (169°)
5	Ph	(Ph)$_2$	—	—	24.9 (178°)
6	CF$_3$	—OCH$_2$CH$_2$O—	17.3	−8.4	20.6 (108°)
7	CF$_3$	—O—C$_6$H$_4$—O—(*ortho*)	14.7	−14.4	20.0 (98°)

ᵃ In CDCl$_3$.

Retention of configuration at the migrating P^{III}-center in reactions of the rearrangement (4.23) and its intramolecular character have been rigorously proved using NMR spectra. As can be seen from Fig. 4.3, the spin coupling of NMe protons with the ^{31}P nucleus is not disturbed even at fairly high rates of the phosphinyl migrations ($k_{40} = 12$ s^{-1}). Under these conditions, diastereotopy of methylene protons in the dioxaphospholane cycle is retained.

As the temperature is raised above 80°, the reactions become intermolecular, and there occurs a transformation of the spectrum of methylene protons into the multiplet A_4X ($X = ^{31}$P) indicating inversion of stereochemical configuration at phosphorus. The ratio between the constants of the intra- and intermolecular processes derived from crossover experiments at 80° (in C_6D_6, $C = 20\%$ for compound No 2 from Table 4.VII) is $1:3 \times 10^{-4}$. Increase in the polarity of the solvent enhances the contribution by the intermolecular mechanism with the **LII** transition state.

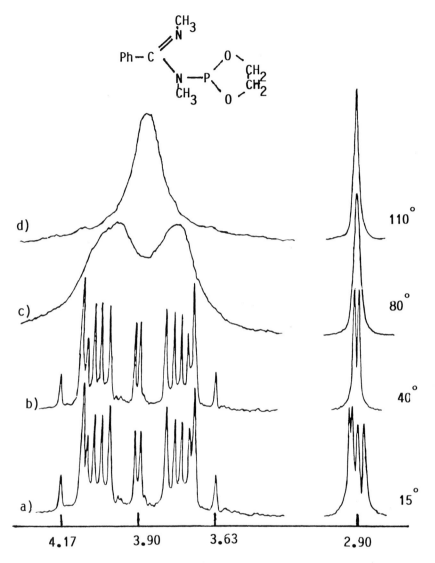

Fig. 4.3. The temperature dependence of the ¹H-NMR spectrum in the region of protons of the dioxaphospholane cycle (³¹P spin decoupled) as well as of protons of the N-methyl groups for compound **XL** (R = Ph, R , R₂ = —OCH₂CH₂O) in nitrobenzene solution: (a) migrations are frozen; (b) intramolecular migration; (c, d) intermolecular migration [90].

LXII

Reversible 1,2-migrations of the phosphinyl groups in diphosphinoxides and sulfides have been found and investigated by Lutsenko and Foss, see the Review [97].

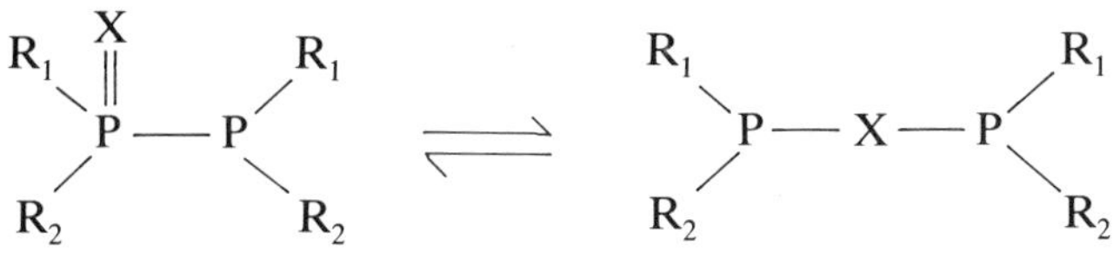

$$R_1, R_2 = Alk, OAlk, Aryl, OAryl, N(Alk)_2$$
$$X = O, S$$

LIIIa   **LIIIb**

3.3. P^V-migrants

Mechanism and stereochemistry of nucleophilic substitution reactions at pentacoordinated phosphorus are still relatively little studied, but data thereon have been accumulating rapidly. According to current views, in the first stage of a nucleophilic attack on phosphorane there is formed a hexacoordinated intermediate whose thermodynamic stability is usually higher than that of the initial reactants [98, 99]. The CNDO/2 and MINDO/3 calculations of the MERP for nucleophilic attack by the hydride ion on simple phosphoranes **LIV** [100] permits the main stages of the reaction to be presented

$$H^- + PH_3R_1R_2 \rightleftharpoons [PH_4R_1R_2]^- \rightleftharpoons PH_3R_1R_2 + H^- \tag{4.24}$$
$$R_1, R_2 = H, CH_3$$

Fig. 4.4 shows the snapshot sequence picture of the hydride ion approaching the pentacoordinated phosphorus atom along the MERP. The distance between the attacking ion and the phosphorus atom has been chosen as the reaction coordinate, while all other geometrical parameters have been optimized. While still at a considerable distance from the center to be attacked, the hydride ion is being oriented in the equatorial plane of phosphorane along the bisector line of the angle between the least electronegative ligands $R_1 = R_2 = CH_3$. This reaction path can be explained by the form of the phosphorane LUMO localized in the equatorial plane (Fig. 4.1). The reaction (4.24) proceeds with no barrier to form the octahedral anion $PH_4R_1R_2^-$.

Since, in the initial phosphorane, the most electronegative groups (X) that are to be substituted occupy an apical position, and for most compounds of the hexacoordinated phosphorus the energy barriers of polytopal rearrangements are quite high [93, 100], the MERP shown in Fig. 4.4 signifies that in the reactions of nucleophilic substitution at the pentacoordinated phosphorus atom a *cis*-configu-

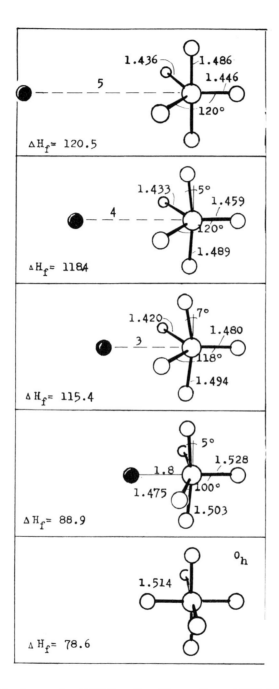

Fig. 4.4. The MERP for the reaction (4.24) $R_1 = R_2 = H$ according to MINDO/3 calculations [100]. Heats of formation are given (in kcal/mol) for various distances (in Å) between the hydride ion (shaded circle) and the molecule being attacked (PH_5).

ration of bonds of the leaving and entering groups is realized. This is consistent with the data from stereochemical studies of nucleophilic substitution reactions in the series of bicyclic phosphoranes [101] as well as with high mobility of the equilibrium (4.25) [102] relating to the σ-valence tautomerism.

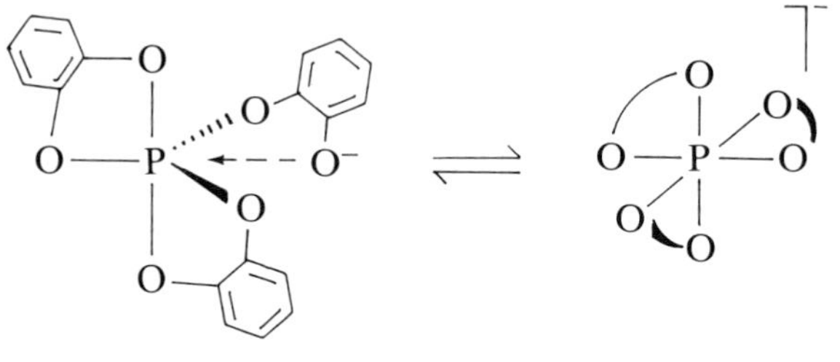

$$(4.25)$$

LIVa   **LIVb**

Thus, the stereochemical course of the degenerate reaction of nucleophilic substitution at the pentacoordinated phosphorus atom in the phosphoranes **XLV** may be represented by the scheme below where X is the most electronegative ligand and the growth in apicophilicity of other ligands is denoted by their growing numbers.

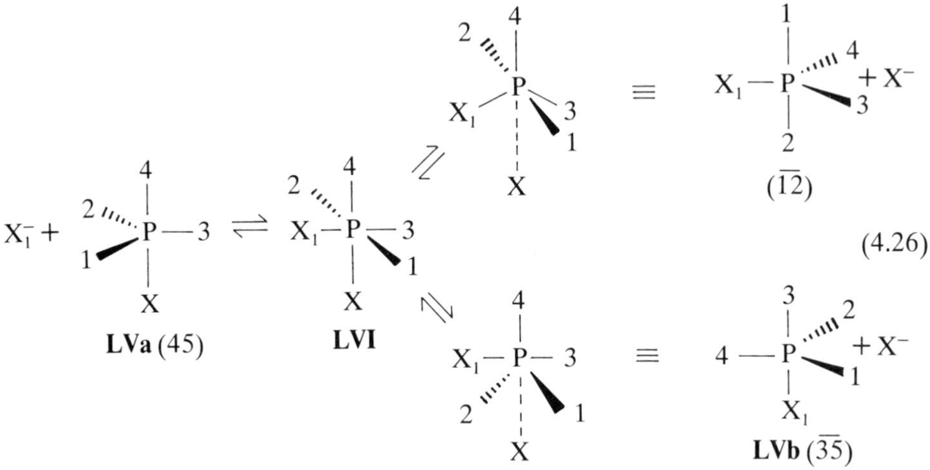

$$(4.26)$$

LVa (45)   **LVI**

After the formation of the octahedral intermediate **LVI** with the *cis*-configuration of the P—X bonds, two stereochemically nonequivalent variants of the *cis*-departure of group X may be envisaged. Only one of these leads to stable phosphorane (35) with the electronegative group X_1 in the apical position. Hence, the most probable stereochemical outcome of the primary reaction is inversion of configuration of the pentacoordinated phosphorus atom. The overall stereochemical result depends on the degree to which phosphorane (35) is subject to secondary permutational isomerizations.

The tautomeric migrations of phosphoranyl groups described by the (4.26) mechanism were first detected by Negrebetski *et al.* [57, 103, 104] in the series of amidines **LVII, LVIII**.

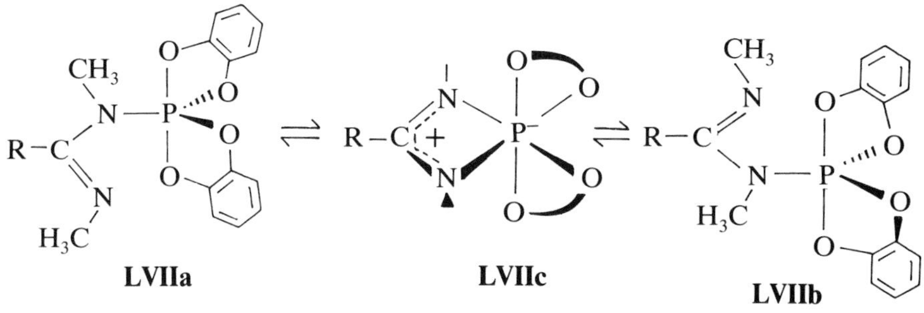

$$ (4.27) $$

LVIIa   **LVIIc**   **LVIIb**

$$ (4.28) $$

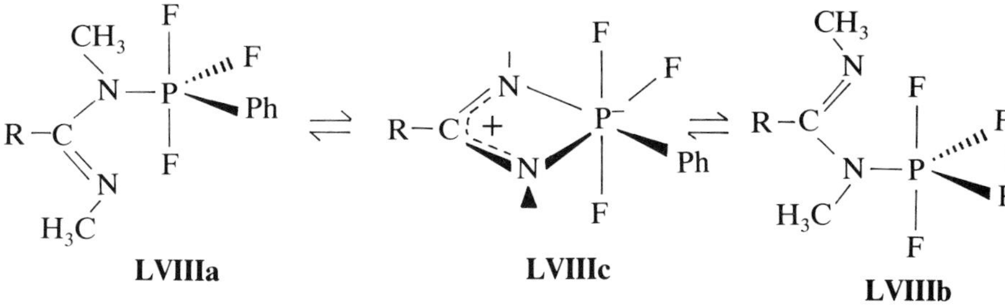

LVIIIa   **LVIIIc**   **LVIIIb**

The position of intermediate phosphorane—phosphorate equilibria strongly depends on the type of the substituent R. When R = Ph, it is completely shifted towards derivatives of the hexacoordinated phosphorus **LVIIc, LVIIIc**. Upon introduction of electron-accepting substituents R, the equilibrium content of phosphorates in solution is lowered and the NMR spectra reveal the degenerate tautomeric migrations of the phosphoranyl groups. Thermodynamic and activational parameters of the reactions (4.27), (4.28) are given in Table 4.VIII.

Intramolecular character of the (4.27), (4.28) tautomeric rearrangements has been proven by observation of averaged $^3J^{PH}$ spin-coupling constants at sufficiently

Table 4.VIII. Thermodynamical Parameters of Phosphorane—Phosphorate Equilibrium and Free Activation Energies of 1,3-Phosphoranyl Group Migration in (4.27) and (4.28) Rearrangements according to Data [103, 104]. (In C_6D_6 solution.)

Compound	Equilibrium content of P^{VI}-isomer(30°)	$\Delta H°$, kcal/mol	$\Delta S°$, e.u.	$\Delta G°_{30}$, kcal/mol	$\Delta G^{\neq}_T(T, °C)$
LVII, R = CCl₃	94%	−3.7	−6.5	−1.7	21.0 (164°)
LVII, R = CF₃	82%	−3.2	−7.6	−0.9	22.0 (164°)
LVIII, R = CCl₃	0	—	—	> 3	21.5 (135°)
LVIII, R = CF₃	0	—	—	> 3	23.4 (135°)

large migration rates. Attachment to the nitrogen atom of amidines **LVII, LVIII** of the prochiral isopropyl, instead of the methyl, group permitted detection of the inversion of configuration at the migrating P^V-center whose rate practically coincides with that of 1,3-migrations of the phosphoranyl group.

Studies of tautomeric migrations of the P^{III}—P^V-containing groups have furnished much important information on the mechanism of these group transfer reactions in the molecular systems with nucleophilic centers in sterically close proximity. Reactions of this type, particularly the phosphorylation and phosphoryl group transfer reactions, lie at the foundations of many vital biochemical processes. Their preparative aspect is also of importance, since they are utilized in obtaining biologically active compounds [89]. Almost all known phosphorus migrations which have been examined in Section 3 follow the *AdRE* mechanism that includes formation of an intermediate and separate stages of breaking the old and making the new bonds. It is noteworthy that all enzymatic reactions at phosphorus, including transfers of the thiophosphoryl moiety from ATP to AMP, from enolpyruvate to ADP, and some others, proceed with inversion [67]. Therefore, they most probably avoid energy-wasting additional pseudorotation steps, preferring a concerted mechanism with one-step phosphoryl and thiophosphoryl transfers. As yet this mechanism has not been implemented in artificial, synthetically-prepared molecular systems.

4. Tautomeric Rearrangements of As-Containing Migrants

Only two well-documented examples of tautomeric migrations of arsenic-containing groups are known to date. They both exhibit substantional acceleration of the As^{III} and As^{IV} group migrations as compared to the P^{III}, P^{IV} migrations in analogous systems. The triphenylarsonio group has been shown to migrate between the nitrogen atoms in *N,N'*-dimethylbenzamidine **LIX** with the energy barrier $G^{\neq}_{-90} <$ 10 kcal/mol [91].

$$(4.29)$$

The reaction (4.29) most probably follows the *AdRE* mechanism. Evidence in favor of this has been found in investigations of the tautomeric rearrangement of (2′-tropolonyl)-1,3,2-benzodioxaarsole **LX** [105]

(4.30)

The ^{13}C-NMR spectrum of this compound contains only seven peaks, even at $-70°C$, which indicates the existence of an effective symmetry plane in **LX** on the NMR time scale and, therefore, very rapid O,O'-arsinyl moiety migrations. Since no line broadening of the peaks corresponding to isochronous nuclei was observed, the frequency of the arsinyl migration at this temperature was estimated to be higher than $10^6 s^{-1}$ and therefore, $\Delta G^{\neq}_{-70} < 6.5$ kcal/mol. This value is lower than the energy barrier of the trialkylsilyl O,O'-migration in tropolone derivatives **XXIXa** ⇌ **XXIXb** (Section 2 of this Chapter). The frequency with which compound **LX** rearranges is similar to that calculated on the basis of the ESR spectrum for the interligand exchange of the odd electron in a stable radical derived from 2'-(O-phenoxyloxy)-1,3,2-benzodioxaarsole in which the arsinyl migration involves an intramolecular homolytic substitition at the tricoordinated As atom [106].

The reason for such fast migrations in a system (4.30) that does not meet steric requirements of concerted rearrangement of the arsinyl migrant (Table 3.III) is made clear by the results of an X-ray investigation of the molecular structure of compound **LX**, which are shown in Fig. 4.5.

Although the distance As----O_4 is longer than the normal covalent bond As—O, it is 1.2 Å shorter than the van der Waals contact (3.4 Å). This indicates a very strong attractive interaction between the centers in question in the molecule **LXa**. Its stable conformation is very close to a pyramidally distorted bisphenoid **LXc$_1$** with a nucleophile (O_4) entering into an axial position. The conformation is thus very well prepared for the steric requirements of the *AdRE*-mechanism (4.30) because the square-pyramidal structures are transition states for the Berry-pseu-dorotation of bisphenoidal compounds [107, 108].

In contrast to the arsenic compounds **LIX**, their antimony analogs exist not as tetracoordinated triphenylantimonio salts but rather as stable symmetrical antimonates **LXI** [91]. This trend is understandable considering the exceptionally high propensity of antimony to form the Sb^{VI}-derivatives.

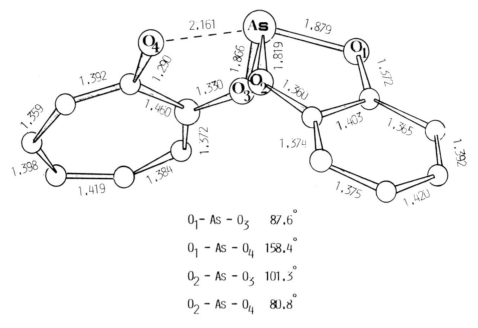

$$O_1 - As - O_3 \quad 87.6°$$

$$O_1 - As - O_4 \quad 158.4°$$

$$O_2 - As - O_3 \quad 101.3°$$

$$O_2 - As - O_4 \quad 80.8°$$

Fig. .4.5. Molecular structure of 2-tropolonyl-1,3,2-benzodioxaarsole **LXa** from X-ray structural data [105].

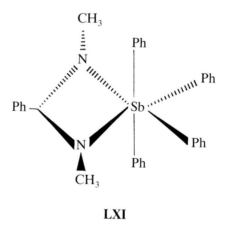

LXI

5. Tautomeric Rearrangements of Sulfur-Containing Groups

Data of Table 3.III show that nucleophilic rearrangements of the S^{II}, S^{III}, S^{IV}-migrants are associated with the formation of trigonal bipyramidal transition state structures or intermediates with the diaxial position of the nucleophilic centers X, Y, as, for example, in the case of the sulfenyl, sulfinyl and sulfonyl groups.

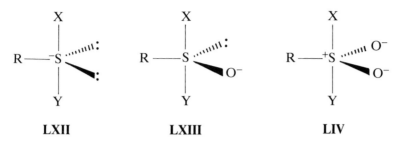

LXII   **LXIII**   **LIV**

The preferability of the **LXII–LXIV** structures over other topomers follows from numerous theoretical and experimental studies on the mechanism of reactions of nucleophilic substitution at sulfur atoms in various coordination states. The order of apicophility of ligands in the sulfur-centered trigonal bipyramids **LXIV** corresponds to that for phosphoranes.

In view of similarities between the **LXII–LXIV** structures and the structures of intermediates of nucleophilic substitution reactions at the P^{III}, P^{IV}-centers, it might be expected that the non-concerted *AdRE* mechanism should be the route for the reaction of nucleophilic substitution at the sulfur atoms. Indeed, this mechanism has been found to operate in the recently studied tautomeric rearrangements of sulfenyl and sulfinyl groups.

5.1. S^{II}-migrants

Reactions of nucleophilic substitution at the bicoordinate sulfur atom play an important part in biological systems [109, 110], they are usually more rapid than such reactions at the S^{III} and S^{IV} atoms [111].

In order to study directions of approach of a nucleophile to the sulfenyl sulfur atom, there have been calculated using the CNDO/2 method optimal trajectories for the reaction of the mixed sulfur dihalogenide with the fluoride ion [112].

$$F_1^-\cdots S-F \rightleftharpoons F_1-\overline{S}-F \rightleftharpoons F_1-S\cdots F^-$$
$$\qquad\quad | \qquad\qquad | \qquad\qquad |$$
$$\qquad\quad Cl \qquad\qquad Cl \qquad\qquad Cl$$

LVIa

$$F_1^- + SClF$$

LXV

$$F_1^-\cdots S-Cl \rightleftharpoons F_1-\overline{S}-Cl \rightleftharpoons F_1-S\cdots Cl^-$$
$$\qquad\quad | \qquad\qquad | \qquad\qquad |$$
$$\qquad\quad F \qquad\qquad F \qquad\qquad F$$

LXVIb

(4.31)

Fig. 4.6 shows sections of the PES of the (4.31) reactions for several distances $F^-\cdots S$ serving as the reaction coordinate in three different directions.

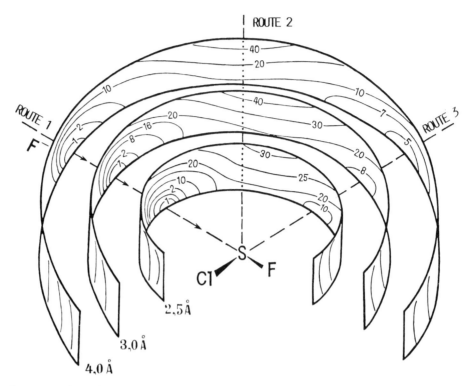

Fig. 4.6. Sections of the potential surfaces of the reaction (4.31) for various distances between F⁻ and the molecule SClF obtained by the CNDO/2 method, according to [112]. In the break-spaces of the curves, the relative energies of the levels are given in kcal/mol.

The orientation of the F_1^- ion along the axial direction relative to a more electronegative ligand (F) in the plane of the molecule being attacked (route 1 in Fig. 4.6) takes place already at quite large distances (4—5 Å) from the sulfur atom. This route is approximately 5 kcal/mol more favorable than the route 2, where the attacking fluoride ion is axially oriented with respect to a less electronegative chlorine atom. A symmetrical approach of the nucleophile (3) does not find the necessary energy valley and is, thus, infeasible. These conclusions are fully consistent with the predictions based on an analysis of a broad series of crystalline structures of compounds of the bicoordinate sulfur atom with intermolecular coordination of nucleophilic centers [113].

The T-shaped structures **LXVI** emerging during the reaction (4.31) are stabilized with respect to the initial reactants. They serve as intermediates rather than transition states of the reaction. The same conclusion could be drawn from the *ab initio* (*DZ*-type basis set) calculations [114] on nucleophilic cleavage of the S—S bond in disulfides, although more recent *ab initio* (6—31 G*) calculations [115], while confirming formation of an intermediate complex between reactants, produced strong evidence that the T-shaped structure possessed properties of a transition state. There are, nevertheless, a number of experimental results indicating stability of the T-shaped sulfuranide anions. In this connection, one may, in addition to the thio-thiophthenes considered earlier (Chapter 1, Section 6.4.2), point to a recently performed synthesis of the salts of the tricoordinated sulfuranide anion [116—118] the structure of which has been confirmed by an X-ray diffraction study.

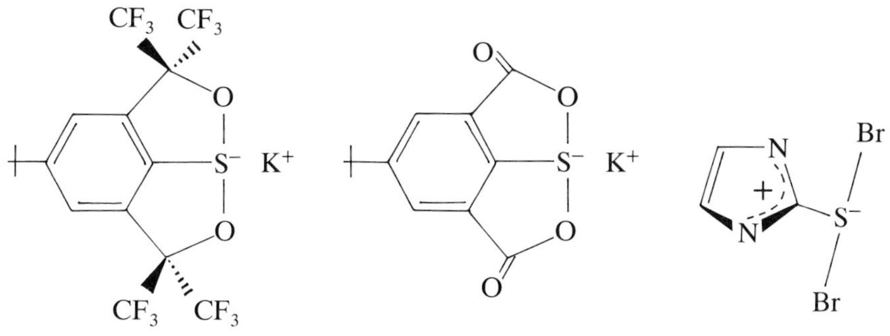

LXVII **LXVIII** **LXIX**

The sterically imposed T-shaped configuration of the breaking and forming bonds leads to fast intramolecular reactions of nucleophilic substitution with the apical groups departing. One may point, for example, to the ready cyclizations of *o*-substituted azobenzenes to thiadiazolium salts [119]

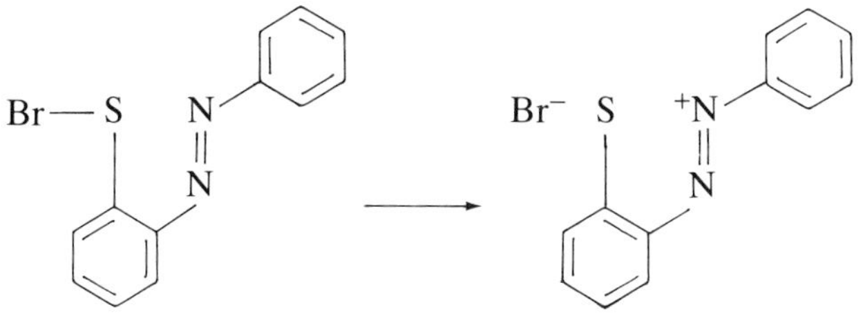

LXXa **LXXb**

and of the *o*-substituted aromatic carboxylic acids [120].

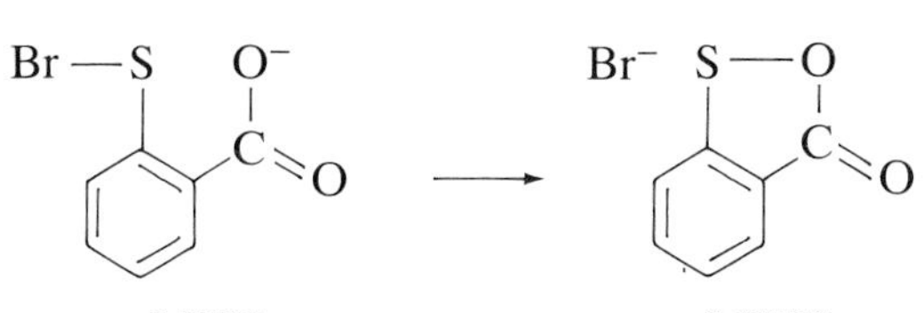

LXXIa **LXXIb**

Also the temperature-variable NMR spectral behavior of the conjugated acid **LXXII** of the sulfuranide anion **LXVII** indicates a rapid ($\Delta G_{25}^{\neq} = 13$ kcal/mol) interconversion (4.32) catalyzed by bases [118].

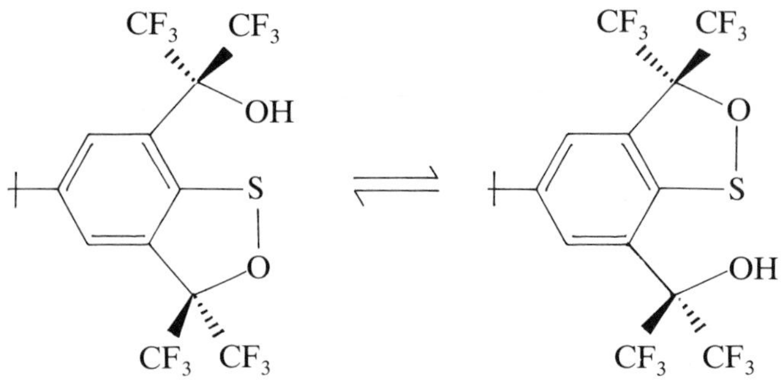

$$\textbf{LXXIIa} \rightleftharpoons \textbf{LXXIIb} \tag{4.32}$$

Since the T-shaped structures may in the course of the nucleophilic substitution reaction at the dicoordinated sulfur atom behave as intermediates with a certain finite lifetime, it will be necessary to examine the possibility and character of their polytopal rearrangements. The topomerization mechanism first devised for the simplest T-shaped structure of the sulfuranide anion $^-SH_3$ [112] and, as has been demonstrated [121, 122], retaining all its principal features for the most diverse T-shaped forms, including tricoordinated complexes of the transition metals, is described by the following scheme (4.33).

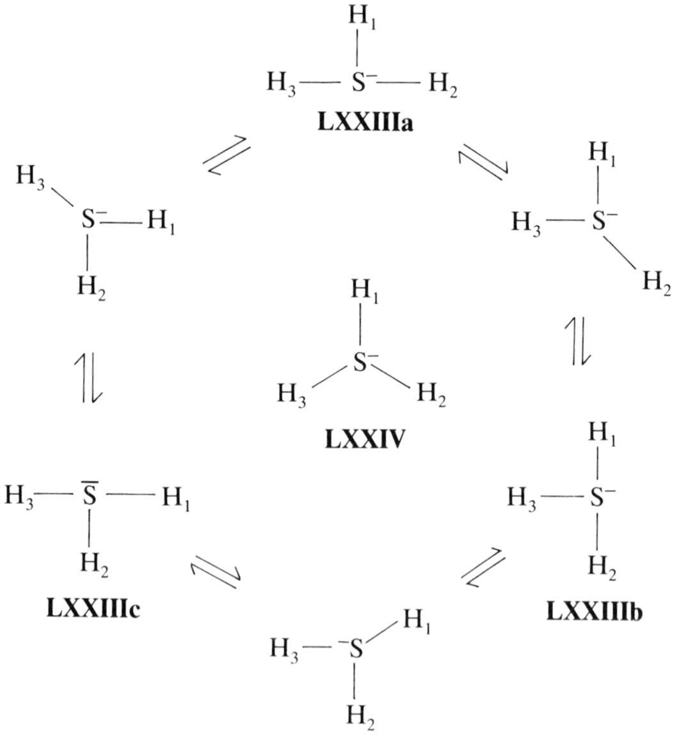

$$\tag{4.33}$$

Interconversions of the three topomeric forms **LXXIIIa—c** occur via the Y-shaped transition state structures formed by moving all the atoms in the common plane. The planar D_3h-form **LXXIV** does not appear in the reaction pathways, since it represents the top of a high hill on the PES.

The presented scheme depicting the exchange of equatorial and axial positions of ligands at the tricoordinated sulfur atom in intermediate sulfuranide structures was employed to describe the *AdRE* mechanism of reactions of tautomeric 1,3-migrations of the sulfenyl groups in amidine derivatives, Scheme (3.17). The structure of amidines **LXXV** does not permit the formation of a four-membered cycle with two axial bonds N—S—N needed for the concerted intramolecular migration of a sulfenyl group. Following the Scheme (4.33) and allowing for the possibility of branching of trajectories of the nucleophilic attack on the sulfenyl sulfur (4.31), a two-step mechanism (4.34) has been suggested [123].

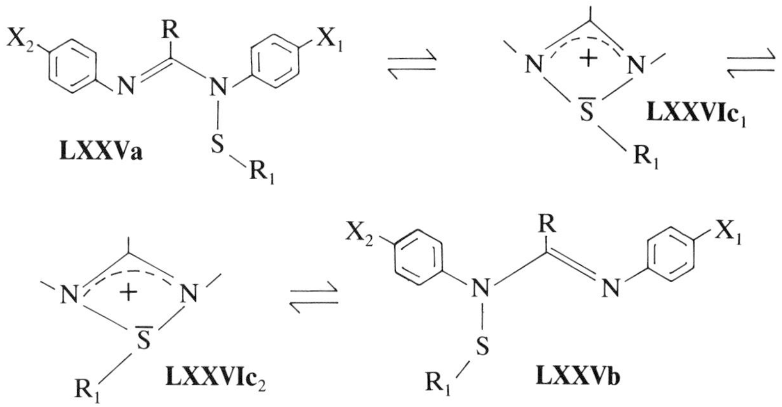

(4.34)

Table 4.IX contains the rate constants and activation parameters of tautomeric rearrangements (4.34) calculated on the basis of the line shape analysis of the dynamic ^{1}H-NMR spectra of compounds **LXXV**.

Similarly to the other reactions considered above of the N,N' transfers of various migrants in amidine systems by the mechanism of associative nucleophilic substitution, the aryl substituents R sharply increase the migration rate due to delocalization of the positive charge in the amidinium triad. High negative values of activation entropy (from -14 to -24 e.u.) are in good agreement with the two-step mechanism of 1,3-migrations of the sulfenyl group in sulfenylamidines which requires substantial reorganization and ordering of the molecular structure in comparison with the initial form **LXXV**.

Stereodynamic processes associated with the hindered rotation about the N—S bond are observed in the NMR spectra of amidines **LXXV** at lower temperatures. Energy barriers against rotation have been estimated from the spectra of compounds **LXXVII** containing prochiral benzyl substituents. They amount to 16 kcal/

Table 4.IX. Kinetic Parameters of Sulfenyl Group Migration in Compounds **LXXV** According to Dynamic NMR ^{1}H Spectral Investigation [123]. (In chlorobenzene solution.)

No.	R	R_1	X_1	X_2	$k_{25} \times 10^2$, s^{-1}	$\Delta H^{\neq}$, kcal/mol	$\Delta S^{\neq}$, e.u.	$\Delta G_{25}^{\neq}$, kcal/mol
1	p-C$_6$H$_4$OMe	2,4-C$_6$H$_3$(NO$_2$)$_2$	p-Me	p-Me	1.7	15.4 ± 0.5	−14.8 ± 1.4	19.8
2	Ph	—"—	—"—	—"—	6.3	15.0 ± 0.6	−13.6 ± 1.2	19.0
3	p-C$_6$H$_4$Br	—"—	—"—	—"—	12.2	13.3 ± 0.5	−18.5 ± 1.4	18.8
4	p-C$_6$H$_4$Cl	—"—	—"—	—"—	15.5	11.9 ± 0.3	−22.2 ± 0.8	18.5
5	p-C$_6$H$_4$NO$_2$	—"—	—"—	—"—	309	9.8 ± 0.3	−23.2 ± 0.9	16.7
6	Ph	2-C$_6$H$_4$NO$_2$	p-OMe	p-OMe	12.2	13.4 ± 0.7	−18.1 ± 1.1	18.8
7	Ph	2,4-C$_6$H$_3$(NO$_2$)$_2$	—"—	—"—	10	13.4 ± 0.4	−17.7 ± 1.3	18.7
8[a]	H	—"—	p-Me	p-Me	~0	—	—	>27
9[b]	Ph	—"—	p-Me	H	22	11.9 ± 0.4	−21.6 ± 1.0	18.3
10[c]	Ph	—"—	p-Me	p-OMe	62	10.7 ± 0.2	−23.6 ± 0.9	17.7
11[d]	Ph	—"—	o-OMe	o-OMe	~0	—	—	>27
12[d]	Ph	2-C$_6$H$_4$NO$_2$	o-Me	o-OMe	~0	—	—	>27
13	Ph	CCl$_3$	p-Me	p-Me	2.2	13.6 ± 0.6	−20.1 ± 1.1	19.6
14[d]	H	2-C$_6$H$_4$NO$_2$	o-OMe	o-OMe	~0	—	—	>27

[a] In hexachlorobutadiene solution;
[b] At 25° the equilibrium content of tautomer **LXXVa** is equal to 45%;
[c] At 25° the equilibrium content of tautomer **LXXVa** is equal to 52%;
[d] In o-dichlorobenzene solution.

mol lying thus within the 11—21 kcal/mol range characteristic of hindered rotations about the N—S bonds in arensulfenylaziridines and arensulfenylamides [124].

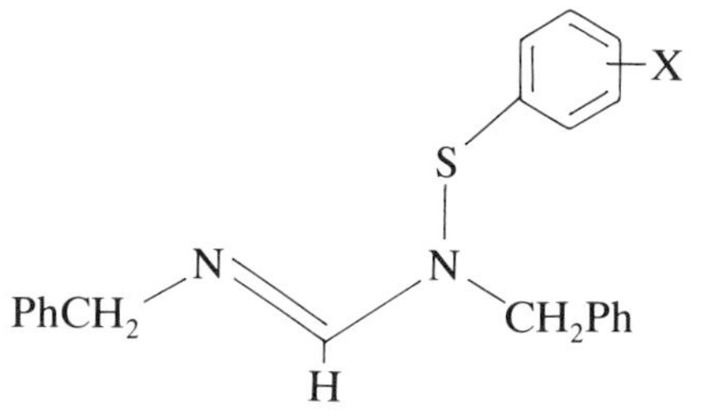

X = 2-NO$_2$, 4-NO$_2$, 2,4-(NO$_2$)$_2$

LXXVII

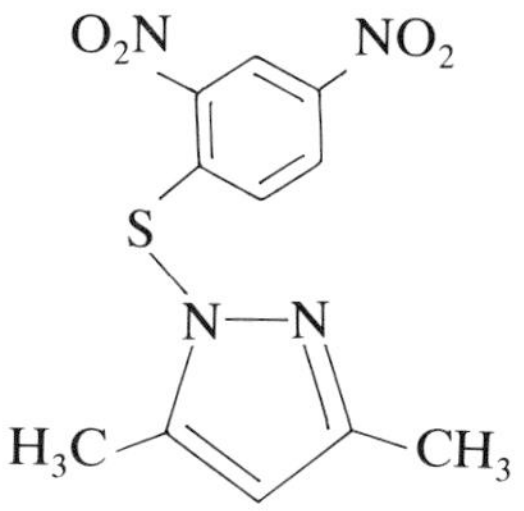

LXXVIII

The formation of the T-shaped intermediates in the *AdRE* reactions of intramolecular transfer of the sulfenyl groups is sterically inhibited in the three-membered cycle. For this reason, the pyrazole derivatives of **LXXVIII** are not susceptible to tautomeric migrations of the arenesulfenyl group even when heated in solution to 200°. On the other hand, it may be assumed that migrations faster than those in the amidines **LXXV** should be observed in the case of 1,4- and 1,5-sigmatropic rearrangements where the formation of intermediates leads to closure of the less strained 5- and 6-membered rings. But no examples of such rearrangements have been found to date.

5.2. S^{III}-migrants

Transition states (intermediates) of the type of sulfuranes **LXIII** in reactions of nucleophilic substitution at the tricoordinated sulfur atom predicted by the VSERP theory as well as by quantum mechanical calculations of the reaction pathways [16, 92] predicate an inversion of configuration at tricoordinated sulfur atom as the stereochemical outcome of the reaction. The vast majority of nucleophilic displacement reactions at the chiral sulfonium or sulfinyl sulfur atoms do occur with inversion of configuration [125, 126].

If **LXIII** is an intermediate, i.e. a compound that can, in principle, be isolated, as for instance the sulfuranide oxide anion **LXXIX**, whose salts have recently been obtained [118, 127], then the substitution reaction is a two-step process (*AdE-*mechanism) leading to inversion of configuration.

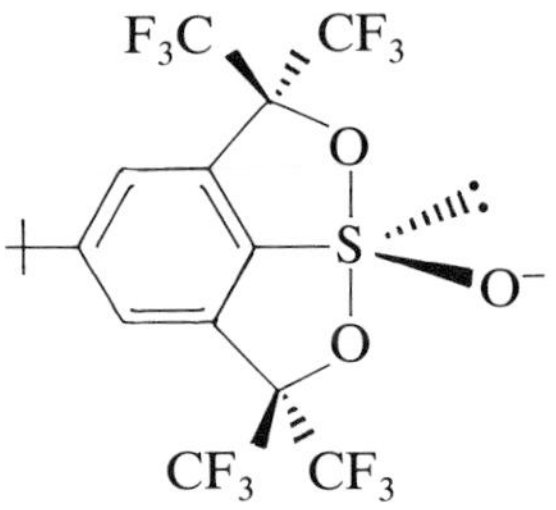

LXXIX

The anion **LXXIX** or its conjugated acid are, apparently, included in fast $(\Delta G_{25}^{\neq} = 13 \text{ kcal/mol})$ tautomeric transformations (4.35) [118].

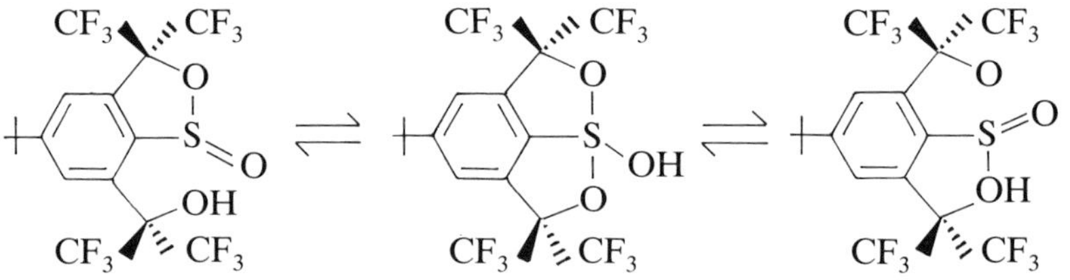

$$\tag{4.35}$$

LXXXa   **LXXXc**   **LXXXb**

Less frequent, but not very rare, are nucleophilic substitution reactions at the tricoordinated sulfur that occur with full or partial retention of configuration. From an analysis of the examples reported [125, 126, 128], it may be concluded that most such cases are due to the formation of a four-membered ring sulfurane intermediate whose ring size excludes any possibility of a diaxial position of entering and departing groups, as in **LXIII**, **LXXIX** and **LXXX**. To explain retention of configuration at sulfur in these reactions, the *AdRE* mechanism including one or several Berry pseudorotations should be invoked. It is completely analogous to the one considered above for the P^{IV} and P^{III} compounds and may be described by the following scheme applied to the sulfinyl group migration:

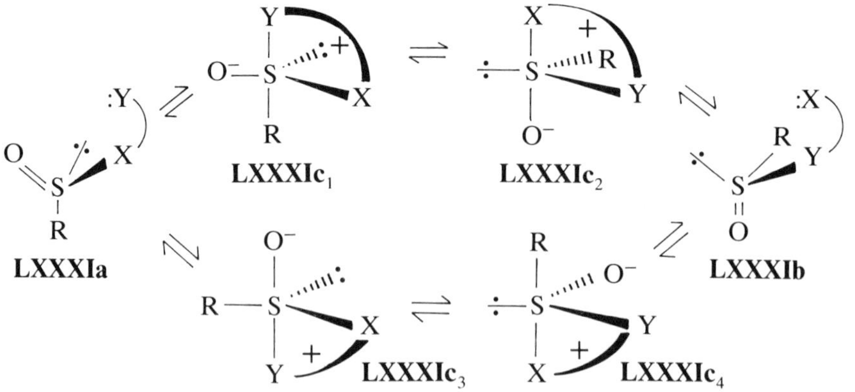

$$\tag{4.36}$$

LXXXIa   **LXXXIc$_1$**   **LXXXIc$_2$**   **LXXXIb**

LXXXIc$_3$ **LXXXIc$_4$**

The distinctive feature of the scheme (4.36) as contrasted to (4.19) is the necessary inclusion in the overall reaction scheme of sulfuranes **LXXXIc₂, c₃** containing in the apical position such an apicophobic ligand as O^- because the lone pair is an even more apicophobic group and ought to be retained in the equatorial position during pseudorotation steps. In spite of that, owing to a considerably larger electrophilicity of sulfur in the sulfinyl and sulfonio groups in comparison with the sulfenyl group, the tautomeric rearrangements of the arenesulfinyl migrants in the amidine system proceed at quite high rates [129].

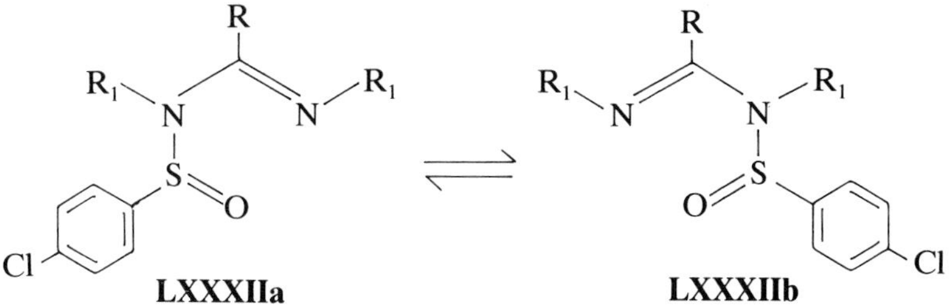

$$\text{(4.37)}$$

LXXXIIa   **LXXXIIb**

$R = H$ $R_1 = C_6H_4CH_3\text{-}p$ $k_{25} = 10^3 \text{ s}^{-1}, \Delta H^{\neq} = 21.5 \text{ kcal/mol}, \Delta S^{\neq} = 0, \Delta G^{\neq}_{25} = 21.5 \text{ kcal/mol}.$
$R = Ph$ $R_1 = C_6H_4Cl\text{-}p$ $k_{25} = 3 \times 10^3 \text{ S}^{-1}, \Delta H^{\neq} = 16.0 \text{ kcal/mol}, \Delta S^{\neq} = -14.3 \text{ e.u.}$ $\Delta G^{\neq}_{25} = 20.9 \text{ kcal/mol}.$

The above kinetic data refer to solutions in chlorobenzene. Similar values for rate constants and activation parameters were also obtained with other solvents including benzonitrile and nitrobenzene. In addition to the absence of concentration dependence, these results indicate intramolecular character of the (4.37) tautomeric rearrangements. In the light of relatively fast migrations (4.37) and also (4.14) (Table 4.VI), somewhat surprising are the rather slow 1,3-migrations of the diphenylsulfonio group in the amidine derivatives **LXXXIII** [91].

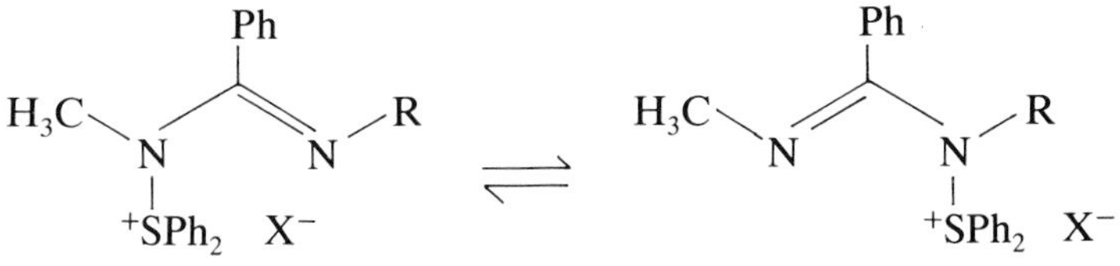

$$\text{(4.38)}$$

LXXXIIIa $(R = CH_3, X = Cl)$ **LXXXIIIb**
              $(R = C_2H_5, X = SbF_6)$
LXXXIVa   **LXXXIVb**

In the *o*-dichlorobenzene or 1,1,2,2-tetrachloroethane solutions of compound **LXXXIII**, the ¹H-NMR spectra exhibit signals of two nonisochronic methyl groups even at 120°C, which points to the freezing of the diphenylsulfonio group 1,3-migration. Nevertheless, such a migration, albeit slow on the NMR time-scale, was observed for compound **LXXXIVa**. When prepared from the corresponding

N-chloroamidine derivative, it exists in the pure **LXXXIV** isomeric form. However, upon standing in the nitromethane solution at 40°C the equilibrium mixture contains 80% of **LXXXIVa** and 20% of **LXXXIVb**.

5.3. *S^{IV}-migrants*

Unlike the (4.37) and (4.38) rearrangements, the 1,3-migrations of sulfuryl (ArSO$_2$) groups in amidine systems do not occur [129]. As a matter of fact, there are to date no reliable data as to intramolecular 1,*j*-migrations of the S^{IV}-centered moieties with the possible exception of the rearrangement (4.39) in specifically constrained structures **LXXXV** designed by Martin *et al.* [118].

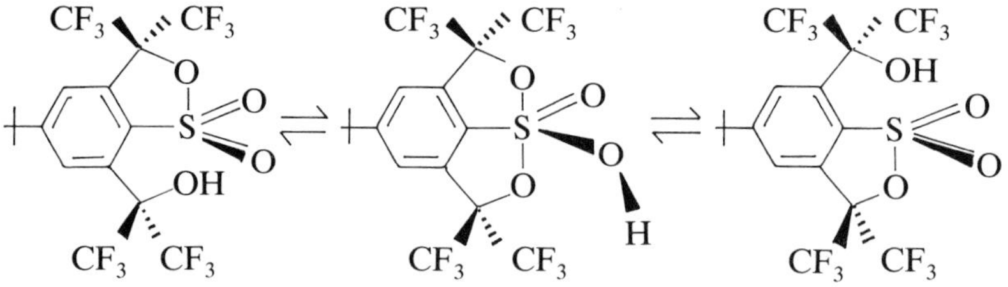

| **LXXXVa** | **LXXXVc** | **LXXXVb** |

$$(4.39)$$

In compounds **LXXXV** the central sulfur atom chemically bound to the rest of the molecular framework is rigidly fixed in a position favorable for the formation of the transition state structure of type **LXIV**. This results in a fairly low energy barrier ($\Delta G_{25}^{\neq} = 14$ kcal/mol) against the degenerate *O,O'*-displacement of the S^{IV}-center in an intermediate sulfuranide dioxide anion or in its conjugated acid **LXXXVc**.

A strong demand for the linear X—S—Y alignment in a single transition state of the reaction without any indication of possibilities involving the *AdRE* mechanism has been evidenced by a thorough study of the sulfonato group (SO$_3$) transfer between pyridine nucleophiles [130] modelling the sulfatase enzyme mechanisms. A possible reason for the hindered intramolecular rearrangements of sulfuryl migrants may lie in the strong polarization of sulfurane dioxide **LXXXVI**, an intermediate in the *AdRE* mechanism. In order to offset unfavorable energy consequences of charge separation in **LXXXVI**, a very careful selection of the molecular system incorporating nucleophilic centers is required.

| **LXXXVI** | **LXXXVII** |

For intramolecular rearrangements of the S^{IV} containing migrants to occur, it appears to be expedient to have recourse to sulfuranes leading to more realistic intermediate forms of the **LXXXVII** type. The general approach to the synthesis of stable sulfuranes and sulfuranide structures **LXXXVII** are currently well developed [118, 131].

Since many of the rearrangement reactions of the main group element migrants considered in this Chapter have been studied in closely related amidine systems, it will be possible to arrange the migrants in a sequence describing their relative migratory aptitudes. Only the principal migrants have been chosen whose kinetical parameters for 1,3-displacements in the benzamidine framework are compared. Though variation of substituents at the migrating center may lead to substantial shifts of the migrant position in the sequence below, it still gives a clear impression of how the origin and coordination number of the central atom influence the rates of migration. These rates vary from more than 10^6 s^{-1} for the triphenylarsonio group on the left side of the sequence to less than 10^{-8} s^{-1} for arenesulfonyl and alkyl groups on the right:

$$^+AsPh_3 > GeMe_3 > SiMe_3 > {}^+PPh_3 > C_6H_2(NO_2)_3 > SC_6H_3(NO_2)_2 >$$
$$S(O)Aryl > {}^+SPh_2 > PF_3Ph > P(NAr)(OAlk)_2 > P(O)(OAlk)_2 > \quad (4.40)$$
$$PPh_2 > COCH_3 > SO_2Aryl \gg Alkyl.$$

Most rearrangements examined in the present Chapter in which the central atom of the migrant is an element located to the right of carbon in the Periodic Table have become known only in the last 5—6 years. Based on reactions of associative nucleophilic substitution, these fast intramolecular rearrangements whose design generally rests on the investigation of the intrinsic reaction mechanism and elucidation of detailed transition state structures not only help uncover the factors responsible for the rate accelerations of the intramolecular reactions over their intermolecular analogs, but also bring us nearer to the understanding of the nature of enzymatic catalysis.

The full potential of rapid and reversible rearrangements resulting in the displacements of C, Si, P, As, S and other main group element migrants between nucleophilic centers as models for a perfect structural organization of chemical processes is only now being realized.

REFERENCES

1. Beak P., *Tetrahedron,* **20**, 831 (1964); *Acc Chem. Res.,* **10**, 186 (1977).
2. Beak P., Bonham J., and Lee J. T., *J. Amer. Chem. Soc.,* **90**, 1569 (1968); *J. Org. Chem.,* **34**, 2125 (1969).
3. Beak P., Woods T. S., and Mueller D. S., *Tetrahedron,* **28**, 5507 (1972).
4. Isaachs N. S., *Reactive Intermediates in Organic Chemistry.* Academic Press. New York. 1972, p. 280.

5. McKillop W. J., Sedor E. A., Calbertson B. M., and Wawzonek S., *Chem. Rev.*, **73**, 255 (1973).
6. Hunter D. H., Stothers J. K., and Warnhoff E. W., in: *Rearrangements in Ground and Excited States* (Ed. P.de Mayo). vol. 1, p. 391. Academic Press. New York. 1980.
7. Schöllkopff und Driessler F., *Liebigs Ann.*, 1521 (1975).
8. Langhals H., Range G., Wistuba E., und Rüchardt C., *Chem. Ber.*, **114**, 3843 (1981).
9. Harrison R. M., Hobson J. D., and Midgley A. W., *J. C. S. Perkin Trans. I*, 2403 (1976).
10. Spangler C. W., *Chem. Rev.*, **76**, 187 (1976).
11. Gajewski J. J., *Hydrocarbon Thermal Isomerizations*. Academic Press. New York. 1981.
12. Ingold C. K., *Structure and Mechanism in Organic Chemistry*. Ch. 7. Cornell University Press. Ithaca and London. 1969.
13. Luckenbach R., *Dynamic Stereochemistry of Pentacoordinated Phosphorus and Related Elements*. G. Thieme Publ. GmbH. Stuttgart. 1973.
14. Cram D. J. and Cram L. M., *Fortschr. Chem. Forsch.*, **31**, 1 (1972).
15. Haake P. C. and Westheimer F. H., *J. Amer. Chem. Soc.*, **83**, 1102 (1961); Dennis E. A. and Westheimer F. H., *J. Amer. Chem. Soc.*, **88**, 3431, 3432 (1966).
16. Minkin V. I., Simkin B. Ya., and Minyaev R. M., *Quantum Chemistry of Organic Compounds. Mechanisms of Reactions*. Khimiya. Moscow. 1986.
17. Alagona G., Chio C., and Tomasi J., *Theor. Chim. Acta*, **60**, 79 (1981).
18. Serre J., *Int. J. Quantum Chem.*, **26**, 593 (1984).
19. Ohta K. and Morokuma K., *J. Phys. Chem.*, **89**, 5845 (1985).
20. Viers J. W., Schug J. C., Stovall M. D., and Seeman J. I., *J. Comput. Chem.*, **5**, 598 (1984).
21. Schlegel H. B., Mislow K., Bernardi F., and Bottoni A., *Theor. Chim. Acta*, **44**, 245 (1977).
22. Dedieu A. and Veillard A., *J. Amer. Chem. Soc.*, **94**, 6730 (1972).
23. Mitchell D. J., *Theoretical Aspects of S_N2 Reactions*. Thesis. Queen's University Kingston. 1981.
24. Wolfe S., Mitchell D. J., and Schlegel H. B., *J. Amer. Chem. Soc.*, **103**, 7692, 7694 (1981).
25. Keil F. and Ahlrichs K., *J. Amer. Chem. Soc.*, **98**, 4787 (1976).
26. Nguyen Trong Anh and Minot C., *J. Amer. Chem. Soc.*, **102**, 103 (1980).
27. Minot C., *Nouv. J. Chim.*, **5**, 319 (1981).
28. Stohrer W.-D., *Chem. Ber.*, **107**, 1795 (1974); **109**, 285 (1976).
29. Maryanoff C. A., Ogura F., and Mislow K., *Tetrahedron Lett.*, 4095 (1975).
30. ElGomati T., Gasteiger I., Lenoir D., and Ugi I., *Chem. Ber.*, **109**, 826 (1976).
31. Vergnani T., Karpf M., Hoesch L., and Dreiding A. S., *Helv. Chim. Acta*, **58**, 2524 (1975).
32. Minkin V. I., Kletskii M. E., Minyaev R. M., Simkin B. Ya., and Pichko V. A., *Zh. Org. Khim.*, **19**, 9 (1983).
33. Paddon-Row M. N., Rondan N. G., and Houk K. N., *J. Amer. Chem. Soc.*, **104**, 7162 (1982).
34. Bach R. D., Badger R. C., and Lang T. J., *J. Amer. Chem. Soc.*, **101**, 2845 (1979).
35. Bach R. D. and Wolber G. H., *J. Amer. Chem. Soc.*, **106**, 1401 (1984).
36. Modena G., *Acc. Chem. Res.*, **4**, 73 (1971).
37. Texier F., Henri-Rousseau O., and Bourgois J., *Bull. Soc. Chim. France*, ser. 2. 86 (1979).
38. Stohrer W.-D., *Tetrahedron Lett.*, 207 (1975).
39. Rappoport Z., *Acc. Chem. Res.*, **14**, 7 (1981).
40. Korobov M. S., Nivorozhkin L. E., and Minkin V. I., *Zh. Org. Khim.*, **9**, 1717 (1973); **14**, 347 (1978).
41. West R., *Adv. Organometal. Chem.*, **16**, 1 (1977).
42. Brook A. G. and Bassindale A. R., in: *Rearrangements in Ground and Excited States*. (Ed. P.de Mayo) Vol. 2, p. 149. Academic Press. New York. 1980.
43. Wilhite D. L. and Spialter J., *J. Amer. Chem. Soc.*, **95**, 2100 (1973).
44. Baybutt P., *Molecular Physics*, **2**, 389 (1975).
45. Brandemark und Siegbahn P. E. M., *Theor. Chim. Acta*, **66**, 233 (1984).
46. Dewar M. J. S. and Healy E., *Organometallics*, **1**, 1705 (1982).
47. Pestunovich V. A., Sidorkin V. F., Dogaev O. B., and Voronkov M. G., *Dokl. Akad Nauk USSR*, **251**, 1440 (1980).

48. Klebe G., *J. Organometal. Chem.*, **293**, 147 (1985).
49. Kanberg F. and Muetterties E. L., *Inorg. Chem.*, **7**, 155 (1968).
50. Stevenson W. H., Wilson S., Martin J. C., and Farnham W. B., *J. Amer. Chem. Soc.*, **107**, 6340 (1985).
51. Hartke K., Kohl A., und Kämpchen T., *Chem. Ber.*, **116**, 2653 (1983).
52. Corriu R. J. P. and Guerin C., *J. Organometal. Chem.*, **198**, 231 (1980).
53. Stevenson W. H. and Martin J. C., *J. Amer. Chem. Soc.*, **107**, 6352 (1985).
54. Wiberg N. und Pracht H. J., *Chem. Ber.*, **105**, 1388 (1972).
55. Scherer O. J. und Hernig P., *Chem. Ber.*, **101**, 2533 (1968).
56. Torocheshnikov V. N., Sergeyev N. M., Viktorov N. A., Goldin G. S., Poddubny V. G., and Koltsova A. N., *J. Organometal. Chem.*, **70**, 347 (1974).
57. Negrebeckii V. V., Kalchenko V. I., Balickaya O. V., Markovskii L. N., and Kornilov M. Yu., *Zh. Obschz. Khim.*, **54**, 2217 (1984); **56**, 114 (1986).
58. Pinnavaia T. J., Collins W. T., and Howe J. J., *J. Amer. Chem. Soc.*, **92**, 4544 (1970).
59. McClarin J. A., Schwartz A., and Pinnavaia T. J., *J. Organometal. Chem.*, **188**, 129 (1980).
60. Reich H. J., Murcia D. A., and Daniel A., *J. Amer. Chem. Soc.*, **97**, 1619 (1975).
61. Kalichman I. D., Vyazankin N. S., Pestunovich V. A., Belousova L. N., and Makarov I. V., *Izv. Akad Nauk USSR. Ser. Khim.* 962 (1977).
62. Winkler T., Philipsborn W., Stroh J., Silhan W., and Zbiral E., *J. C. S. Chem. Commun.*, 1645 (1970).
63. Minkin V. I., Olekhnovich L. P., and Zhdanov Yu. A., *Molecular Design of Tautomeric Systems*. Rostov-on-Don University Publ. House. Rostov-on-Don. 1977.
64. Minyaev R. M., Minkin V. I., and Kletskii M. E., *Zh. Org. Khim.*, **14**, 449 (1978).
65. Minyaev R. M. and Minkin V. I., *Zh. Struct. Khim.*, **20**, 842 (1979).
66. Jencks W. P., *Catalysis in Chemistry and Enzymology*. McGraw-Hill. New York. 1969.
67. Westheimer F. H., in: *Rearrangements in Ground and Excited States*. (Ed. P.de Mayo). Vol. 2, p. 229. Academic Press. 1980.
68. Frey P. A., *Tetrahedron* **38**, 1541 (1982).
69. Kolind — Andersen H. and Lawesson S. O., *Acta Chem. Scand.*, **B29**, 430 (1975).
70. Westheimer F. H., *Acc. Chem. Res.*, **1**, 70 (1968).
71. Gorenstein D. G., Luxon B. A., Findlay J. K., and Momil R., *J. Amer. Chem. Soc.*, **99**, 4170, 8048 (1977).
72. Gorenstein D. G., Luxon B. A., and Findlay J. K., *J. Am. Chem. Soc.*, **101**, 5869 (1979).
73. Holmes R. R., *Pentacoordinated Phosphorus*. Vols 1,2. American Chemical Society: Washington, DC. 1980. ACS Monogr. 175, 176.
74. Granoth J. and Martin J. C., *J. Amer. Chem. Soc.*, **100**, 5229 (1978).
75. Granoth J., Segal Y., Waysbort D., Shirin E., and Leader H., *J. Amer. Chem. Soc.*, **102**, 4253 (1980).
76. Holmes R. R., *Acc. Chem. Res.*, **12**, 257 (1979).
77. Furmanova N. G. and Kompan O. E., *Problems of Crystallochemistry*. Nauka. Moscow. p. 122. 1985.
78. Berry R. S., *J. Chem. Phys.*, **32**, 933 (1960).
79. Ugi I., Marguarding D., Klusacek H., Gillespie P., and Ramirez F., *Acc. Chem. Res.*, **4**, 288 (1971).
80. Strich A. and Veillard A., *J. Am. Chem. Soc.*, **95**, 5574 (1973).
81. Altman J. A., Yates K., and Csizmadia I. G., *J. Amer. Chem. Soc.*, **98**, 1450 (1976).
82. Kutzelnigg W. and Wasilewski J., *J. Amer. Chem. Soc.*, **104**, 953 (1982).
83. Muetterties E. L., Mahler W., and Schmutzler R., *Inorg. Chem.*, **2**, 613 (1963).
84. Hoffmann R., Howell J. M., and Muetterties E. L., *J. Amer. Chem. Soc.*, **94**; 3047 (1972).
85. Trippett S., *Pure Appl. Chem.*, **40**, 595 (1974); *Phosphorus and Sulfur*, **1**, 89 (1976).
86. McDowell R. S. and Streitwieser A., *J. Amer. Chem. Soc.*, **107**, 5849 (1985).
87. Mislow K., *Acc. Chem. Res.*, **3**, 321 (1970).
88. Zimin M. G., Afanasyev V. M., and Pudovik A. N., *Zh. Obschz. Khim.*, **49**, 2621(1979).
89. Pudovik A. N. and Zimin M. G., *Usp. Khim.*, **52**, 1803 (1983).

90. Negrebeckii V. V., Bogelfer L. Ya., Sinitza A. D., Kryshtal V. S., Kalchenko V. I., and Markovskii L. N., *Zh. Obschz. Khim.*, **50**; 2133, 2420, 2806 (1980). **52**, 40, 1496 (1982).
91. Hartke K. and Wolff H.-M., *Chem. Ber.*, **113**, 1394 (1980).
92. Minyaev R. M., Minkin V. I., and Kletskii M. E., *Zh. Org. Khim.*, **14**, 449 (1978).
93. Trinquier G., Daudey J.-P., Caruana G., and Madaule Y., *J. Amer. Chem. Soc.*, **106**, 4794 (1984).
94. Kyba E. O., *J. Amer. Chem. Soc.*, **98**, 4805 (1976).
95. Mikolajczyk M., *Pure Appl. Chem.*, **52**, 959 (1980).
96. Nielsen J. and Dahl O., *J. C. S. Perkin Trans. II*, 553 (1984).
97. Lutsenko I. F. and Foss V. L., *Pure Appl. Chem.*, **52**, 917 (1980).
98. Lerman C. L. and Westheimer F. H., *J. Amer. Chem. Soc.*, **98**, 179 (1976).
99. Wolf R., *Pure Appl. Chem.*, **52**, 1141 (1980).
100. Minyaev R. M. and Minkin V. I., *Zh. Struct. Khim.*, **20**, 842 (1979).
101. Trippett S. and Wadding R. E., *Tetrahedron Lett.*, 193 (1979).
102. Pickering M., Jurado B., and Springer C. S., *J. Amer. Chem. Soc.*, **98**, 4503 (1976).
103. Negrebeckii V. V., Bogelfer L. Ya., Sinitza A. D., Kryshtal V. S., and Markovskii L. N., *Zh. Obschz. Khim.*, **51**, 956 (1981).
104. Kalchenko V. I., Negrebeckii V. V., Rudyi R. B., Atamas L. I., Povolockii M. I., and Markovskii L. N., *Zh. Obschz. Khim.*, **53**, 932 (1983).
105. Minkin V. I., Olekhnovich L. P., Metlushenko V. P., Furmanova N. G., Bally I., and Balaban A. T., *Tetrahedron* **37** (Supplement Number 1), 421 (1981).
106. Prokof'ev A. I. Malysheva N. A., Bubnov N. N., Solodovnikov S. P., and Kabachnik M. I., *Izv. Akad. Nauk USSR. Ser. Khim.*, 1969 (1978).
107. Minkin V. I. and Minyaev R. M., *Zh. Org. Khim.*, **11**, 1993 (1975); *Zh. Struct. Khim.*, **18**, 274 (1977).
108. Chen M. M. L. and Hoffmann R., *J. Amer. Chem. Soc.*, **98**, 1647 (1976).
109. Allison W. S., *Acc. Chem. Res.*, **9**, 293 (1976).
110. Zefirov N. S. and Makhonkov D. I., *Chem. Rev.*, **82**, 615 (1982).
111. Kice J. L., *Progr. Inorg. Chem.*, **17**; 147 (1972).
112. Minkin V. I. and Minyaev R. M., *Zh. Org Khim.*, **13**, 1129 (1977).
113. Rosenfield R. E., Pathasarathy R., and Dunitz J., *J. Amer. Chem. Soc.*, **99**, 4860 (1977).
114. Pappas J. A., *J. Amer. Chem. Soc.*, **99**, 2926 (1977).
115. Aida M. and Nagata C., *Chem. Phys. Lett.*, **112**, 129 (1984).
116. Lau P. H. W. and Martin J. C., *J. Amer. Chem. Soc.*, **100**, 7077 (1978).
117. Arduengo A. J. and Burgess E. M., **99**, 2376 (1977).
118. Perkins C. W., Wilson S. R., and Martin J. C., *J. Amer. Chem. Soc.*, **107**, 3209 (1985).
119. Burawoy A., in: *Organic Sulfur Compounds*, vol. 1. Ch. 25. Ed. N. Kharash. Pergamon Press. Oxford. 1961.
120. Block E., *Reactions of Organosulfur Compounds.* Academic Press. New York. 1978.
121. Hoffmann R., in: *IUPAC Frontiers of Chemistry*, (Ed. K. J. Laidler) p. 247. Pergamon Press. Oxford. 1982.
122. Albright T. A., Burdett J. K., and Whangbo M. H., *Orbital Interactions in Chemistry.* Chs. 14, 18. J. Wiley. New York. 1985.
123. Olekhnovich L. P., Minkin V. I., Mikhailov I. E., Ivanchenko N. M., and Zhdanov Yu. A., *Dokl. Akad. Nauk USSR*, **233**, 874 (1977); *Zh. Org. Khim.*, **15**, 1355 (1979).
124. Kost D. and Raban M., *J. Amer. Chem. Soc.*, **98**, 8333 (1976); Raban M. and Yamamoto G., *J. Amer. Chem. Soc.*, **99**, 4160 (1977).
125. Tillett J. G., *Chem. Rev.*, **76**, 747 (1976).
126. Mikolajczyk M. and Drabowicz J., *Topics in Stereochem.*, **13**, 333 (1982).
127. Perkins C. W. and Martin J. C., *J. Amer. Chem. Soc.*, **105**, 1377 (1983).
128. Mikolajczyk M., Drabowicz J., and Bujnicki B., *Tetrahedron Lett.*, **26**, 5699 (1985).
129. Mikhailov I. E., Ivanchenko N. M., Olekhnovich L. P., Minkin V. I., and Zhdanov Yu. A., *Dokl. Akad. Nauk USSR*, **263**, 366 (1982).
130. Bourne N., Hopkins A., and Williams A., *J. Amer. Chem. Soc.*, **107**, 4325 (1985).
131. Martin J. C., *Science* **191**, 154 (1976).

5. Dyotropic and Polytropic Tautomeric Systems

1. Scope and Definition

The tautomeric systems examined in Chapters 2—4 are based on reactions in which one sigma-bond migrates between two reaction centers. Conceivable are also more complex tautomeric systems, in which simultaneous or consecutive intramolecular migrations of two and more σ-bonds take place. They will be referred to as dyotropic and polytropic, respectively.*

The concept of a dyotropic rearrangement was introduced by Reetz [2, 3] who singled out two classes of such reactions.

In the simplest case, the dyotropic rearrangement is, in essence, a mutual exchange of positions between the migrants linked to a common σ-bond.

$$
\underset{\textbf{Ia}}{\overset{R_2}{\underset{R_1}{\diagup}}X\!\!-\!\!Y} \quad \rightleftharpoons \quad \underset{\textbf{Ib}}{\overset{R_2}{\underset{R_1}{\diagdown}}X\!\!-\!\!Y} \tag{5.1}
$$

The rearrangements (5.1) have been known for quite a long time. They include, for example, a rearrangement of nitrile oxides to isocyanates [4] detected 100 years ago.

$$
\underset{\textbf{IIa}}{R\!-\!C\!\equiv\!N\!\longrightarrow\!O} \rightleftharpoons \underset{\textbf{IIc}}{\overset{R}{\underset{O}{C\!=\!\!=\!N}}} \rightleftharpoons \underset{\textbf{IIb}}{O\!=\!C\!=\!N\!-\!R} \tag{5.2}
$$

Among the best-studied reactions of type (5.1) are the thermal rearrangements of silyl silylmethyl ethers **III** [3, 5, 6];

---

* Earlier we proposed an additional requirement for the assignment of molecular systems to the polytropic class, namely the concerted character of the reaction, i.e. transfer of both migrants in the course of one kinetical step [1]. The research over the last few years has, however, shown this condition to be too stringent as it can be met only for a very limited number of rearrangements.

IIIa IIIb

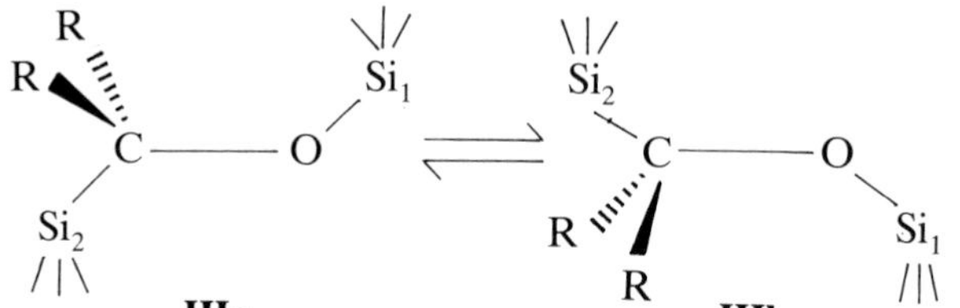

(5.3)

and of bis- and tris(silyl)hydroxylamines **IV** [7].

IVa IVb

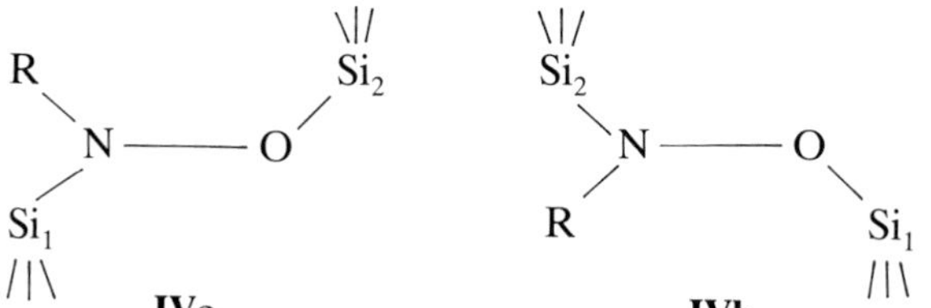

(5.4)

The dyotropic rearrangements of class **II** also involve double migration but do not include positional exchange. The degenerate double 1,4-silyl migrations in the derivatives of squaric acid **V** [8] and bis-(2,2-imidazoline) **VI** [9] studied by the method of dynamic NMR may serve as examples of such dyotropic reactions.

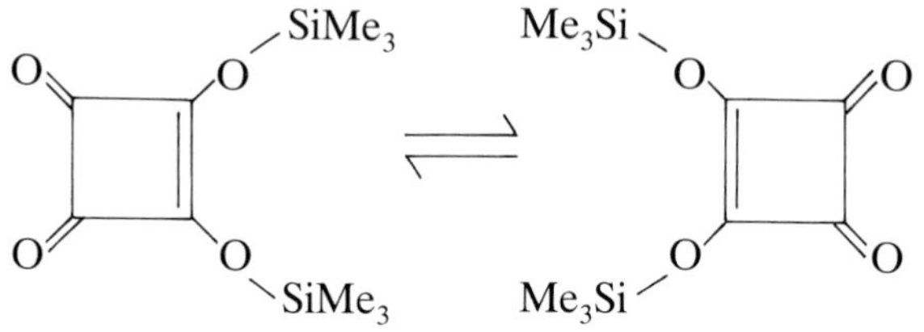

Va Vb

(5.5)

VIa VIb

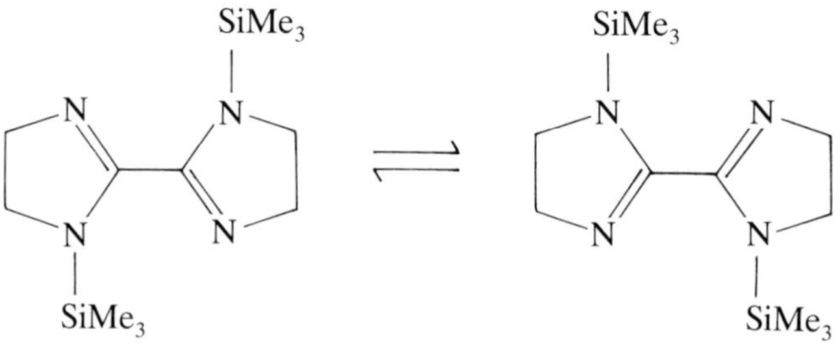

(5.6)

An important case of dyotropic reactions are the double prototropic rearrangements that are intramolecular analogs of a number of significant biochemical reactions. The derivatives of 2,5-dihydroxy-1,4-benzoquinones [10—14] are a well-studied example of tautomeric dyotropic systems where the protons act as migrants.

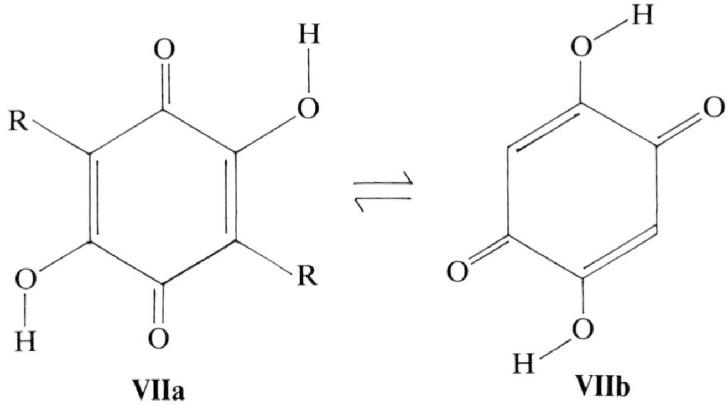

$$(5.7)$$

VIIa   **VIIb**

2. The Mechanism and Energetics of the Dyotropic Reactions

When studying the mechanisms of dyotropic reactions, there arise the following important questions: (1) Are these intramolecular reactions concerted, i.e. proceeding in a single kinetic step without formation of intermediates? (2) If they are, do the processes of the bond-breaking and bond-making progress uniformly in both reactive sites of the molecule? (3) What is the influence exerted by the nature of a migrant and by the shape of the rest of the molecular system on the energetics of the dyotropic reactions? These questions may be answered at different levels. The ascertainment of the concerted character of a reaction or its assignment to a stepwise process are based on the theoretical and experimental studies of the kinetics of the reaction, isotopic effects and, primarily, on the fixation of intermediates. Whether or not the processes of making and breaking of two or more bonds are synchronous is a question of the timing of these events, which may be properly investigated only by calculation of the pertinent reaction paths for dyotropic rearrangements.

2.1. Concerted and Nonconcerted Dyotropic Rearrangements

Prior to solving the question of whether a dyotropic rearrangement is concerted, its intramolecular character has to be demonstrated. The type (5.1) rearrangements belong to thermally forbidden ($\sigma_s^2 + \sigma_s^2$) reactions [2, 15], consequently, they are associated with quite high activation energies and usually require for their realization rather severe conditions (thermolysis in inert solvent at $140-235°C$). In spite of these conditions, the dyotropic rearrangements of silyl silylmethyl ethers (5.3), where the substituents R_1, R_2 are represented by a polycyclic moiety, e.g. **VIII**, **IX**, were shown by crossover experiments to be 100% intramolecular [3].

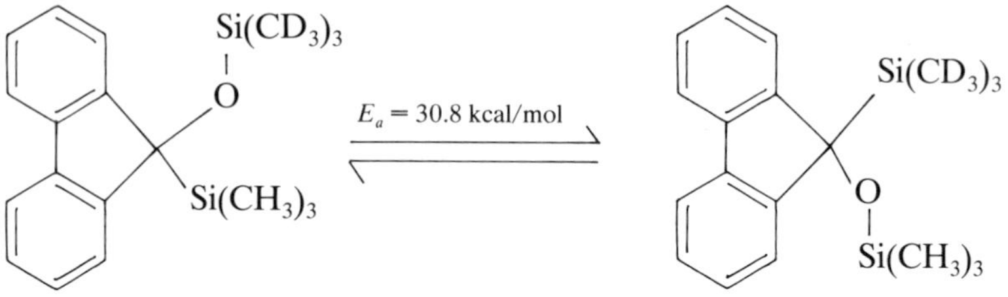

VIIIa   **VIIIb**

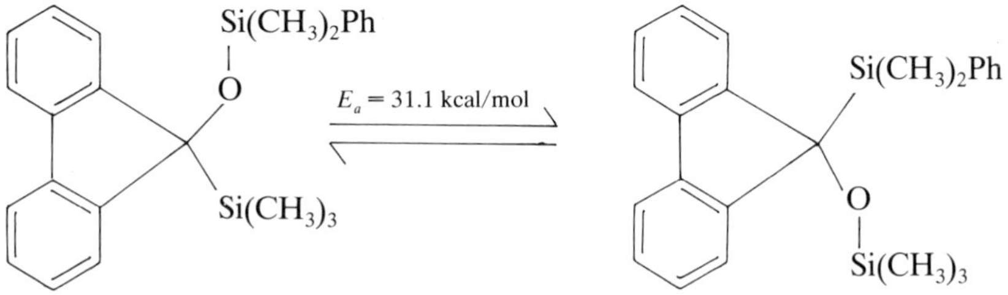

$E_a = 30.8$ kcal/mol

IXa   **IXb**

$E_a = 31.1$ kcal/mol

On the other hand, crossover experiments clearly indicate the intermolecular character of the dyotropic rearrangement (5.5) even though, according to data from dynamic ^{13}C-NMR spectroscopy, this reaction develop fairly rapidly: $\Delta G_{T_c}^{\neq} = 16.9$ kcal/mol ($T_c = 83°C$). Upon standing, the $1:1$ molar mixture of **Va** and its D_{18}-derivative **Vc** at room temperature, a statistical $1:1:2$ distribution of isotopomeric compounds **Va, c, d** was rapidly achieved, as has been demonstrated by mass spectroscopy [8].

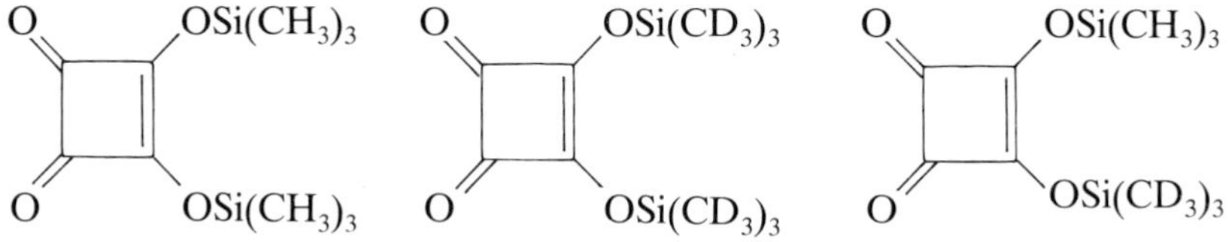

Va   **Vc**   **Vd**

This result does not rule out a slower intramolecular dyotropic rearrangement of type (5.5), but it clearly indicates predominance of the rapid intermolecular reaction. Obviously, for an intermolecular reaction to occur, homolysis or heterolysis of the bonds formed by migrants is required. Reactions of this type are strongly solvent dependent, polar and coordinating solvents facilitating bond scissions. This makes it possible to effectively control, in some cases, the degree of the intramolecular character of a reaction. For instance, the double proton migrations in dihydroxy benzoquinones **VII** (5.7) in triethyl phosphate ester—

diethyl ether $(1:1)$ solution are predominantly intermolecular [12, 14]. In the ^{13}C-NMR spectra of a mixture of compound **VIIa** (R = Cl) with its D_2-derivative (R = Cl), the splitting of the signals of the ^{13}C nuclei due to an isotopic shift, as well as the spin coupling with the constant $^3J_{C(3)H} = 7.6$ Hz are retained only at low temperatures (below −60°C). When the temperature of the solution is raised, the intermolecular exchange becomes predominant. The rate constants and free activation energies of this reaction are given in Table 5.I.

Table 5.I. Rate Constants and Free Energy of Activation for Intermolecular Double Proton Transfers (5.7) According to ^{13}C-NMR($f = 25.18$ MHz). Investigation [12]. (Solvent: $1:1$ mixture of diethyl phosphate and diethyl ether).

Compound **VII**	T_c, °C	$k_{T_c}\, 10^{-3}, \text{s}^{-1}$	$\Delta G^{\neq}_{T_c}$, kcal/mol
R = H	−36°	1.4	10.4
R = Cl	−50°	1.1	9.8
R = Cl(OD)$_2$	−38°	1.1	10.3
R = OH	+158°	2.6	18.8

When one goes to less polar and low-basicity solvents (CDCl$_3$, CCl$_4$, perfluoro-toluene), the reactions of the double proton exchange (5.7) proceed entirely by the intramolecular mechanism. Unambiguous confirmation of this fact in regard to compounds **VII** is furnished by the two-dimensional ^{1}H-NMR spectra of a mixture of quinones **VII** (R = i-C$_3$H$_7$) and **VII** (R = C$_7$H$_{15}$) shown in Fig. 5.1.

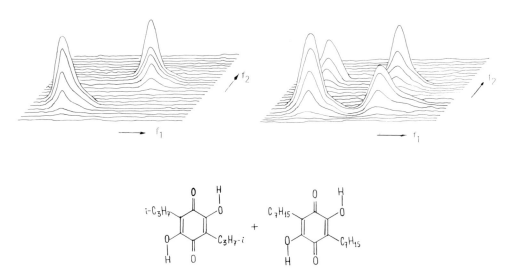

Fig. 5.1. Two-dimensional ^{1}H-NMR spectra of the $1:1$ mixture of 2,5-dihydroxy-3,6-dialkyl-1,4-quinones at 60°C in (a) CDCl$_3$; (b) CDCl$_3$ with additions of (CD$_3$O)$_3$PO [14].

In deuterochloroform solutions, the protons of hydroxyl groups of each compound, for a wide range of concentrations and temperatures, appear as separate auto-peaks (chemical shifts of 7.80 and 7.56 ppm.), which clearly points to the absence of an intermolecular exchange. When basical polar solvents (trimethyl phosphate, dimethyl sulfoxide, tetrahydrofurane) are added to the solution of the above mixture of quinones, there appear in the two-dimensional ^{1}H-NMR spectrum, along with the signals of hydroxyl protons lying on the main diagonal, off-diagonal exchange cross-peaks (Fig. 5.1b). The cross-peaks indicate occurrence of a chemical exchange of the intermolecular proton transfer type. Similar results were obtained for solutions of the above compounds and their mixtures in the presence of proton-donating additives, such as water, methanol, acetic acid, the last substance most effectively catalyzing the intermolecular proton exchange.

Thus, in order to realize the intramolecular dyotropic rearrangement of type (5.7), both base and acid catalysis must be excluded.

Let us first examine the question concerning the concertedness of type (5.1) dyotropic rearrangements. The rearrangement of nitrile oxides to isocyanates (5.2) has been found to be intramolecular and to proceed with complete retention of the stereochemical configuration of the migrating group R [16]. This has been deduced from the premise that the reaction develops concertedly with the doubly bridged transition state structure **IIc** (the migration of the oxygen atom has been assumed to proceed further than that of R), however, a detailed quantum mechanical *ab initio* (4—31 G) search for a reaction path of such a rearrangement (R = H [17], R = CH$_3$ [18]) has led to the conclusion that here a more compicated mechanism appears to be operative including passage through oxazirine **IId** and singlet acylnitrene **IIe**.

$$R-C\equiv N\longrightarrow O \quad\longrightarrow\quad R-C\overset{\displaystyle O}{=\!=}N \quad\longrightarrow\quad R-C\overset{\displaystyle O}{\underset{\displaystyle N:}{\big\langle}} \quad\longrightarrow\quad R-N=C=O \tag{5.8}$$

IIa	**IId**	**IIe**	**IIb**

However, at the level of approximation being employed, **IId** and **IIe** are not true intermediate structures belonging to the local minima of the respective PESs. Therefore, a concerted mechanism cannot be disproved by calculations [17, 18], even though they clearly present evidence in favor of nonsynchronous migration of two atomic groups exchanging their places in the molecule **IIa**.

A nonconcerted, asynchronous mechanism is also assumed for the dyotropic rearrangement of silyl silylmethyl ethers (5.3) [3]. Short-lived intermediates **IIIc, d** are invoked whose formation in the rate-determining step of the reaction is consistent with the large negative value of the activation entropy.

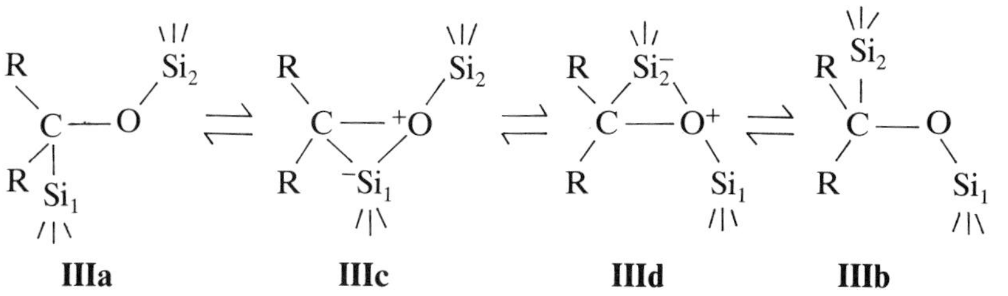

$$
\begin{array}{cccc}
\underset{\textbf{IIIa}}{\overset{\displaystyle R}{\underset{R}{\overset{\displaystyle \diagdown}{\underset{\diagup}{C}}}}\!\!-\!\!O} &
\rightleftharpoons &
\underset{\textbf{IIIc}}{} &
\underset{\textbf{IIId}}{} &
\underset{\textbf{IIIb}}{}
\end{array} \tag{5.9}
$$

The main reasons for the nonconcertedness of these and other dyotropic rearrangements of type (5.1) lie in the unfavorable conditions for an overlap of the electron lone pair orbitals of the heteroatom and the lower valence orbital of the migrant and then in the forbiddenness of these reactions in the ground electronic state. The first of these factors is hard to improve in a molecular system rigidly constrained for the 1,2-shift. By contrast the thermal forbiddenness can be overcome by using migrants with vacant low-lying *d*-orbitals or *p*-orbitals. How significantly the electronic nature of a migrant can be reflected in the energetics and, consequently, in the mechanism of a dyotropic rearrangement may be judged from the data of quantum mechanical (MP2/6—31 G*) calculations [19] on the relative energies of the symmetrical doubly-bridged structures of ethane and 1,2-dilithioethane.

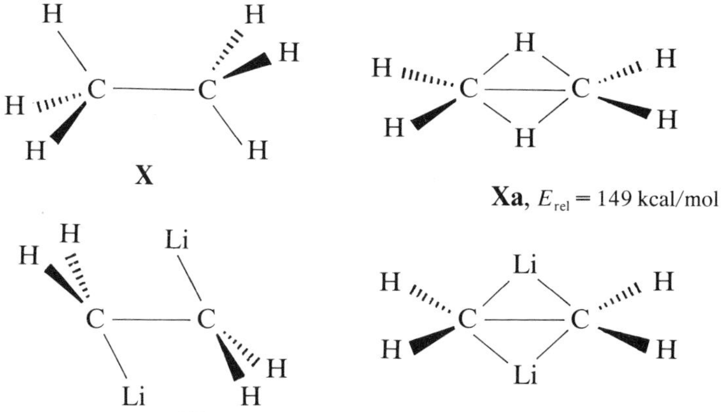

X

Xa, $E_{rel} = 149$ kcal/mol

XI

XI, $E_{rel} = 1.9$ kcal/mol

The structures **X** and **XI** represent ground states of these molecules. The total energy of the D_{2h}-structure **Xa**, a possible candidate for a concerted dyotropic proton shift, is 149 kcal/mol higher than that of X, which violates this mechanism because the C—H bond dissociation energy in ethane (~ 100 kcal/mol) is considerably smaller. At the same time, the symmetrically *trans* doubly-bridged structure **Xa** of dilithioethane that represents a transition state structure for the dyotropic rearrangement of **Xi** is only 1.9 kcal/mol higher in energy. The decisive role of steric factors in determining favorable conditions for the orientation of electron lone pair orbitals of heteroatoms in the reaction sites becomes apparent when studying the reaction mechanisms of the dyotropic systems assigned, according to Reetz, to class II.

As regards the derivatives of naphthazarin **XII**, their dyotropic rearrangement may be conceived as a process based on two independent 1,5-shifts **XII** (a → d, e → b) or as a synchronous shift of two migrants.

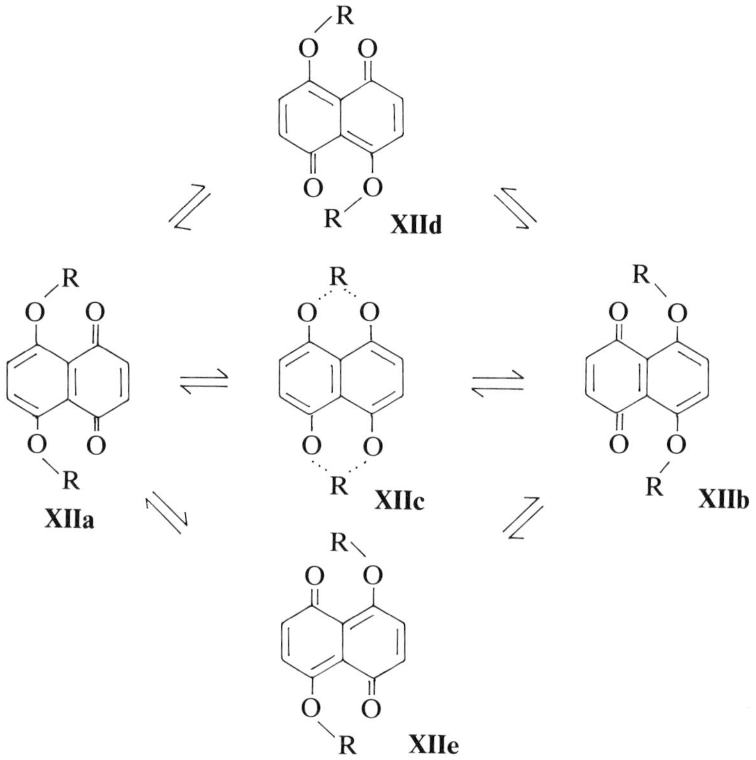

$$ \text{(5.10)} $$

As was noted in Chapter 2, the data of the NMR spectra show that in the acetyl transfer reactions (R = COCH$_3$) the compounds **XIId, e** are not observed as stable enough intermediates, and the reaction (2.27) is accomplished, most probably, in a single kinetic step.

The reaction (5.10) was studied particularly fully in the case of proton transfer (R = H). According to the data obtained from the IR [20] and NMR [21] spectra, a very fast intramolecular exchange **XIIa** ⇌ **XIIb** takes place in solutions of naphthazarin with the structures **XIId, e** not being fixed as intermediates. This finding is in accordance with the data of the quantum mechanical (STO–3G) calculation of the reaction mechanism [22]. The structures **XIId, e** are identified as transition states for the proton migrations in **XII** with the calculated energy barrier totalling merely 2.5 kcal/mol. The symmetrical structure **XIIc** that must emerge in the reaction path in the case of the concerted and synchronous mechanism of the rearrangement (5.10) corresponds to the hilltop on the PES of **XII** which lies 28 kcal/mol above the minima of the stable structures **XIIa, b**. Thus. the favorable steric conditions for proton and acyl group transfer in the naphthazarin system are conducive to a concerted mechanism of the dyotropic rearrangement but do not ensure a synchronous shift of two migrants.

The question as to the feasibility of a synchronous shift of two protons is still quite controversial in the well-studied intramolecular tautomeric transformations of porphyrins whose free activation energies, determined by dynamic ^{1}H and ^{15}N-NMR, amount to 10—15 kcal/mol [23—27].

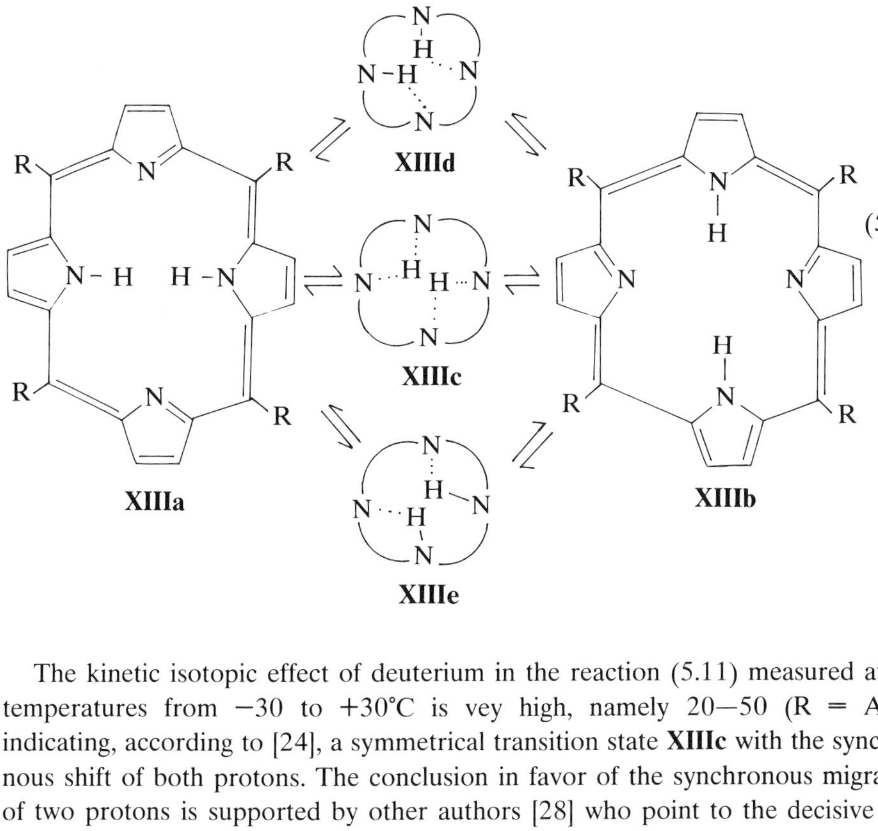

$$(5.11)$$

The kinetic isotopic effect of deuterium in the reaction (5.11) measured at the temperatures from −30 to +30°C is vey high, namely 20—50 (R = Aryl), indicating, according to [24], a symmetrical transition state **XIIIc** with the synchronous shift of both protons. The conclusion in favor of the synchronous migration of two protons is supported by other authors [28] who point to the decisive role of tunnelling in controlling the rate of the proton transfer. The last-mentioned view is, however, disputed [29] on the ground that, at low temperatures, the kinetics of a slow proton transfer of the type (5.11) readily obeys the Arrhenius relationship.

The magnitude of the kinetical HH/DD isotopic effect in reactions (5.11) depends strongly on the type of the substituent R and may, in some cases, be relatively low [23]. Therefore, some authors [23, 25] are inclined to prefer a two-step mechanism **XIIIa** ⇌ **XIIId, e** ⇌ **XIIIb** for the proton transfer. Their judgement conforms also to the data of quantum mechanical calculations of the

PES of reactions (5.11). The CNDO/2 calculated energy barriers of the two-step proton migration for the **XIIIa → XIIId, e** step are 40—60 kcal/mol lower than those of the synchronous reaction [30—32].

The results of the CNDO/2 calculations [33] of the nondegenerate intramolecular double transfer in bispyrrolindigo also agree with the two-step nonconcerted mechanism that includes the formation of an intermediate **XIVc** corresponding to the local minimum on the PES of the reaction (5.12).

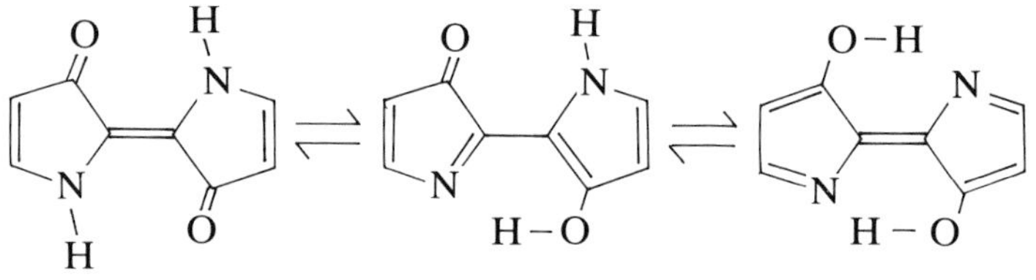

$$\tag{5.12}$$

| **XIVa** | **XIVc** | **XIVb** |

However, the spectral studies on tautomerization of 3,6-bis(benzoxazolyl-2)catechole **XV** [34] and 2,2′-dipyridyl-3,3′-dione **XVI** [35] indicate likelihood of a concerted mechanism for the shift of both protons in the singlet electronic excited states.

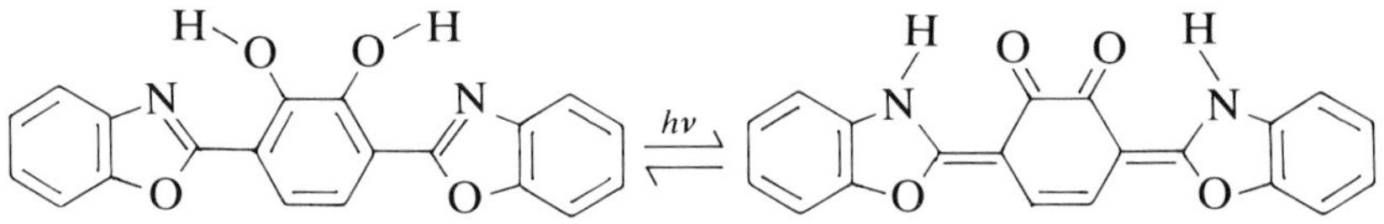

$$\tag{5.13}$$

| **XVa** | **XVb** |

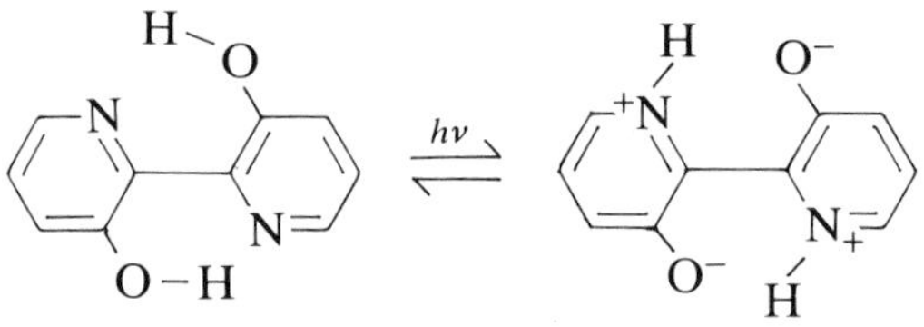

$$\tag{5.14}$$

| **XVIa** | **XVIb** |

Thus, the data regarding the concertedness of the mechanism of double migrations in the dyotropic reactions of class II are contradictory. In spite of the fact that in all reactions considered, associated with the proton transfer (5.10)—(5.14), there exist, in the initial molecules, favorable steric conditions for the motion of every proton along the hydrogen bond bridges, some of these reactions proceed concertedly while others include intermediates. A distinctive feature of all these reactions is the possibility to represent such intermediates, which are products of the transfer of one migrant only, by the normal classical structural formulas, such as **XIIc—XIVc**. This fact alone almost invariably points to the existence of a local minimum on the PES corresponding to this structure, although it naturally cannot guarantee that the MERP of the dyotropic reaction should unavoidably have to pass through this minimum. However, in the case of some other dyotropic reactions such as, for instance, (5.5)—(5.7), (5.14), it is not possible to write the classical structural formula for a potential intermediate corresponding to the product of the transfer of one migrant. Whenever the transfer of a single silyl group in the derivative of squaric acid **V** leads to the formation of an intermediate compound, its structure is described either as a bipolar ion **Vc** or as a biradical **Vd**.

Structures of this type, in which energy-consuming processes of intramolecular charge separation (**Vc**) or of homolysis (**Vd**) have occurred, may be assumed to be unstable and lie outside the MERP of the dyotropic reaction.

2.2. *Multigraphs of Dyotropic and Polytropic Systems*

Provided the above assumption is valid, it will be admitted that the intramolecular double migrations in compounds of type **V**—**VII** must proceed only in a concerted manner with a fairly high degree of synchronization between two migratory processes. This conclusion is a simple consequence following from the classical structural theory.

Within the framework of this theory, there may be developed a simple formal scheme for the design of dyotropic and, more generally, polytropic systems where the migration of n-σ-bonds ($n \geqslant 2$) will be subject to the condition that only their simultaneous transfer can lead to a product with the classical structure. The products of a stepwise transfer of m bonds ($1 \leqslant m \leqslant n$) cannot be described by the classical structural formula (in this case, the formula in which all atomic centers retain a constant valence).

The displacement of one σ-bond necessarily entails redistribution of the formal order of all other bonds in a conjugated system (the so-called canonical bond redistribution [36]); likewise, the simultaneous migration of several bonds is subject to certain structural restrictions.

In the following, rules of formal construction of the graphs for polytropic, mainly, dyotropic systems are described. They are based on combinations of graphs for sigmatropic tautomeric structures.

'*Sewing*' *together of graphs of 1,j-sigmatropic systems.* The simplest technique for obtaining the graph of a polytropic system consists in the 'sewing' together of the graphs denoted in Table 3.2 as the 'bridge' —X— (see footnote in Table 3.2). This 'bridge' represents either a conjugated chain $+\!\!\!\!-\!\!\!\!-\!\!\!\!-\!\!\!\!)_n$ or any nonconjugated system. In the simplest case it is an ordinary bond.

Below, some examples are given illustrating construction of dyotropic graphs and the corresponding symbolic reactions using ordinary bond as a model of the 'bridge' —X—. As we are concerned with tautomeric systems, the examples are restricted to degenerate transformation.

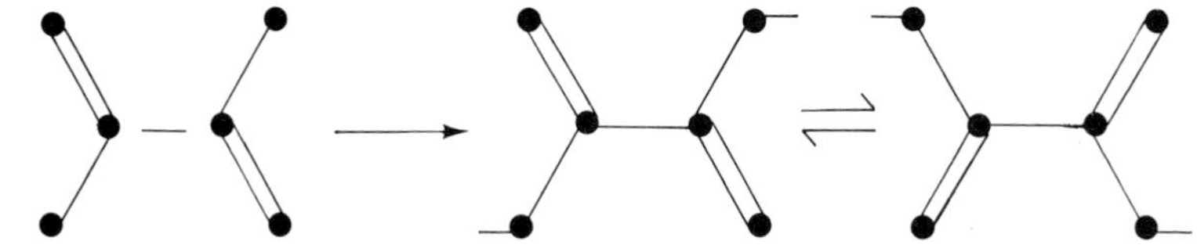

$$\hspace{11cm}(5.15)$$

$$\hspace{11cm}(5.16)$$

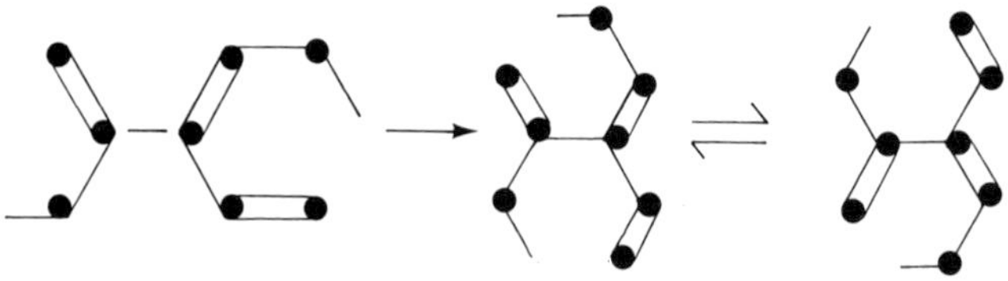

$$\hspace{11cm}(5.17)$$

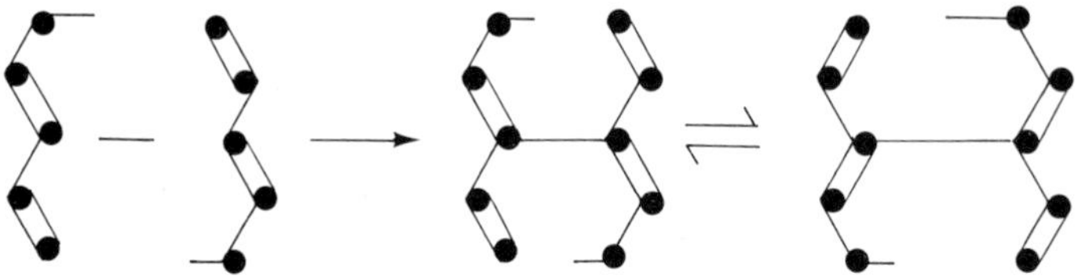

The symbolic equations (5.15)—(5.17) describe dyotropic reactions in dyotropic systems in a generalized form, which can be made concrete when certain atomic groups are substituted for corresponding centers (see footnote in Table 3.2). The transfer of the σ-bonds in a dyotropic system occurs from the donor center of the σ-bond in the structure isomorphic with the first component of the graph being sewn together to the acceptor center of the structure corresponding to the second component of the graph. As in this process the chain of bonds between the donor and the acceptor center is elongated by one bond (or an odd number of multiple bonds), there arises the possibility of $1,2k$-sigmatropic reactions in an uncharged dyotropic system ($k = 1, 2 \ldots$).

It is not hard to pass from the graphs to real molecular structures. *N,N'-*Disubstituted amides and oxamidines **XVII** are the compounds in which dyotropic tautomerism may conceivably take place due to two synchronous 1,4-sigmatropic shifts. This process seems to be favored because of the preferability of the transoid conformation in various derivatives of 1,2-dicarbonyl compounds [37] which is just necessary for the aforementioned transfers.

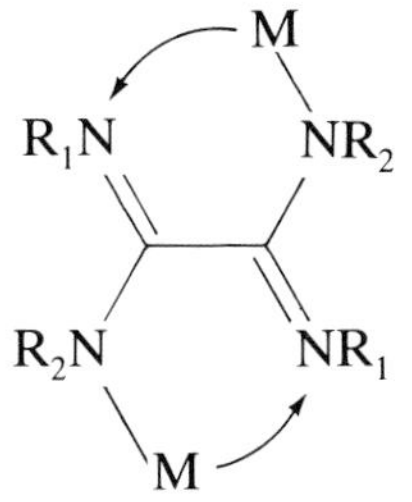

XVIIa 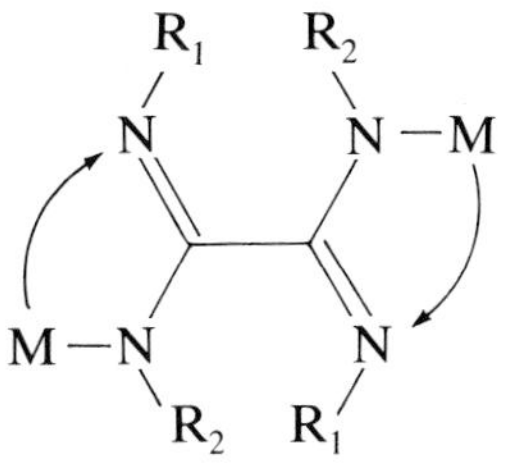   **XVIIb**

Fast intramolecular double proton migrations were observed in oxamidines by Limbach *et al.* [28] by the method of dynamic ^{15}N-NMR spectroscopy. The observed averaging of the signals was attributed to the type **XVIIa** (M = H) double 1,4-shifts; however, the same result might also be achieved in a different way, i.e. by means of type **XVIIb** coupled 1,3-shifts. By contrast, the latter possibility does not exist in the cases of intramolecular rearrangements of 2,2′-bis-Δ^2-imidazoline and its bis-silyl derivative **VI**. The dyotropic 1,4-migrations of the trimethylsilyl group **VIa** $\rightleftharpoons$ **VIb** have been shown [9] to occur at a fairly high rate ($\Delta G_{25}^{\neq} = 17.8$ kcal/mol, $\Delta H^{\neq} = 10.9$ kcal/mol, $\Delta S^{\neq} = -23.3$ in *o*-dichlorobenzene solution) exceeding the rate of transfer reactions of one silyl group in pyrazole derivatives (Chapter 4, Section 2). This is probably due to more favorable steric conditions for the formation of the Si—N bond in a 1,4-sigmatropic reaction as contrasted with the 1,2-transfer reaction. In the search for molecular systems originating from a given structural graph, the geometry factors must be meticulously taken into account. For instance, molecules isomorphic to the (5.16), (5.17) graphs can be readily synthesized as derivatives of 1,3-diketones. However, the compounds of both types exist as *cis*-(*Z*) isomers with respect to the double bond in the enol forms, even with M = H **XVIII** [38]. Clearly, this is connected with the steric constraints for the *trans*-form, in which the necessary proximity of the reactive centers is ensured. In the derivatives of methylene-bis-dimedone **XIX**, these constraints are removed and the required *trans*-configuration is rigidly fixed. The structure of this compound rules out any possibility of independent 1,5-sigmatropic shifts in each 1,3-diketone fragment.

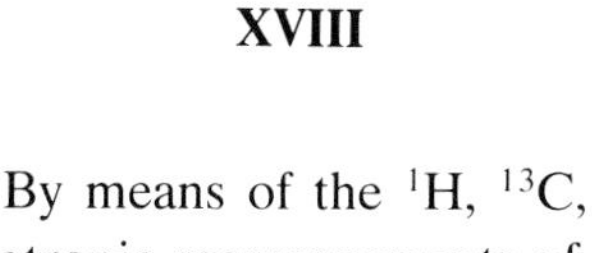

XVIII 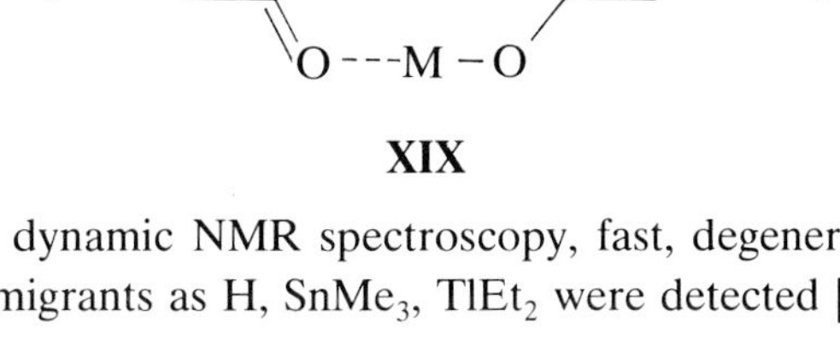   **XIX**

By means of the ^{1}H, ^{13}C, ^{119}Sn dynamic NMR spectroscopy, fast, degenerate, dyotropic rearrangements of such migrants as H, SnMe$_3$, TlEt$_2$ were detected [39] in compounds **XIX**. On the other hand, with the $COCH_3$, SiR_3 migrants, no such rearrangements could be observed on the NMR time scale. A fast dyotropic tautomeric rearrangement takes place also in the case of the nitrogen-substituted derivative of **XIX** — N^1, N^2, N^3, N^4-tetraphenyl-bis(5,5-dimethylcyclohex-2-yl-1,3-diimino)methane **XX**.

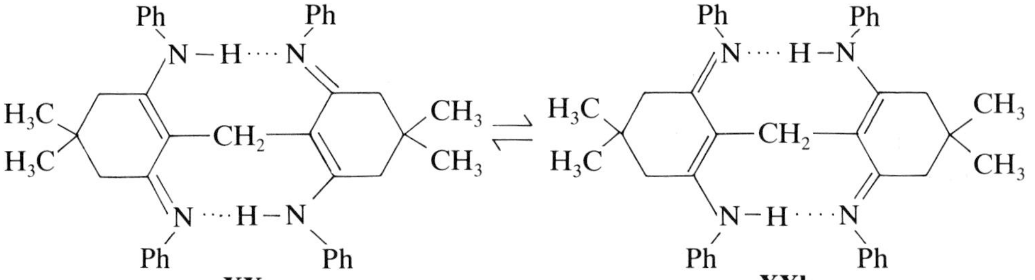

In the solutions of this compound, even when such high-polarity solvents as DMSO are applied, a very strong intramolecular hydrogen bond is established. Fast intramolecular shifts of two protons along the H-bridges are particularly clearly observed in the ^{15}N-labelled compounds **XX**, in which, under the conditions of fast exchange, the NH proton gives a triplet signal with the spin—spin coupling constant of 43 Hz being equal to one half of the normal $^1J_{^{15}NH}$ value. The kinetic parameters for the reaction **XXa** $\rightleftharpoons$ **XXb**, determined in a deuterochloroform solution by means of the total line shape analysis of ^1H-NMR spectra are as follows: $k_{-100} = 1300$ s^{-1}, $\Delta G^{\neq}_{-100} = 7.5$ kcal/mol, $\Delta H^{\neq} = 3.0$ kcal/mol, $\Delta S^{\neq} = 26.0$ e.u.

The same technique may be applied for obtaining a graph of the polytropic system of a higher than dyotropic order, requiring only that the graph of the 'bridge' —X— be made up of nonconjugated or conjugated bonds with more than two branches for the linkage with the tautomeric 1,*j*-sigmatropic graph. As applied to the last-mentioned case, this is possible because of the closure of the double bonds of a conjugated bridge into a cycle, e.g. the benzene cycle (closure of three double bonds).

The derivatives of the mesidinic acid so produced, for instance, amidines, are promising candidates for studying the triotropic tautomerism of the following type:

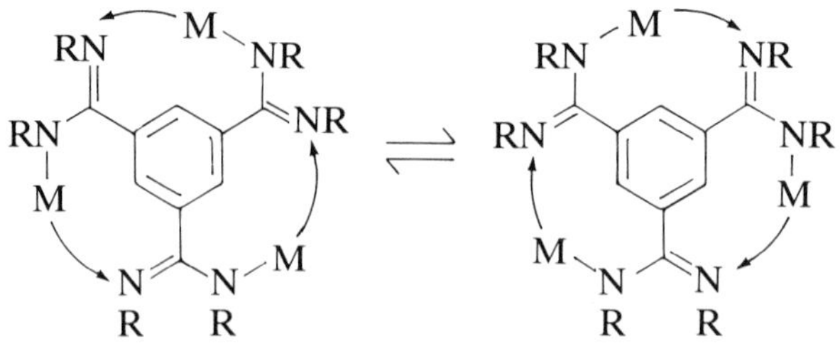

In this system, favorable steric conditions are afforded for the transfer of the groups such as methyl, sulphenyl, sulphinyl and others requiring linear orientation of the reactive centers of the forming and breaking bonds in a transition structure.

The systems of type (5.1) studied by Reetz are also derived in accordance with the sewing-together-rule. But in this case the components of the dyotropic graph comprise only the carrying center and the migrating bond

$$\tag{5.18}$$

When this graph is sewn to the graphs of the 1,*j*-sigmatropic series, it leads to prototypes of some interesting systems with an unequal number of the centers donating and accepting the migrating σ-bonds, e.g.

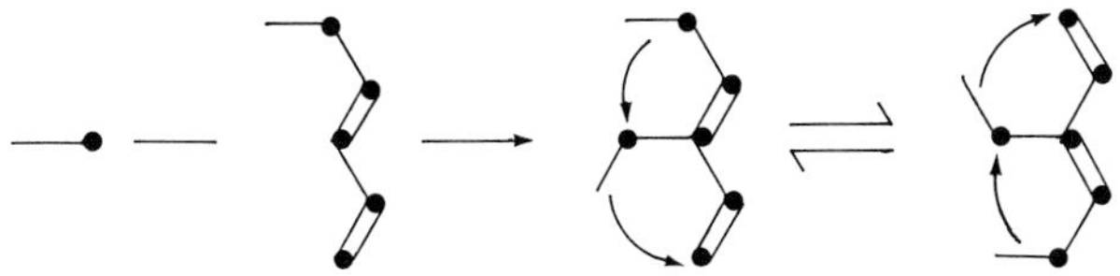

$$\tag{5.19}$$

The guanidine derivatives **XXII** may serve as an example of the systems where it may be expected to find such dyotropic rearrangement.

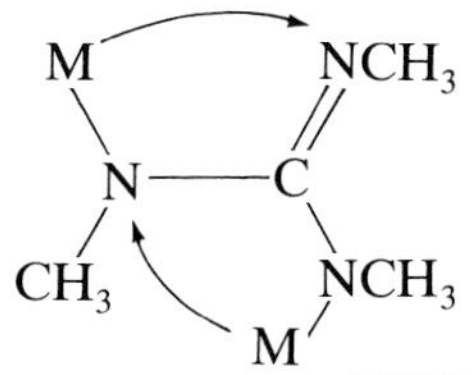

XXII   M = H, Acyl, Aryl

Sewing of the graph of the ordinary bond to the 1,5-sigmatropic graph gives another system.

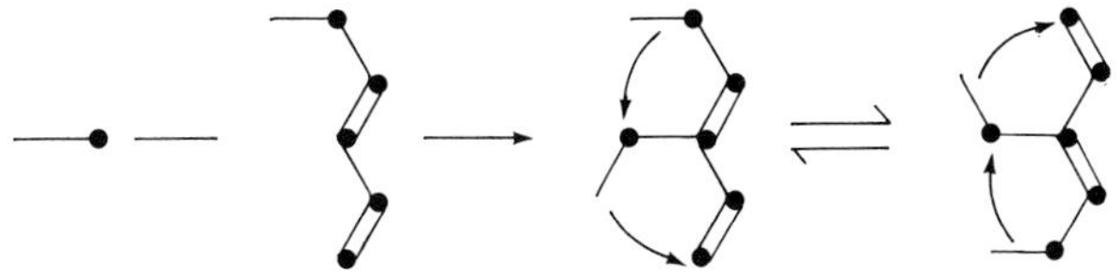

$$\tag{5.20}$$

Triose reductone (2-propen-2,3-diol-1-one) **XXIII** is one of the compounds in which dyotropic rearrangements of type (5.20) actually take place. According to a ^{1}H-NMR study [40], in solution it undergoes tautomeric transformations as a result of two synchronized 1,4-proton shifts

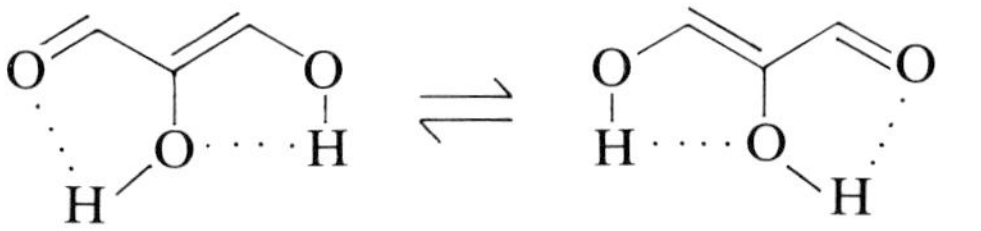

XXIIIa   **XXIIIb**

In reductic acid (2-cyclopentene-2,3-diol-1-one) **XXIV**, the hydrogen bonding scheme is totally similar to that found in the crystal structure of **XXIII** [41].

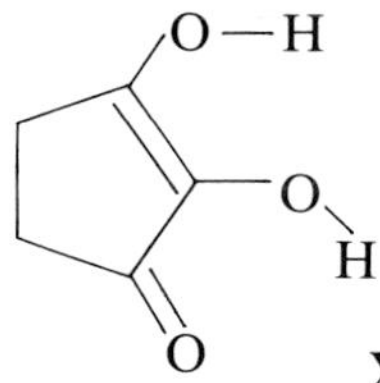

XXIV

Fusion of the graphs of 1,j-sigmatropic systems with the formation of a common edge. The second way of producing a graph of a polytropic system consists in the linkage of two graphs of 1,*j*-sigmatropic systems in such a manner that the edge of an ordinary bond of one graph is 'fused' with the edge of a double bond of another. This may be illustrated by an example with the graphs of 1,5-sigmatropic systems. Two variants are conceivable.

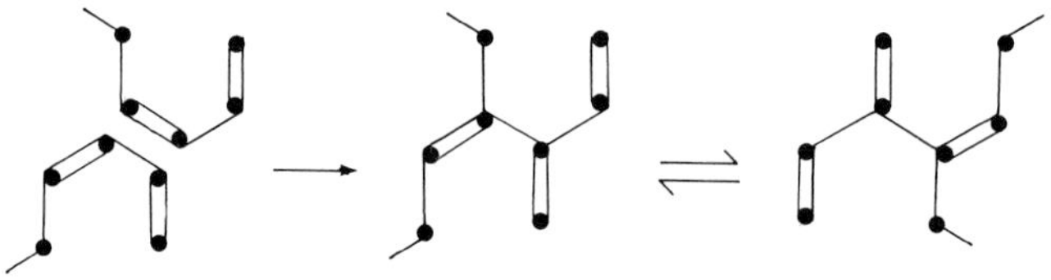

(5.21)

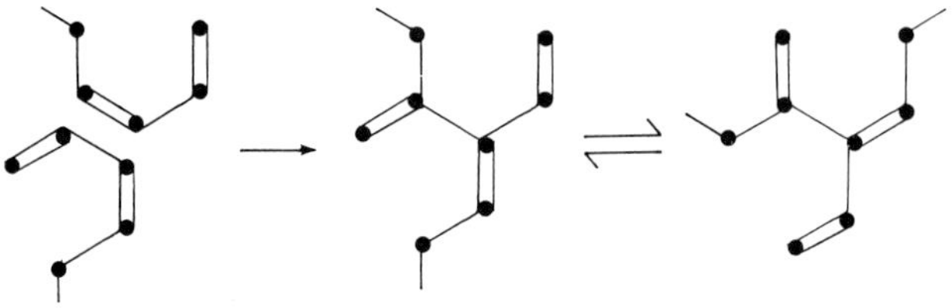

(5.22)

The (5.22) variant duplicates the dyotropic graph (5.16) obtained by another method. The graph (5.21) cannot be produced by the sewing procedure based on bridging as described earlier.

Graphs of tri-, tetra- and higher polytropic systems can be constructed in a like manner, e.g.

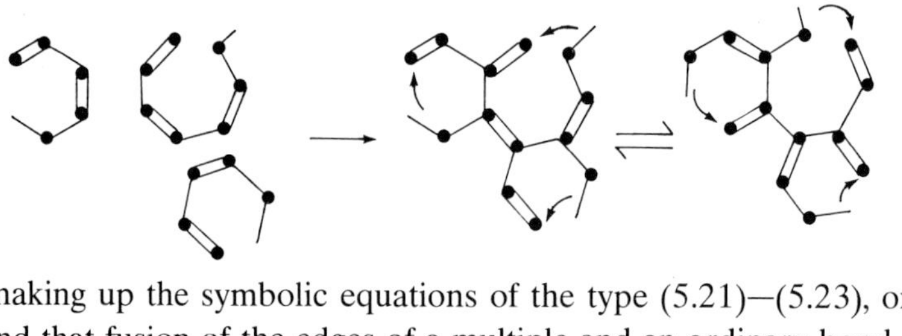

(5.23)

When making up the symbolic equations of the type (5.21)–(5.23), one has to bear in mind that fusion of the edges of a multiple and an ordinary bond produces an ordinary bond edge.

Combination of the graphs of dyotropic systems. The assemblage of a polytropic graph can also be performed by combining the graphs of dyotropic systems. This technique may be explained by the following equation:

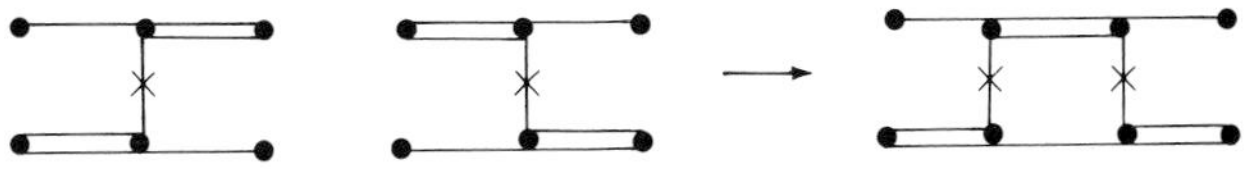

$$(5.24)$$

The double edge of one graph is laid on the double edge of another graph of the dyotropic system producing the common double edge of the resultant graph. In the same manner, the edges of ordinary bonds are joined.

Below two examples are given illustrating dyotropic graphs constructed by Scheme (5.24) as well as their respective molecular systems.

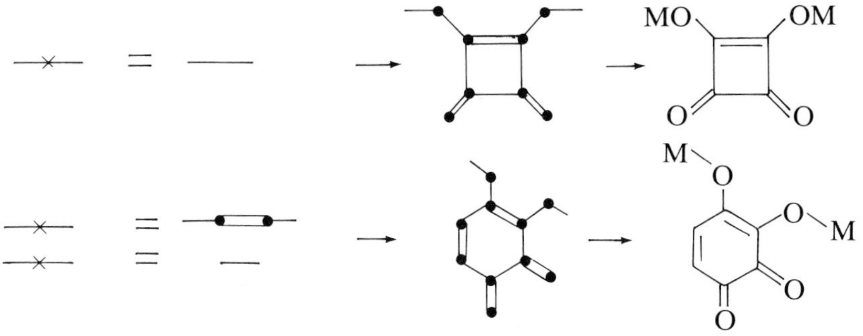

By combining several dyotropic graphs according to the rule (5.24), one may obtain the graphs of polytropic systems.

To be able to change the order in which the edges of dyotropic graphs are combined, a unit consisting of one apex and two different edges —X— has to be included in the procedure of the assemblage of a polytropic graph. Then, via the apex noted, the ordinary and the multiple edges of the dyotropic graphs can be combined as follows

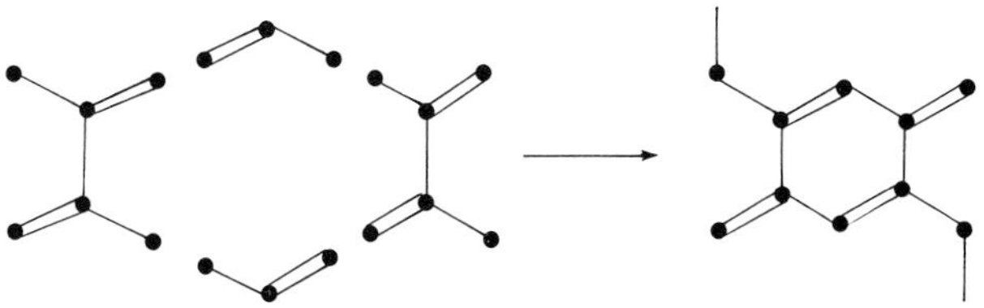

$$(5.25)$$

This graph represents the molecular system of 2,5-dihydroxy-1,4-benzoquinone **VII**.

2.3. *Quantum Mechanical Studies on the Mechanism of Dyotropic Reactions*

Construction of multigraphs and symbolic equations in accordance with the rules

(5.15)–(5.25) is a convenient means for classification and molecular design of dyotropic and polytropic systems. Can, however, the correspondence of an actual molecular system to the graphs discussed be regarded as a reliable test for the concertedness of a polytropic reaction? The data of quantum mechanical calculations available to date testify that the classical structural selection rule (absence of classical structural formula for a possible intermediate — the product of the migration of one bond) is not rigid.

The calculations of the MERP for the dyotropic reactions of 1,4-proton transfer between the oxygen atoms in oxalic and squaric acids, triose reductone and 1,5-dihydroxy-2,4-benzoquinone made by the MINDO/3 [42] method and between the nitrogen atoms in azophenine **XXVII** by the AM 1 [43] method revealed in each case the formation of stable bipolar intermediates indicating a nonconcerted mechanism of these reactions.

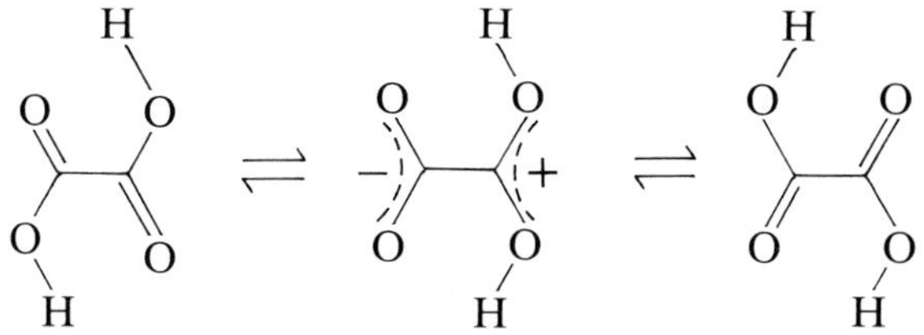

$$E_{rel} = 16.6 \text{ kcal/mol} \qquad (5.15a)$$

XXVa   XXVc   XXVb

$$E_{rel} = 8.7 \text{ kcal/mol} \qquad (5.24a)$$

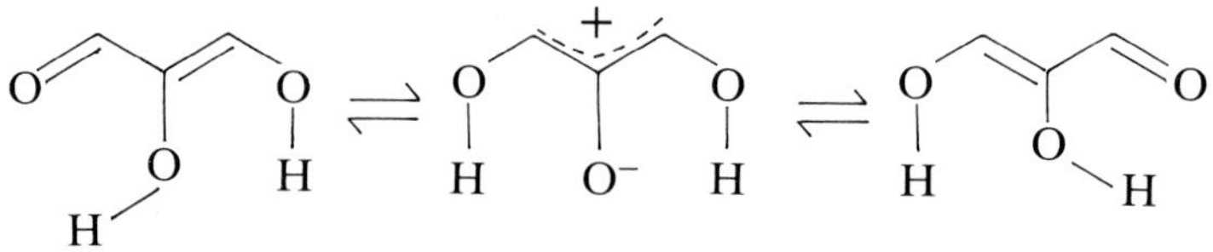

XXVIa   XXVIc   XXVIb

$$E_{rel} = 35.9 \text{ kcal/mol} \qquad (5.20a)$$

XXIIIa   XXIIIc   XXIIIb

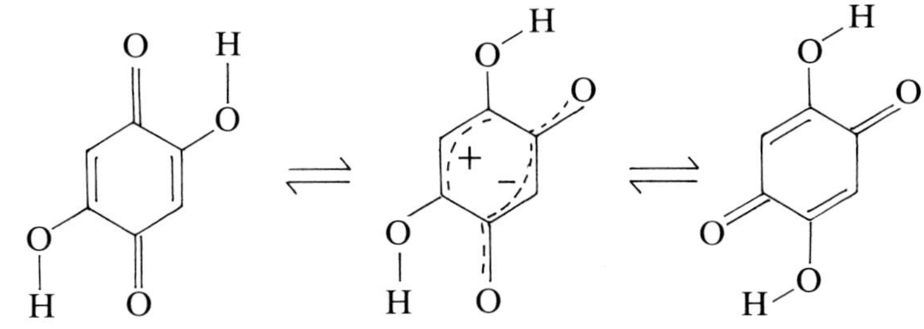

VIIa (R= H)   **VIIc**   **VIIb**

$E_{rel} = 17.6$ kcal/mol

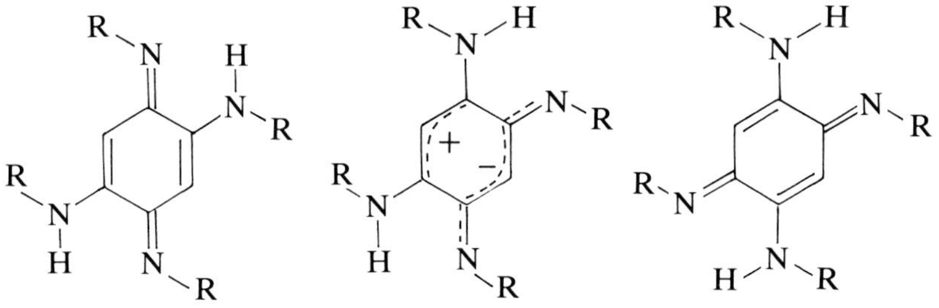

XXVIIa (R = H, Ph)   **XXVIIc**   **XXVIIb**

$E_{rel} = 22.4$ kcal/mol

(5.25a)

In all these cases, the symmetrical structures emerging in the course of the synchronous 1,4-shift of two protons between the heteroatomic centers are hilltops on the respective PESs, i.e. they are not transition-state structures. The typical pattern of the PES of a dyotropic 1,4-shift reaction is given in Fig. 5.2.

The calculated electron distributions and geometry characteristics of nonclassical bipolar structures of intermediates in the reactions considered are presented in Fig. 5.3. Their relative energies (in relation to the stable initial form) are appended to the structural formulas.

The activation energies predicted by calculations [42, 43] for a single proton 1,4-transfer (40—70 kcal/mol) are significantly higher than the experimental values measured for these reactions (about 10 kcal/mol for 5.25 a, b) [14, 25]. The difference seems far too large to be attributed to the deficiencies of semiempirical MINDO/3 and closely related AM 1 procedures. In view of the very small widths of the energy barriers for single proton transfers, especially in excited vibrational states of the molecules, a decisive role of tunnelling from vibrationally excited states of the stable form is suggested [43]. The effective activation energy should then be equal to the endothermicity of the single proton transfer steps. It may be seen that the calculated relative energies of the intermediates of dyotropic reactions are indeed very close in value to experimental activation energies.

Calculations [42] show that the conclusion as to the nonconcerted mechanism of dyotropic 1,4-rearrangements holds for quite diverse systems such as hydro-

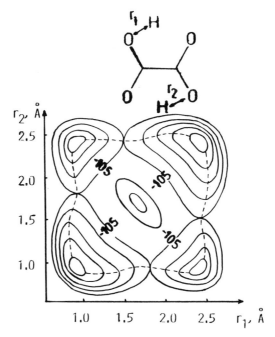

Fig. 5.2. The PES of the dyotropic rearrangement **XXVa** ⇌ **XXVb** calculated by the MINDO/3 method [42]. The equipotential curves are spaced 14 kcal/mol apart. The MERP is denoted by the dashed line.

carbons, compounds with heavy migrants (BR_2, NO, $COCH_3$) and nondegenerate species. It is, however, still unclear whether this conclusion is general, i.e. whether it embraces all dyotropic processes, or depends on steric factors of the reaction site. Some light is thrown on the matter thanks to calculations of the energy profile of the reaction (5.26) at various values of the parameters R and θ determining the stereochemistry of the proton transfer.

$$\text{XXVIIIc} \longrightarrow \text{XXVIIIa} \tag{5.26}$$

XXVIIIc   **XXVIIIa**

The hydroxyl and carbonyl oxygen atoms are models of respective nonequivalent centers in dyotropic systems. The fragments **XXVIIIa** correspond to classical structures while the **XXVIIIc** fragments to the structures of intermediates in the products of the single proton transfer (one of the possible resonance forms). If at certain values of R and θ there is a minimum on the PES corresponding to the

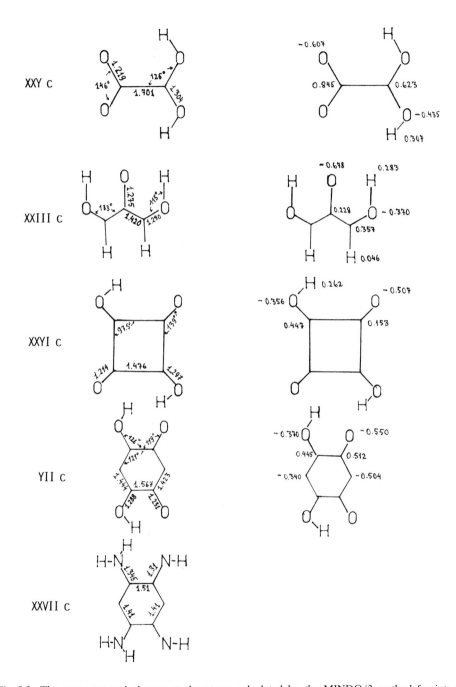

Fig. 5.3. The geometry and charges at the atoms calculated by the MINDO/3 method for intermediates emerging in the course of the single proton 1,4-transfer in dyotropic reactions [42]. In the case of the intermediate **XXVIIc**, the geometry has been calculated using the AM 1 method [43].

structure **XXVIIIc**, then it may be assumed that the dyotropic reaction with the analogous stereochemistry of proton transfer would proceed in a nonconcerted way. By contrast, when the minimum on the PES corresponding to this structure does not exist, the reaction should be concerted.

Calculations [42] have shown that the whole region of variation of the parameters R and θ divides into two subregions, in one of which there is a double-well potential corresponding to the reaction (5.26) whereas in the other only one minimum of **XXVIIIa** is present (Fig. 5.4).

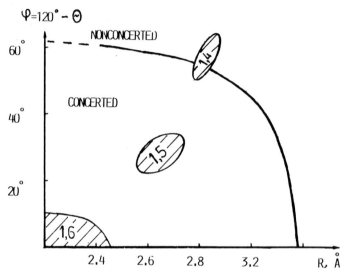

Fig. 5.4. Regions of concerted and nonconcerted mechanisms in 1,*j*-dyotropic reactions. Shaded areas represent 1,*j*-dyotropic systems of the appropriate structure.

The steric parameters of 1,4-dyotropic systems ($R \simeq 2.8\,\text{Å}, \theta \simeq 60°$) fall in the subregion of the double-well potential, while the concerted mechanism of a dyotropic reaction is expected in the systems with the short linear hydrogen bond, i.e., in the case when there exist optimal steric conditions of the reactive site for the proton transfer step (Table 3.III). Such conditions can be created in 1,5- and 1,6-dyotropic systems ($R = 2.5-2.6\,\text{Å}, \theta \simeq 90°$ and $R \leqslant 2.4\,\text{Å}, \theta \simeq 120°$, respectively).

Indeed, calculations of the MERP for 1,5- and 1,6-dyotropic reactions

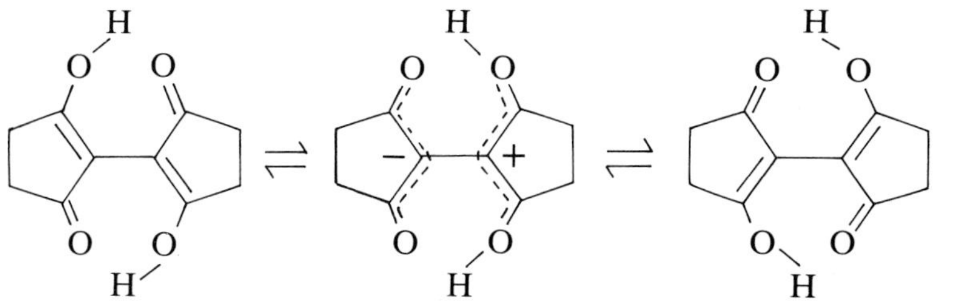

$$\text{XXXa} \qquad \text{XXXc (TS)} \qquad \text{XXXb} \qquad (5.28)$$

made be the MINDO/3 method have shown that the intermediate structures **XXIXc, XXXc**, incorporating two fragments **XXVIIIc** each, represent, not the local minima, but the transition state structures similar to the structures **XIId, e** in the reaction (5.10). The relative energies of **XXIXc, XXXc** are 47 and 10 kcal/mol, respectively.

The concerted reactions of double proton transfer in dyotropic systems model most important intermolecular processes with an analogous mechanism of proton motion along the hydrogen bridges. The double concerted proton transfers determine the mechanism of the bifunctional acid-base catalysis [44]. As has been shown by *ab initio* calculations [45, 46], the coupled proton migrations in DNA base pairs, in particular, in the H_1-H_2 and H_1-H_3 pairs in the guanine—cytosine dimer proceed in a concerted manner without formation of intermediates.

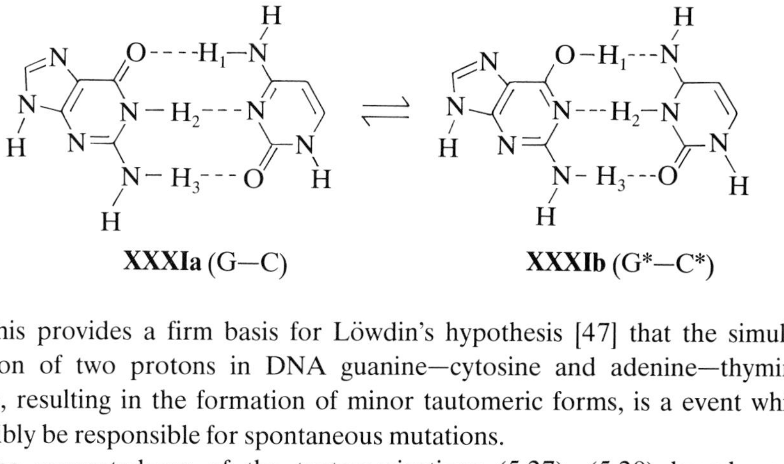

$$\text{XXXIa (G—C)} \qquad\qquad \text{XXXIb (G*—C*)} \qquad (5.29)$$

This provides a firm basis for Löwdin's hypothesis [47] that the simultaneous motion of two protons in DNA guanine—cytosine and adenine—thymine base pairs, resulting in the formation of minor tautomeric forms, is a event which may possibly be responsible for spontaneous mutations.

The concertedness of the tautomerizations (5.27)—(5.29) based on double proton migrations by no means signifies that the motion of both protons along hydrogen bond bridges is a synchronous process. By a careful scanning of the entire PES of the reaction (5.29) it was found that the optimum path for the double proton transfer corresponded to moving first one proton, and moving the second one only after very significant stretching of the $N-H_1$ bond [46]. Synchronous motion of both protons in **XXXIa** towards **XXXIb** warrants the prediction of an unreasonably high barrier for the double proton transfer (61.6 kcal/mol), which is only partly due to the deficiencies of the approximation being employed (PRDDO) [48]. Thus, in all the dyotropic reactions considered above, for which sufficiently full computations of critical portions of the potential surfaces (5.7), (5.10), (5.15), and (5.27)—(5.29) have been performed, the motion of protons in hydrogen bond bridges is nonsynchronous. The symmetrical structures corresponding to a synchronous double proton transfer are energy-rich and belong to the hilltops on the respective PESs (stationary points of the second and higher order).

This conclusion is in agreement with the recent generalizing statement by Dewar [49] that "synchronous multibond mechanisms are normally prohibited", which means that no bond-making and bond-breaking processes during an elementary step of a reaction can take place in unison. This rule, apparently, pretends to eliminate strictly synchronous mechanisms for virtually all the pericyclic reactions, including the Diels—Alder reactions between symmetric dienes and dienophiles.

And yet a very probable exception to this rule has already been found in experimental and theoretical investigations of the butadiene—ethylene reaction whose mechanism has been shown to be both concerted and synchronous [50, 51]. Some other important examples of synchronous concerted multibond reactions have been found in the case of double proton transfers in the hydrogen bonded dimers of carboxylic acids. By a series of infrared spectroscopic measurements [52], it has been demonstrated that in various crystalline carboxylic acids there coexist two molecular forms which are convertible to each other by double proton transfer in the dimers.

$$
\underset{\textbf{XXXIIa}}{
\begin{array}{c}
\mathrm{O-H\cdots O} \\
R-C \qquad\qquad C-R \\
\mathrm{O\cdots\cdots H-O}
\end{array}}
\;\rightleftharpoons\;
\underset{\textbf{XXXIIc (TS)}}{
\begin{array}{c}
\mathrm{O\cdots H\cdots O} \\
R-C \qquad\qquad C-R \\
\mathrm{O\cdots H\cdots O}
\end{array}}
\;\rightleftharpoons\;
\underset{\textbf{XXXIIb}}{
\begin{array}{c}
\mathrm{O\cdots H-O} \\
R-C \qquad\qquad C-R \\
\mathrm{O-H\cdots O}
\end{array}}
\tag{5.30}
$$

The height of the potential barrier for this conversion has been estimated using the solid-state NMR technique and found to drop to about 1 kcal/mol in the case of compounds **XXXII** with R = $C_6H_4CH_3$-p [53]. Detailed *ab initio* calculations [54, 55] of the MERP for the reaction (5.30) testify to a concerted and rigorously synchronous mechanism of the shift of two protons in the dimers **XXXII**. In all portions of the MERP, the system **XXXII** remains centrosymmetrical; the symmetrical D_{2h}-structure **XXXIIc** is a true transition state of the reaction. In the case of the dimer of the formic acid (R = H), the activation energy totals at the highest level of approximation (4—31 G + CI [55]) 7.1 kcal/mol. The characteristic feature of the transition state structure **XXXIIc** is a very significant shortening of the O·····O distance as compared to the ground state structure **XXXIIa**.

Studies of the detailed mechanism of multibond reactions, in general, as well as dyotropic and polytropic reactions, in particular, are still in their early stages. All information on mechanisms of these reactions is of great importance both for a target-oriented design of relevant molecular systems and for gaining better insight into mechanisms of some intricate cooperative processes such as simultaneous redistribution of bonds in complex molecular systems or reactions of polymers and solids.

REFERENCES

1. Minkin V. I., Olekhnovich L. P., and Zhdanov Yu. A., *Molecular Design of Tautomeric Systems*. Rostov-on-Don University Publ. House. Rostov-on-Don. 1977.
2. Reetz M. T., *Tetrahedron, 29*, 2189 (1973).
3. Reetz M. T., *Adv. Organometal. Chem., 16*, 34 (1977).
4. Gabriel S. and Hoppe M., *Ber., 19*, 1145 (1886); Wieland H., *Ber., 42*, 4207 (1909).
5. Reetz M. T., Kliment M., and Plachky M., *Chem. Ber., 109*, 2716 (1976).
6. Reetz M. T., Greif N., and Kliment M., *Chem. Ber., 111*, 1083, 1095 (1978).
7. Nowakowski P. and West R., *J. Amer. Chem. Soc., 98*, 5616 (1976).
8. Reetz M. T., Neumeier G., and Kashuke M., *Tetrahedron Lett.*, 1295 (1975); Reetz M. T. and Neumeier G., *Lieb. Ann. Chem., 1234* (1981).
9. Bren V. A., Chernoivanov V. A., Konstantinovskii L. E., Borisenko N. I., Bren Zh. V., and Minkin V. I., *Dokl. Akad. Nauk USSR, 255*, 107 (1980).
10. Phillisborn W., *Pure Appl. Chem., 40*, 159 (1974).
11. Graf F., *Chem. Phys. Lett., 62*, 291 (1979).
12. Bren V. A., Chernoivanov V. A., Konstantinovskii L. E., Nivorozhkin L. E., Zhdanov Yu. A., and Minkin V. I., *Dokl. Akad. Nauk USSR, 251*, 1129 (1980).
13. Joseph-Nathan P., Abramo-Bruno D., and Ortega D. A., *Org. Mag. Res., 15*, 311 (1981).
14. Chernoivanov V. A., Bren V. A., Chernysh Yu. E., Borodkin G. S., Minkin V. I., and Zhdanov Yu. A., *Dokl. Akad. Nauk USSR, 283*, 885 (1985).
15. Hoffmann R. and Williams J. E., *Helv. Chim. Acta, 55*, 67 (1972).
16. Grundmann C., Kochs P., and Boal R., *Lieb. Ann. Chem., 761*, 162 (1972).
17. Poppinger D., Radom L., and Pople J. A., *J. Amer. Chem. Soc., 99*, 7806 (1977).
18. Poppinger D. and Radom L., *J. Amer. Chem. Soc., 100*, 3674 (1978).
19. Kos A. J., Jemmis E. D., Schleyer P. v. R., Gleiter R., Fischbach U., and Pople J. A., *J. Amer. Chem. Soc., 103*, 4996 (1981).
20. Bratan S. and Strohbush F., *J. Mol. Struct., 61*, 409 (1980).
21. Lyerla J. R., Yannoni C. S., and Fyfe C. A., *Acc. Chem. Res., 15*, 208 (1982).
22. De la Vega J. R., *J. Amer. Chem. Soc., 104*; 3295 (1982).
23. Abraham R. J., Hawkes G. E., and Smith K. M., *Tetrahedron Lett.*, 1483 (1974).
24. Eaton S. S. and Eaton G. R., *J. Amer. Chem. Soc., 99*, 1601 (1977).
25. Irving C. S. and Lapidot A., *J. C. S. Chem. Commun.*, 184 (1977).
26. Grishin Yu. A., *Zh. Struct. Khim., 20*, 352 (1979).
27. Yeh H. J. C., Sato M., and Morishima J., *J. Mag. Resonance, 26*, 365 (1977).
28. Limbach H. H., Hennig J., Gerritzen D., and Rumpel H., *Faraday Discuss. Chem. Soc., 74*, 822 (1982).
29. Stilbs P., *J. Mag. Resonance, 58*, 152 (1984).
30. Kuzmitsky V. A., Sevchenko A. I., and Solovyov K. N., *Dokl. Akad. Nauk USSR, 239*, 308 (1978).
31. Kusmitsky V. A. and Solovyov K. N., *J. Mol. Struct., 65*, 219 (1980).
32. Dvornikov S. S., Kujzmitsky V. A., Knyushto V. I., Solovyov K. N., Gurkova A. E., and Ivanov E. G., *Khimitzeskaya Fizika, 4*, 889 (1985).

33. Sühnel J. and Gustav K., *Zeitschr. Chemie,* **17**, 297 (1977).
34. Mordzinski A. and Grabowska A., *J. Mol. Struct.,* **114**, 337 (1984).
35. Bulska H., *Chem. Phys. Lett.,* **98**, 398 (1983).
36. Trach S. S. and Zefirov N. S., *Zh. Org. Khim.,* **18**, 1581 (1982).
37. Minkin V. I., Osipov O. A., and Zhdanov Yu. A., *Dipole Moments in Organic Chemistry.* Ch. 4. Plenum Press. New York. 1970.
38. Krongaus E. S. and Berlin A. A., *Usp. Khim.,* **38**, 1479 (1969).
39. Minkin V. I., Chernoivanov V. A., Bren V. A., and Borisenko N. I., *Zh. Org. Khim.,* **23**, (in press) (1987).
40. Frimmel F., Fritz H. P., and Kreiter C. G., *J. Cryst. Mol. Struct.,* **1**, 25 (1971).
41. Semingsen D., *Acta Chem. Scand.,* **B28**, 141 (1974): **B31**, 81 (1977).
42. Simkin B. Ya., Golyanskii B. V., and Minkin V. I., *Zh. Org. Khim.,* **17**, 3 (1981).
43. Dewar M. J. S. and Mertz K. M., *J. Mol. Struct. THEOCHEM.,* **124**, 183 (1985).
44. Jencks W. P., *Acc. Chem. Res.,* **9**, 425 (1976); **13**, 161 (1980).
45. Clementi E., Mehle J. and von Niessen W., *J. Chem. Phys.,* **54**, 508 (1971).
46. Clementi E., Corongiu G., Detrich J., Chin S., and Domingo L., *Intern. J. Quantum Chem. Quantum Chem. Symp.* **18**, 601 (1984).
47. Löwdin P.-O., *Rev. Mod. Phys.,* **35**, 724 (1963); *Adv. Quant. Chem.,* **2**, 213 (1965).
48. Scheiner S. and Kern C. W., *J. Amer. Chem. Soc.,* **101**, 4081 (1979); Scheiner S., *J. Amer. Chem. Soc.,* **103**, 315 (1981).
49. Dewar M. J. S., *J. Amer. Chem. Soc.,* **106**, 209 (1984).
50. Bernardi F., Bottini A., Robb M. A., Field M., Hillier I. H., and Guest M. F., *J. C. S. Chem. Commun.,* 1051 (1985).
51. Houk K. N., Lin Y.-T., and Brown F. K., *J. Amer. Chem. Soc.,* **108**, 554 (1986).
52. Umemura J., *J. Mol. Struct.,* **36**, 35 (1977); *J. Chem. Phys.,* **68**, 42 (1978).
53. Meier B. H., Graf F., and Ernst R. R., *J. Chem. Phys.,* **76**, 767 (1982).
54. Graf F., Meyer R., Ha T.-K., and Ernst R. R., *J. Chem. Phys.,* **75**, 2914 (1981).
55. Hayashi S., Umemura J., Kato S., and Morokuma K., *J. Phys. Chem.,* **88**, 1330 (1984).

6. Dissociative and Photochemical Mechanisms of Intramolecular Tautomerism

The tautomeric rearrangements described in the previous Chapters of this book are based on associative mechanisms of nucleophilic substitution. Fitting of a molecular structure to the structure of a transition state or an intermediate of a rearrangement reaction is an effective approach to the design of an intramolecular tautomeric system.

This general approach appears promising not only in regard to associative but also to dissociative mechanisms of intramolecular substitution. As yet few data are available on intramolecular sigmatropic tautomeric systems in which migratory processes are nonconcerted, proceeding via the stages of breaking the bond with the migrant, forming an intermediate ionic or radical pair and their subsequent rearrangement. Nonetheless, a brief analysis of data from the literature given below shows that dissociative mechanisms of both heterolytic and homolytic type should not be discarded when designing tautomeric systems. Furthermore, these mechanisms may play an important role in the photochemical reversible rearrangements.

The stereochemistry of dissociative mechanisms is radically different from that of the associative substitution mechanisms, hence, the steric demands of the molecular system carrying a migrant are different with these two mechanisms, and general conclusions concerning the structure of a system ensuring a tautomeric reaction presented in Chapters 2—4 have to be reexamined in the case of the dissociative type reactions.

1. Heterolytic Dissociative Mechanism

The basic scheme of this mechanism of intramolecular tautomeric migrations consists of:

(a) heterolytic rupture of the bond between a migrant and the nucleophilic center;
(b) formation of a tight ionic pair **Ic**;
(c) transfer of the migrant within the ionic pair to a new nucleophilic center in the anion

$$
R_nM \underset{Y}{\diagdown} \underset{X}{\diagup} \rightleftharpoons \overset{+}{R_nM} \underset{Y}{\diagdown} \underset{X}{\diagup} \rightleftharpoons MR_n \underset{Y}{\diagup} \underset{X}{\diagdown} \tag{6.1}
$$

Ia   **Ic**   **Ib**

Unlike transition structures of the associative mechanisms of tautomeric migrations (transition states or covalent intermediates) considered earlier, the electron pair shared by the X—M and Y—M bonds in **Ia** and **Ib** is completely withdrawn from the migrant in **Ic**. At the same time, the cation M^+ is held in close proximity to the related anion meeting thus the condition of a tight ionic pair where the counterions are separated, if at all, by not more than one molecule of the solvent. Only the tight ionic pairs **Ic**, apparently, ensure the intramolecular character of the process. The feasibility of such a process of the (6.1) type has already been proven by remarkable studies of Goering and co-workers [1, 2] who showed that the solvolysis of optically active *p*-nitrobenzoate labelled on the oxygen is accompanied by the randomization of the isotopic label. Inclusion of an independent stage (6.2) in the overall reaction scheme has been demonstrated to be the reason for this effect.

$$
\underset{C_6H_4NO_2\text{-}p}{\overset{^{18}O}{\diagup}} C \overset{O-^+CH}{\underset{C_6H_4Cl\text{-}p}{\diagup}} \rightleftharpoons \underset{C_6H_4NO_2\text{-}p}{\overset{^{18}O\text{-}\text{-}\text{-}O}{\diagup}} C \overset{^+CH}{\underset{}{\diagup}} \rightleftharpoons \underset{C_6H_4NO_2\text{-}p}{\overset{^+CH-^{18}O}{\diagup}} C \overset{O}{\diagup} \tag{6.2}
$$

IIa   **IIc**   **IIb**

At this stage, the configuration of the tetrahedral carbon of the migrating alkyl group is fully retained. This means that the carbocation is not a kinetically independent particle and its structure in the ionic pair is rigidly fixed.

In the detailed Review [3] some more examples are described of nucleophilic substitution reactions or kinetically confirmed separate steps of a complete reaction which follow the general scheme (6.1) of internal return in an ionic pair. They all belong to 1,3-sigmatropic shifts in carboxyl, thiocarboxyl and thiocyanato systems with the discrete rearrangement step revealed by means of an anylysis of the kinetical scheme of reaction.

More straightforward evidence for the (6.1) type mechanism being operative was derived by studying some circumambulation rearrangements in cyclic polyenes in saturation transfer experiments or line shape simulations of the ^{1}H and ^{13}C-NMR spectra. Due to the high sensitivity of the spectral line shape to the migration pathway, it became possible to specify the NCS shift in tropyl isothiocyanate **III** as an unambiguously random process (1.23), see Chapter 1, Section 6.3.

5-(1-X-4-nitrophenylazo)-1,2,3,4,5-penta(methoxycarbonyl)cyclopentadienes exist as azo forms **IVa, b** in solution. Using the dynamic ^{1}H-NMR method, one can observe fast and reversible intramolecular migrations of the arylazo group along the cyclopentadiene ring proceeding by the randomization mechanism via formation of the diazonium salt **IIIc** [4]

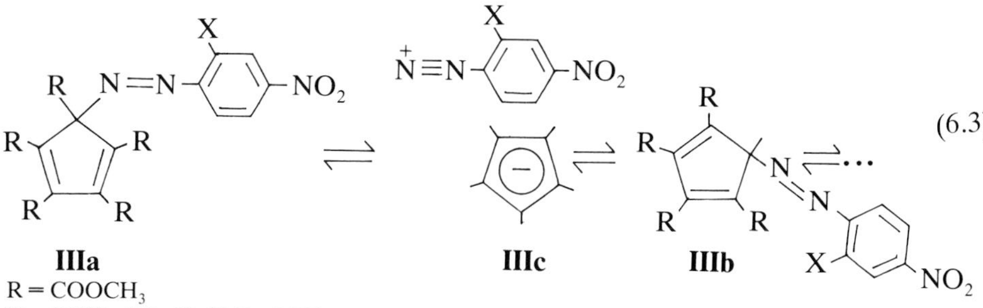

$$(6.3)$$

IIIa   **IIIc**   **IIIb**

R = COOCH$_3$
X = OCH$_3$, CH$_3$, H, Cl, Br, I, NO$_2$

The propensity of compounds **III** to rearrange in accordance with the dissociative mechanism (6.3) is manifested in the strong dependence of their molecular structure in the crystalline state on electron-acceptor properties of the substituent X in the arylazo migrant. So, according to an X-ray structural investigation [5], the 2,4-dinitro derivative exists in crystal as an azo form (**IIIa**, X = NO$_2$) while the 4-nitro derivative has the ionic structure of a diazonium salt (**IIIc**, X = H).

An examination of the deuterium kinetic isotopic effects at the β- and γ-carbons of α-methylphenyl sulfoxide **IV** has led to the conclusion that the rearrangement (6.4) is highly dissociative with the ion-pair intermediate **IVc** being included [6].

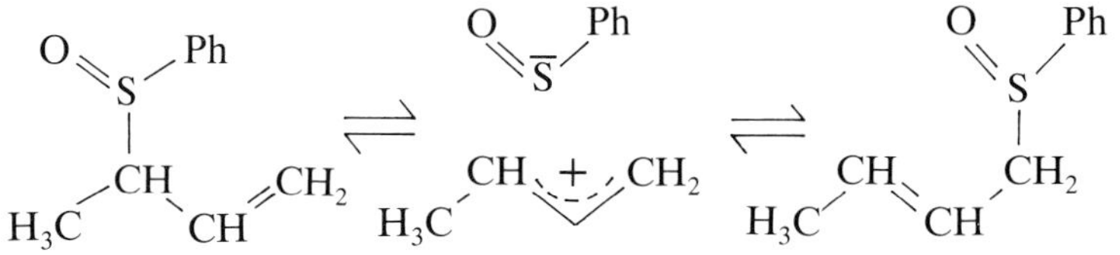

$$(6.4)$$

IVa   **IVc**   **IVb**

The heterolytic dissociative mechanism may eventually be employed in the design of tautomeric systems with such migrants (CR$_1$R$_2$R$_3$, POR$_1$R$_2$ and others) in regard to which the concerted associative mechanism is connected with rigorous steric demands on the molecular structure. Factors favoring the mechanism (6.1) would have to be taken into consideration. The stability of the cation MR$_n^+$ can be achieved through a selection of electron-donating substituents R. The stabilization of the anion in the ionic pair **Ic** is facilitated by electronegative heteroatoms X, Y. The most important and subtle point in the procedure is the correct choice of a solvent.

A distinctive feature of the (6.1) mechanism is actually its multicomponent character. Indeed, only because of the specific solvation of the initial compounds and the ion pair, can there occur the heterolysis of bonds with the migrant in the former and the appropriate stabilization of the latter, though a solvent is not explicitly present in the equations of reactions (6.1) and (6.2). Moreover, it is precisely a stabilization of the tight, but not of the solvent-divided, ion pair which has to be established. In other words, solvation must be achieved in the outer, but by no means in the internal, coordination sphere.

Theoretical approaches that would allow calculation and prediction of such subtle solvation effects are as yet in the process of development; however, a considerable amount of experimental data concerning the influence of the solvent on the structure and properties of ion pairs has already been gathered [7, 8]. The use of these data makes it possible to effectively choose an environment favoring the dissociation mechanism (6.1). Thus, employing organic solvents in the reactions of monosubstituted phosphates has led to the detection of phosphoryl transfer reactions through a facile dissociative decomposition to give a metaphosphate intermediate [9].

2. Homolytic Dissociative Mechanism

When a homolytic — as opposed to a heterolytic (as in (6.1)) — rupture of a bond with the migrating center is energetically more favorable, the scheme of a tautomeric reaction may be represented by the following dissociation-recombination mechanism

$$(6.5)$$

In this mechanism, the intermediate **Vc** is a stabilized radical pair susceptible to recombination into either of the alternative tautomeric forms with approximately equal probability. The favored direction for the rearrangement is determined by the thermodynamic stabilities of the species **Ia**, **Ib**. If the migrant MR_n is linked with the rest of the molecular framework by more than one σ-bond, then the intermediate **Vc** corresponds to a biradical and the Scheme (6.5) in this case describes a biradical route for the 1,*j*-sigmatropic shift, which is typical of some circumambulatory rearrangements [10].

The effects of chemically-induced dynamic polarization of nuclei observed in the NMR spectra [11] have shown the mechanism (6.5) to be operative in allyl systems of the following type:

$$
\begin{array}{ccc}
\underset{\textbf{VIa}}{\overset{\displaystyle X}{\underset{\displaystyle H_2C}{\overset{\displaystyle \diagdown CH_2}{\diagup}}}\,\overset{\displaystyle CH_2}{\underset{\displaystyle CH}{\diagup}}}
&\rightleftharpoons&
\underset{\textbf{VIc}}{\overset{\displaystyle X}{\underset{\displaystyle H_2C}{\overset{\displaystyle CH_2}{}}}\,CH_2\,CH}
&\rightleftharpoons&
\underset{\textbf{VIb}}{\overset{\displaystyle X}{\underset{\displaystyle CH_2}{\overset{\displaystyle CH_2}{}}}\,CH_2\,CH}
\end{array}
\qquad (6.6)
$$

Some other examples of the dissociation–recombination mechanism may be found among ylide rearrangements similar to those of Stevens, Meisenheimer and Wittig [12] proceeding in accordance with the mechanism of a 1,2-shift of the migrating group. Usually, the (6.5)-type radical mechanism is interfered with by other competing mechanisms. For instance, the Wittig rearrangement of benzohydryl ethers,

$$
\begin{array}{ccc}
\underset{\textbf{VIIa}}{\overset{\displaystyle Ar \quad\; R}{\underset{Ar'\; Li^+}{C-O}}}
&\rightleftharpoons&
\underset{\textbf{VIIc}}{\overset{\displaystyle \dot{R}}{\underset{Ar'\; Li^+}{\overset{Ar}{C}-O^-}}}
&\rightleftharpoons&
\underset{\textbf{VIIb}}{\overset{\displaystyle R}{\underset{Ar'}{Ar-C-O^-}}\; Li^+}
\end{array}
\qquad (6.7)
$$

which follows mechanism (6.7) in tetrahydrofuran (for $R = C_6H_5$), drastically changes its mechanism when diethyl ether is used as a solvent. In the case of an allyl-type migrant ($R = CH_2-CH=CH_2$), the mechanism (6.7) competes with the concerted 1,2-shift, and the vinyl ethers (**VII**, $R = -CH=CH_2$) rearrange, probably by the mechanism (6.1) [13, 14].

This example shows what a delicate balance is required regarding the structural factors in the system (6.1) and the parameters of the medium in order to realize a pure radical dissociation–recombination mechanism of intramolecular tautomeric migrations. The chief structural modifications should be intended to stabilize radical moieties in the radical pairs. Exactly this purpose is served when a heteroatom X is introduced into the migrating group CH_2X in compounds **VI**. Another factor of crucial importance is the choice of a solvent which has to ensure an 'in-cage' lifetime of the radical pair sufficient for a recombination to occur.

The radical mechanisms of the type (6.5) play an important part in many important photochemical processes.

3. Ion-radical Mechanism

This (still hypothetical) mechanism is a combination of dissociative and associative steps. In the preliminary activation stage it includes an intramolecular redox reaction leading to an ion—radical pair.

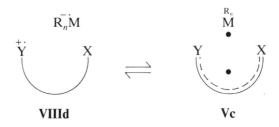

$$(6.8)$$

The probability of realizing an electron-transfer-catalyzed reaction of this type increases with:

(a)   the increase in the oxidative ability (lowering of the ionization potential) of the nucleophilic centers X, Y; and

(b)   the enhancement of the electron affinity of the migrant, its ability to stabilize the anion-radical form.

The latter ability is most clearly exhibited by aromatic nitro-compounds for which an $S_{RN}1$ nucleophilic substitution mechanism of Russel and Bunnett is known [15]. For this reason, the aromatic nucleophilic substitution reactions may serve as prototypes for the intramolecular tautomeric system (6.8). In case the reaction (6.8) exactly follows the $S_{RN}1$ mechanism, a dissociation stage preceding the formation of the major intermediate **VIIIc** enters the reaction scheme.

The structural factors facilitating the realization of the stages of intramolecular electron transfer in donor—acceptor interactions have been analyzed in detail [16—18]. A special role belongs in these steps to the geometric deformation of the initial compound **Ia** accompanied by a reorganization of the solvate shell. This ensures achievement of the structural region which is isoenergetic and isostructural with the **VIIId** system with the transferred electron. A considerable overlap of frontier orbitals of the donor (Y) and acceptor (MR$_n$) centers in the reaction zone is the second important condition for the occurence of a thermal rearrangement with sufficiently low activation energy. Semiempirical calculations [19] show that the optimal route for a nucleophilic attack on the aryl substrate leading to the formation of an ion—radical pair is different from the standard pathway of the reaction giving the σ-complex. Thus, although both the concerted tautomeric rearrangements by the mechanism of aromatic nucleophilic substitution (Chapter 2) and the hypothetical mechanism (6.8) are passing through identical transition states (intermediates) **VIIIc**, the pathways leading to them are different. This means that the structural requirements of initial compounds that would guarantee sufficiently low activation barriers of tautomeric migrations are also different.

4. Photo-initiated Carbonotropic Rearrangements

Radical and ion—radical mechanisms considered in Sections 2 and 3 are relatively rare in the case of rearrangements in the ground electron states of the molecules because the homolysis of the bond formed by a migrant requires considerable energy expenditure. On the other hand, in electron-excited states where a molecule possesses redundant energy of the absorbed light quantum, the dissociative mechanisms of rearrangement are much more frequent. Photoinitiation of rearrangement reactions also creates fundamentally new possibilities for the realization of the associative rearrangement mechanisms. This is due to the fact that the transition of the molecule, as a result of photoexciation, to an energetically higher electronic state causes a significant redistribution of the electron density and a change in the orders of the bonds between atoms and activates rotations about double bonds, which are hindered in the ground state. This provides additional possibilities for a reorganization of the geometry of the system, and for adapting it to the requirements of the transition state of the reaction. Furthermore, the energy imparted to the molecule by photoexcitation is known to be higher than the energy of thermal initiation, which may lead to the occurence of strongly non-degenerate conversions and to the formation of isomers that are energetically 'richer' than the starting compounds.

4.1. *Photochromic Transformations*

In the light of the problem of tautomerism, which is central to this book, reversible photorearrangements of the following type are extremely interesting:

$$A \underset{\Delta(h\nu_2)}{\overset{h\nu_1}{\rightleftarrows}} B. \tag{6.9}$$

Here the forward and reverse reactions take place in different electronic states. The most important is the case in which the conversion A → B proceeds upon photoexcitation while B → A occurs as a dark reaction in the electronic ground state. Such rearrangements are characterized by the phenomenon of photochromism [20, 21].

Fig. 6.1. shows the basic scheme of photochromic transformation exemplified by a proton transfer reaction [22].

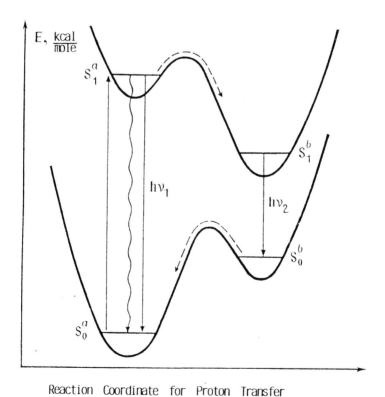

$$\tag{6.10}$$

Reaction Coordinate for Proton Transfer

Fig. 6.1. Typical scheme of the mechanism of a photochromic reaction in the interconversion (6.10).

In the electronic ground state, the energy barrier for the reaction **IXa** → **IXb** is large, due to a considerable energy difference between the stable **IXa** and unstable **IXb** isomers.

In the excited singlet state, however, the PES section of the system (6.10) along the coordinate of the migrating group transfer reaction has a different shape. Upon absorption of (hv_1) in the long-wave-band region of the electronic spectrum of **IXa** (340 nm), the molecule comes to be in the vibrationally excited region of the first single state of **IXb**, and on emission of the quantum hv_2, by which the presence of the tautomer **IXb** can be established, or through a radiationless transition, this molecule, after overcoming an insignificant activation barrier $\Delta E^{\neq}_{a \to b}$, returns to the stable form **IXa**. An analogous mechanism with only singlet excited states is, apparently, operative also in the photoinitiated reaction of the O ⇌ N transfer of the acetyl group [23].

$$\text{(6.11)}$$

Xa   **Xb**

In frozen solutions of the compound **Xa** at 77 K, irradiation in the spectral region of 365 nm gives rise to two concurrent processes; namely, fluorescence at 520 nm with an anomalously large Stokes shift, and the formation of a colored short-lived isomeric form **Xb**.

In general, the mechanisms of photochromic reactions may be much more complex, including triplet excited states, intermediate structural isomerizations, etc. Let us now consider the principal data on photoinitiated reactions associated with the transfers of carbon-containing migrants giving particular attention to the photochromic transformations.

4.2. *Photoacylotropic Rearrangements*

The light-induced acyl group migrations have been studied quite extensively. These reactions may be divided into two types: the first type comprises rearrangements in which the energy of the excitation is chiefly spent on breaking the bond between the acyl migrant and the rest of the molecule and, in a similar manner to reaction (6.11), the acyl group transfer occurs in the excited electronic state.

In the photo-Fries rearrangement of aryl esters and their analogs, the limiting stage is the formation of the radical pair **XIc** [24, 25] in accordance with the Scheme (6.5).

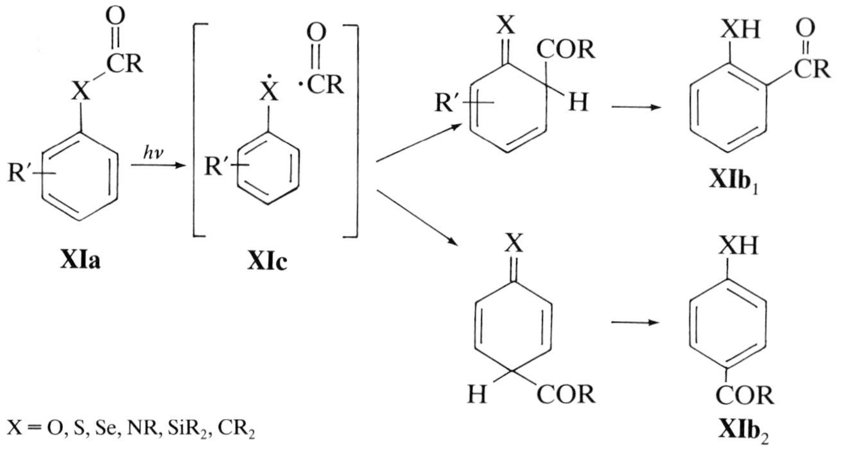

(6.12)

X = O, S, Se, NR, SiR$_2$, CR$_2$

The radical migration proceeds primarily within the solvent cage and the *ortho*-rearrangement usually serves as a predominant route. In order to avoid a cage escape, which leads to a considerable amount of concomitant Ar—XH products, various improvements in synthetic procedures, including variation of the solvent, two-phase photochemistry etc. were invoked [26].

An interesting example of the sterically enforced *meta*-photo-Fries rearrangement has been found [27].

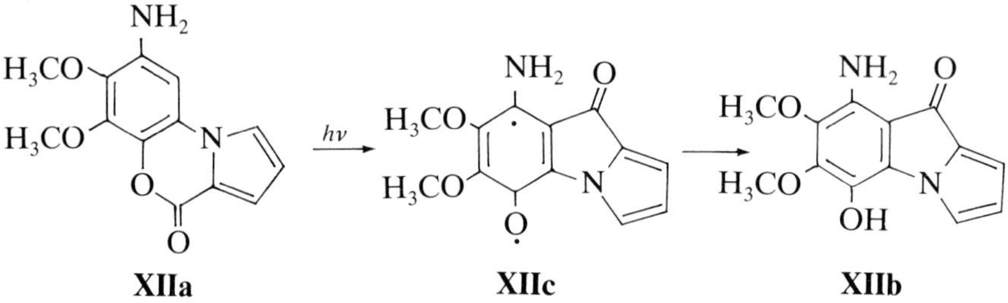

(6.13)

The reaction mechanism includes an intermediate formation of the biradical **XIIc**. A radical mechanism also forms the basis for 1,3-acyl photorearrangements of enamides.

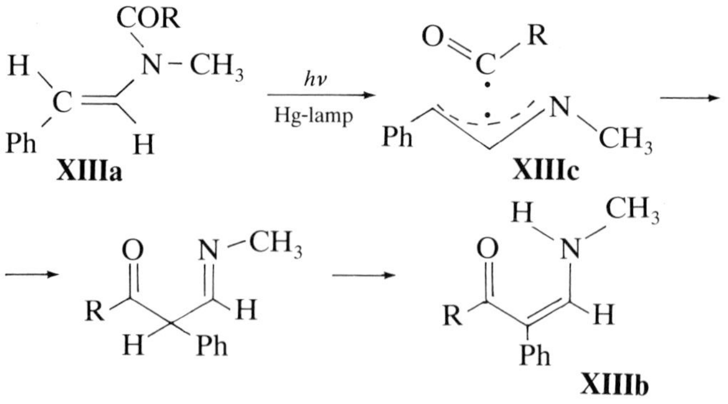

(6.14)

The formation of a radical pair **XIIIc** is, according to [28], achieved only in the higher (S_2 or T_2) electron-excited states since irradiation in the long-wave absorption band of **XIIIa** ($\lambda_{max} > 300$ nm) results merely in a $Z-E$-isomerization with respect to the double bond. Full light of a high pressure mercury lamp initiates the reaction (6.15) with the reaction product **XIIIb** being formed at an equal rate from both Z- and E-enamides of **XIIIa**. The study of reactions of isotopomeric compounds **XIIIa** has shown that this reaction proceeds more than to 90% by an intramolecular mechanism.

The radical mechanism is not, however, the only possible route for reactions of the 1,3-photorearrangements of acyl groups. For instance, intramolecular photo-rearrangements of acylated enoles which model the photo-Fries rearrangements, proceed via singlet excited states of the molecule with a high quantum yield, achieved only in the case of those compounds in which the excitation energy is not localized in the aryl group of the migrant (R = Aryl) [29].

For other compounds of this series, the α-cleavage mechanism with the formation of a radical pair retains its importance in the forward as well as reverse reactions [30, 31].

A photoinitiated thermally reversible intramolecular transfer of acyl groups corresponding to the photochromic reaction (6.9) was detected in the series of *peri*-acyloxy-9, 10- and 1,4-anthraquinones [32, 33]. Upon irradiation of ethanolic solutions of these compounds frozen at 77 K, or under impulse photolysis at room temperature, spectra are observed of photoisomeric *ana*-quinonoid compounds **XIVb** which are the products of a light-induced transfer of acyl groups. The reverse reaction **XIVb** → **XIVa** does not occur at 77 K under conditions of photochemical initiation (irradiation at 400 nm), but can be achieved only by heating the photoisomer. Type **XIV** compounds are quite stable against destruction and the cycle (6.15) may be reproduced many times.

(6.15)

Table 6.I contains some spectral characteristics of photochromic compounds **XIV** and rate constants of the dark reaction of O,O'-acyl rearrangements. The effect of the substituents R in an acyl group on the rate of a thermal rearrangement is qualitatively analogous to the effect found earlier in regard to reactions of acylotropic rearrangements of acetylacetone enolacylates and O-acyl derivatives of tropolone (Chapter 2): the electron-donating groups lower, while the electron-accepting groups raise, the rate of migrations.

Table 6.I. Maxima of long-wave absorption bands in the electronic spectra of 1-acyloxy-9, 10-anthraquinones **XIVa** and their photoisomers **XIVb** (in ethanol, 77 K) and rate constants of thermal isomerization **XIYb** ∞ **XIYa** (in benzene, 285 K) according to data [33].

No.	Migratory group C(O)R	λ_{max}, nm		$k_{12} \times 10^{-5}$, s^{-1}
		XIVa	**XIVb**	
1	—CH$_3$	374	526	1.5
2	—CH$_2$Ph	369	526	2.8
3	—(CH$_2$)$_{10}$CH$_3$	370	532	2.0
4	—OC$_2$H$_5$	368	526	0.03
5	—C$_6$F$_5$	366	530	2.6
6	—2,4-C$_6$H$_3$Cl$_2$	370	532	2.0
7	—4-C$_6$H$_4$NO$_2$	369	532	1.4
8	—3-C$_6$H$_4$F	372	532	8.2
9	—2-C$_6$H$_4$OCH$_3$	370	532	0.67
10	—C$_6$H$_5$	369	526	0.46
11	—4-C$_6$H$_4$F	368	532	0.42
12	—4-C$_6$H$_4$OCH$_3$	372	526	0.18

A theoretical analysis of the mechanism of the photoreaction (6.15), performed on the basis of state correlation diagrams, points to an adiabatic mechanism of the acyl group migration via the lowest excited single state [34]. In this case, as is demonstrated by a calculation of electron distributions, it becomes probable that the acyl group migrates in the form of the acyl cation RCO$^+$ in conformity with the heterolytic dissociative mechanism (6.1).

The second type of photoacylotropic rearrangements comprises those reactions in which photoexcitation energy is spent on the activation of structural isomerizations, not associated with the acyl group transfer itself, but providing favorable steric conditions, absent in the initial structure, for the subsequent stage of the thermal migration.

The rearrangement of O-acylisoimide derivatives **XVa** considered above (Chapter 2, Section 2.1) may serve as an example of the photoinduced acyl 1,3-transfers. The O-acylisoimides **XV** (R = Ar) obtained by acyloxylation of nitriles exclusively in the form of the Z-isomer **XVa** show no capacity for rearrangement to the stable amide form **XVb** [35, 36]. This is explained by an unfavored orientation of the interacting orbitals of the electron lone pair in nitrogen and the carbonyl group [36, 37].

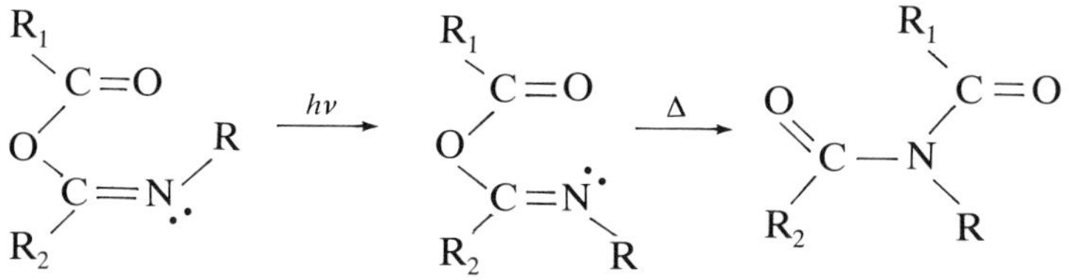

$$(6.16)$$

XVa(*Z*)   **XVc** (*E*)   **XVb**

The *Z*-configuration **XVa** is the reason for the extremely low rates of formation of the amide isomer also in the case of derivatives of oximes (R = OCH$_3$) and hydrazones (R = NCH$_3$Ar) [38]. However, the rearrangement **XVa** ⇌ **XVb** proceeds rapidly and completely upon UV irradiation of solutions of **XVa**, leading to the intermediate formation of the *E*-isomer **XVc**, whose structure meets the steric requirements of the 1,3-acyl transfer reaction. An analogous method for the initiation of acyl transfers has been used by Hartke and coworkers [39—42] for 1,5-rearrangements of *O*-acylenols of cyclic 1,3-diketones.

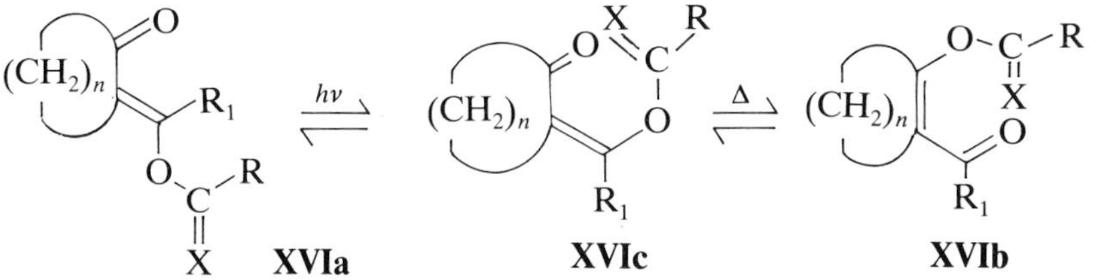

$$(6.17)$$

XVIa   **XVIc**   **XVIb**

X = O, S
R = Alk, Ar, N(CH$_3$)$_2$; *n* = 2—5

The same basic reaction scheme was found [23, 43, 44] to operate in photo-initiated acylotropic rearrangements of *N*-acylaminomethylene derivatives of 3(2*H*)benzo(*b*)thiophenones and their analogs **XVIIa**.

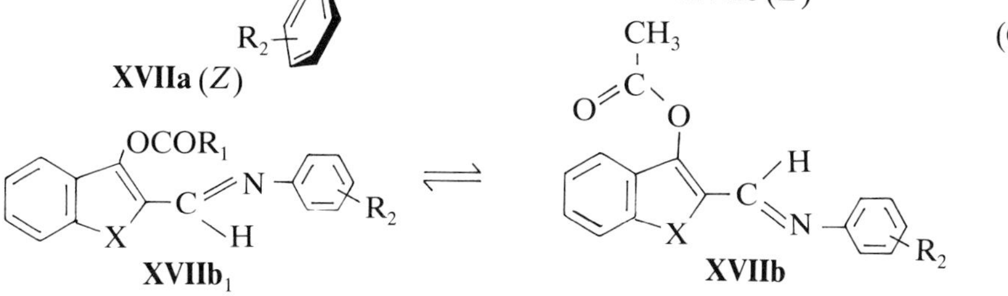

$$(6.18)$$

XVIIa (*Z*)   **XVIIc** (*E*)

XVIIb$_1$ **XVIIb**

X = S, Se; R$_1$ = CH$_3$, Ph, OCH$_3$, CH(CF$_3$)$_2$; R$_2$ = H, Hal, CH$_3$, NO$_2$.

The *Z*-configuration **XVIIa** of *N*-acyl derivatives of arylaminomethylene compounds, obtained by acylation by different methods, has been verified spectroscopically and by direct X-ray measurements.

A characteristic structural feature of compounds **XVIIa** is the *s-cis* conformation of the aryl group with respect to the C—N bond where the bulky heteroatom and the C(N) atom have been brought together to a distance less than the van der Waals contacts. The N—C(O)R$_1$ bonds in all compounds studied are elongated in comparison with the typical amide bonds (1.325 Å). Especially important is the sizeable lengthening of the C=C bonds relative to which there occurs geometric isomerization in the excited and ground electron states. Table 6.II lists some parameters of type **XVIIa** compounds derived from X-ray structural investigations.

In view of the unsuitability of the structure of the compounds **XVIIa** (*Z*) for the steric requirements of a 1,5-acyl transfer reaction, the thermal rearrangement **XVIIa** → **XVIIb** is not observed even at high temperatures. At the same time, irradiation of solutions of compounds **XVIIa** at room temperature in the region of the long-wavelength absorption band (400—430 nm) with a mercury lamp or sunlight leads to the rapid formation of the *O*-acyl isomer **XVIIb** with a long-wavelength absorption band in the region of 340—360 nm.

Fig. 6.2 shows some spectral changes characterizing the forward and reverse reactions (6.18).

The compounds **XVIIb** are the only products of the photoreaction and are isolated preparatively in a yield of up to 95%. Their structure is confirmed by spectral data and X-ray analysis. They can also be obtained preparatively, for example, from 3-acyloxybenzo(*b*)thiophene-2-aldehyde and the corresponding amines. The photoreaction is insensitive to triplet-state sensitizers. It is also monomolecular in nature and is described by exponential kinetic equations. The reaction constants under the experimental conditions have a value of 0.1—1 sec^{-1}. The quantum yields are high and amount to approximately 0.5—0.7. The rate of the photoreaction is practically independent of the solvent and of the solution concentration, indicating its intramolecular nature.

The intermediate formation of the *E*-isomer **XVIIc**, which precedes the step of 1,5-acyl transfer, is recorded by the flash photolysis method. The lifetime of **XVIIc** (X = S, R$_1$ = CH$_3$) has a value of 0.05 s; λ_{max} = 450 nm. The compound **XVIIc** rapidly relaxes to the isomer **XVIIb** with an *O*-transferred acyl group. The only possible mechanism is a non-adiabatic transfer mechanism, because of the significantly higher energy of the first excited singlet state of **XVIIb**, as compared with the excited singlet level of **XVIIc** (Fig. 6.2).

Irradiation of solutions of compounds **XVIIb** at the absorption band or with the light of a mercury lamp does not cause a reverse reaction. However, upon heating solutions of compounds **XVIIb** in high-boiling inert solvents to 100—180°C, a reverse dark O → N acyl-group transfer reaction is observed, which can be readily recorded by means of electronic absorption spectra. The reaction products may

Table 6.II   Structural Parameters of Compounds **XVII**, **XVIII** as determined by X-ray investigations [45—47].

No.	X	Z	R_2	R	$l(C_1=Z)$, Å	$l(Z-N)$, Å	$l(N-COCH_3)$, Å	$\varphi°$	Photochromism
1	S	CH	$m\text{-}NO_2$	CH_3	1.326	1.387	1.425	82.2	+
2	S	CH	H	$CH(CF_3)_2$	1.341	1.385	1.410	85.7	+
3	O	CH	H	CH_3	1.336	1.368	1.415	87.2	−
4	$-^a$	CH	H	CH_3	1.328	1.391	1.392	76.0	−
5	Se	CH	H	CH_3	1.343	1.380	1.406	87.6	+
6	Se	N	H	CH_3	1.271	1.368	1.382	87.1	+
7	S	N	H	CH_3	1.288	1.352	1.395	87.0	+
8	NCH_3	CH	H	H	1.338	1.405	1.389	41.9	−

^{a} 1-Benzoyl-2-(*N*-acetyl-*N*-phenylamino)ethylene possesses the same configuration relative to the C=C bond and conformation as do compounds **XVIIa**.

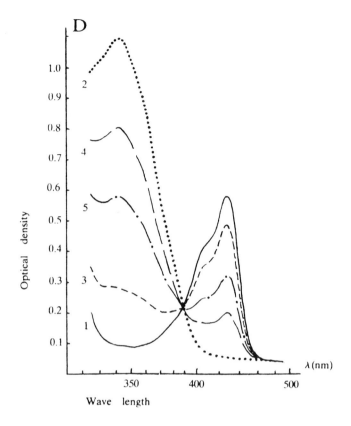

Wave   length

Fig. 6.2. Changes in the absorption spectrum of the compound **XVIIa** (R_1 = CH_3, R_2 = 3—NO_2) in hexachlorobutadiene (C_0 = 5 × 10^{-5} M) with consecutive photochemical and thermal actions according to data from [44]: (1) Starting solution; (2) Irradiation at λ = 436 nm for 2 min; (3) Heating of the irradiated solution at 160°C for 2 min and cooling to 20°C; (4) Irradiation for 30 s; (5) Heating at 160°C for 1 min and cooling to 200°C.

be isolated in preparative yields, indicating the complete reversibility of the cycle **XVIIa** ⇌ **XVIIb**, including a forward photoreaction and a reverse dark reaction. The reaction cycle of N → O and O → N transfer can be repeated many times.

The values of the free activation energy of the dark reaction **XVIIb** → **XVIIa** are approximately 25 kcal/mol [23, 48]. The metastable isomers **XVIIb** can be stored for years in the crystalline state and for weeks in solution. The reaction is, however, subject to acid catalysis. At the ratio of **XVIIb** to CCl_3COOH = 1 : 1, the isomerization **XVIIb** → **XVIIa** is complete within 15—20 minutes at room temperature, which corresponds to a significant lowering (to approximately 10 kcal/ mol) of the activation energy barrier. The rate of the reverse reaction is increased by the electron-donor substituents R_2. The thermal effect of the reverse reaction (6.18) has been determined calorimetrically for compound **XVIIa** (X = S, R = *p*-OCH_3). It totals 5.2 and 5.7 kcal/mol in toluene and acetonitrile solutions,

respectively. Spectral and kinetical data obtained make it possible to construct a
scheme of energy levels of the photochromic system (6.18) shown in Fig. 6.3.

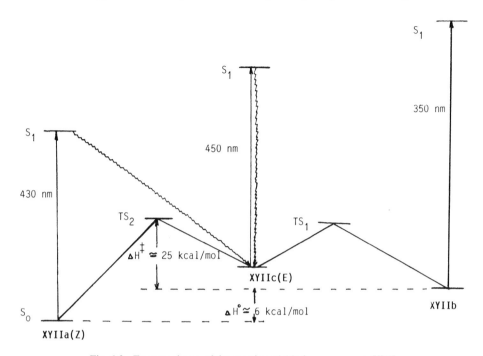

Fig. 6.3. Energy scheme of the reaction (6.18) for compounds **XVII**.

By varying structural fragments of type **XVII** compounds it is possible to
substantially change the relative order of the energy levels corresponding to
individual isomers, thereby changing the character of reactive transformations
(6.18). Acylated heterocyclic aminoenones can be classified by their capacities for
photo- and thermal rearrangements and isomerizations into three groups. The
principle for such division is sketched in Fig. 6.4.

The first group comprises compounds whose energy level order is represented
by the scheme in Fig. 6.3. There are the *N*-acyl derivatives of *N*-arylaminomethy-
lene-3(2*H*)benzo(*b*)thiophenone. The order of the energy levels

$$E(A - Z) < E(A - E) > E(B) \qquad (6.19)$$

constitutes the condition for the occurence of a type (6.18) photochromic reaction.

The same order of the energy levels is retained in acyl hydrazones **XVIIIa**.
These compounds are also capable of an N $\rightarrow$ O phototransfer of the acyl groups
accompanied by a reverse dark reaction whose much lower activation barrier is
the main point of difference from compounds **XVII**.

Furthermore, upon initiation of transformations (6.20) by irradiating the solutions of compounds **XVIII** (X = S, Se) with light at a wavelength $\lambda_1 = 405$ nm or

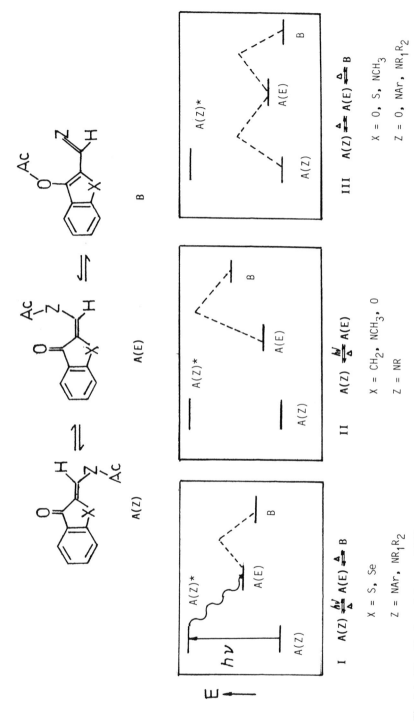

Fig. 6.4.   Classification of the type **XVII** compounds in three groups by the relative positions of the ground state energy levels of the Z, E isomers of the N-acylaminomethylene and O-acyleneimine (B) forms.

λ_2 = 436 nm, the formation of a mixture of *Z*- and *E*-isomers **XVIIIb**, **d** is observed, which are capable of the interconversions characteristic for azobenzenes. Irradiation of the solution with light at a wavelength of 365 nm leads to an accumulation of the *Z*-isomer **XVIIIb** which is completely converted to the form **XVIIId** in 15—20 minutes at room temperature. The rate constants of the photoreaction **XVIIIa** → **XVIIIb, d** have values of the order of 10^{-2} s^{-1}.

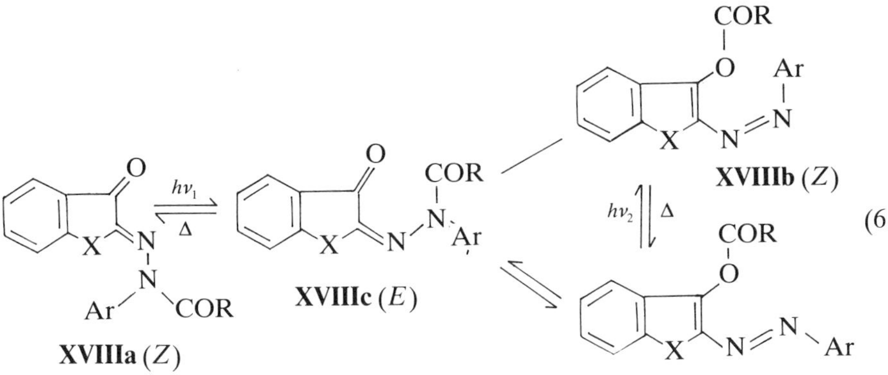

XVIIIa (*Z*) XVIIIc (*E*) XVIIIb (*Z*) XVIIId (*E*) (6.20)

The second group consists of compounds which do not exhibit acylotropic migrations. This is due to strong destabilization of the *O*-acyl enamine isomer **XVIIb**. Hence, the *E*-isomer **XVIIc** formed by irradiation of the *Z*-isomer **XVIIa** does not suffer subsequent thermal rearrangement with the N → O acyl migration, and the reaction (6.18) stops at the stage of the $Z \rightleftharpoons E$ isomerization. Such a character of the photoreaction is particularly typical of the indoxyl derivatives (X = NCH_3). In view of considerable steric hindrances caused by the bulky heterocyclic methylimine group, the molecules of these compounds have a substantially coplanar structure as has been demonstrated by an X-ray diffraction study [47, 49].

Interestingly, not the *Z* but the *E*-diastereomeric structure is favored in the case of the *N*-formyl derivative **XX**.

XIX

XX

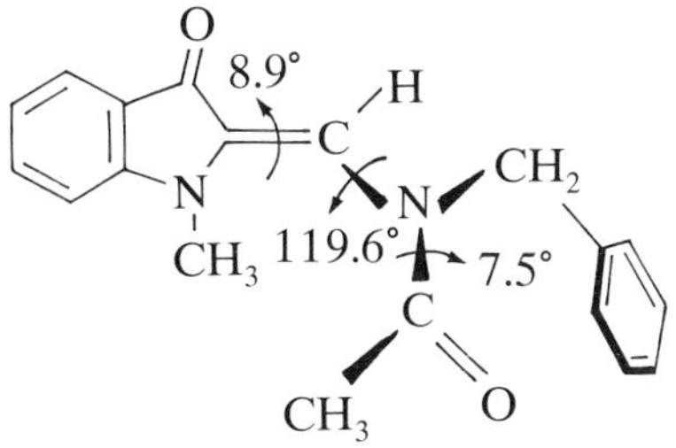

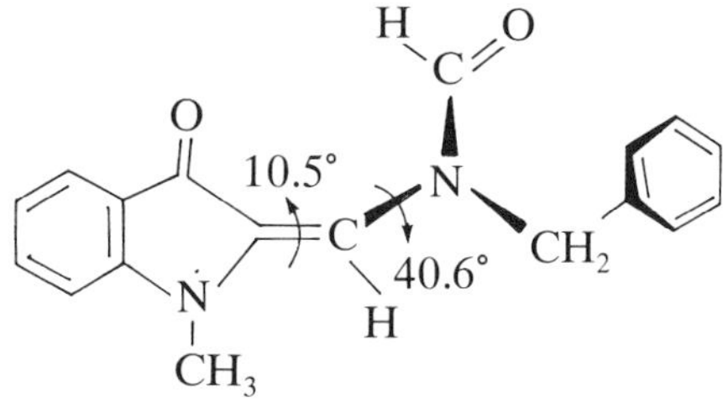

Experimental data as well as semiempirical quantum mechanical calculations [44, 48] indicate that the relative destabilization of *O*-acyl isomers **XVIIb** as contrasted with the *N*-acyl forms **XVIIa** grows in the following order

$$X = CH_2, O, S, Se, NCH_3 \tag{6.21}$$

Regarding the compounds classified as the third group, the ground states of the isomeric forms **XVIIa** and **XVIIb** are energetically so close that, in addition to photochemical, the purely thermal N $\rightleftharpoons$ O acylotropic rearrangements become possible. In this case, the energy level of the *E*-isomer **XVIIc** is usually close to the energy levels of **XVIIa** and **XVIIb**, and activation barriers of the thermal forward and reverse reactions do not exceed the value of 18—22 kcal/mol. Compounds with these properties include acetyl derivatives of heterocyclic hydroxyaldehydes, of their *N,N*-disubstituted hydrazones, as well as compounds **XVII** (R$_1$ = OCH$_3$) [48—50]. In the last case, sufficiently fast thermal rearrangements **XVIIa** $\rightleftharpoons$ **XVIIb** occur in solution at room temperature even without catalysts.

4.3. *Photoacylotropic Compounds as Abiotic Photochemical Solar Energy Storage Systems*

Photochromic conversions (6.18) of compounds **XVIIa** belonging to the first group and possessing a system of energy levels shown in Figs. 6.3 and 6.4a provide for a possible transformation and storage of light, including solar energy in the form of strain energy of the metastable isomer **XVIIb**. A part of the energy stored may be released as the thermal energy of the reverse dark reaction. Most promising in regard to solar energy storage are the *N*-aryl derivatives of benzothiophenones and benzoselenophenones **XVII** (X = S, Se). The properties of these compounds meet the necessary requirements of the photochemical systems needed to convert the Sun's energy into low-temperature thermal energy [48, 51—54], namely

(1) The initial substance (**XVIIa**) possesses intensive absorption band with the wavelength λ_{max} = 425 nm, which ensures fairly effective utilization of the solar radiation. The portion of the solar spectrum useful for organic photo-chemistry is known to lie in the ultraviolet and visible range of 300—700 nm which corresponds to about 50% of the total energy falling on the earth.

(2) The barrier of the reverse thermal reaction **XVIIa** $\rightleftharpoons$ **XVIIb** amounts to 25—30 kcal/mol, permitting storage of energy in the photoisomer **XVIIb** over several weeks at ambient temperature.

(3) The reverse reaction **XVIIb** $\rightarrow$ **XVIIa** releasing energy accumulated in the

system is susceptible to catalytic acceleration and thermal initiation, thus allowing this process to be controlled.

(4) While the initial substance **XVIIa** is colored, the photoisomer **XVIIb** is colorless (λ_{max} = 350 nm). Therefore, the reverse reaction **XVIIb** → **XVIIa** is practically not subject to photochemical initiation.

(5) The forward intramolecular photoreaction **XVIIa** → **XVIIb** does not need sensitization; there is no luminiscence of solutions of **XVIIa** at room temperature, which provides for a high quantum yield of the photoreaction (φ = 0.6—0.7).

(6) The system (6.18) can withstand, without noticeable destruction, several dozens of cycles (irradiation with high pressure Hg lamp, thermal initiation of the reverse reaction by heating to 140°C).

(7) Compounds **XVII** are nontoxic. Products of industrial organic synthesis serve as starting materials for the synthesis of compounds of the **XVII** benzo(b)thiophene series.

Table 6.III lists main characteristics of the system **XVII** as exemplified by 2-(N-phenyl-N-acetylaminomethylene)-3($2H$)-benzo(b)thiophene as well as data on the best photochemical systems of Solar energy storage found so far, see [51—53]. These include derivatives of norbornadiene whose action is based on an intramolecular photochemical (2 + 2) cycloaddition reaction and N,N'-diacylindigo (Z—E-isomerizations). As the effectiveness of a photochemical system is defined by many parameters, special functions are employed for its determination [48, 51].

The Calvert factor is given by

$$Q = \frac{100\Delta H\varphi}{E_{h\nu}} \tag{6.22}$$

where ΔH is the heat of the dark reaction, φ is the quantum yield of the photoreaction, $E_{h\nu}$ is the energy of 1 einstein, i.e. 1 mol of photons of the light absorbed (usually for the maximum of the long-wave absorption band of the initial substance).

Clearly, Q is equivalent to an efficiency ratio and would equal 100% only on condition that $\Delta H = E_{h\nu}$ (with $\varphi = 1$), which is physically infeasible.

Unfortunately, the factor Q does not reflect some of the features of photochromic processes affecting the effectiveness of the use of the solar energy. Therefore, an additional coefficient η_A is introduced which takes into consideration the degree of the utilization of the solar radiation in an ideal photochromic substance fully absorbing the sun light in the region $\lambda \leqslant \lambda_{gr}$. The quantity η_A is the ratio of the energy of one einstein (i.e. of one mol of photons) with the wavelength of λ_{gr} multiplied by the number of einsteins in the solar spectrum

Table 6.III Solar Energy Storage Systems

$A \underset{\Delta}{\overset{h\nu}{\rightleftarrows}} B$	λ, nm	φ	ΔH, kcal/mol	$\tau^{1/2}$ (B)	Q, %	η_A	η_w, %
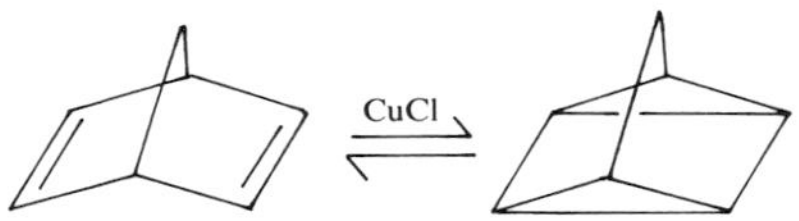 (quadricyclane ⇌ norbornadiene)	388	0.4	26.3	14h (140°)	14	0.015	0.21
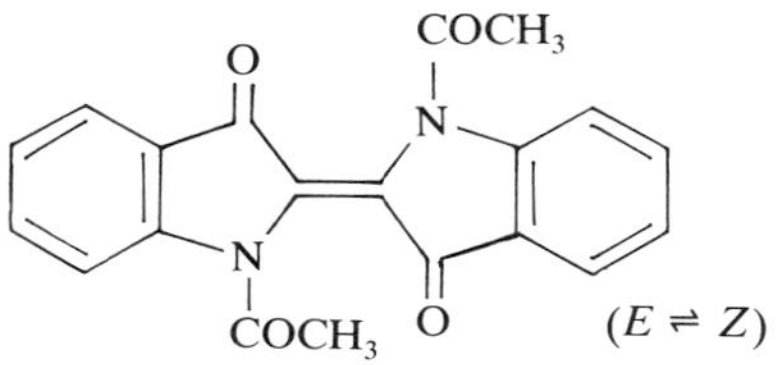 $\xrightarrow{\text{CuCl}}$ ($E \rightleftharpoons Z$)	610	0.2	7.2	0.05h (25°)	3.5	0.27	0.95
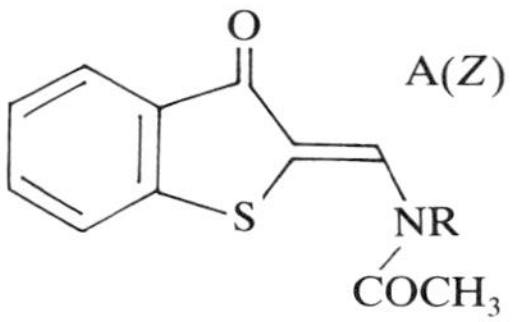 $A(Z) \rightleftarrows A(E) \rightleftarrows B$	460	0.6	5.7	1/2 year (25°)	6.3	0.11	0.7
$CO_2 + H_2O \xrightarrow{\text{Photosynthesis}} \frac{1}{6}C_6H_{12}O_6 + O_2$	700	0.1	111	2h (70°)	26	0.36	8.0

(observed near the surface of the earth) at $\lambda \leqslant \lambda_{gr}$ to the total energy of solar radiation

$$\eta_A = \frac{\displaystyle\int_0^{\lambda_{gr}} J_\lambda(\lambda/\lambda_{gr})\, d\lambda}{\displaystyle\int_0^\infty J_\lambda\, d\lambda} \tag{6.23}$$

where J_λ is the distribution of the power of solar radiation over the wavelengths.

The coefficient η_A depends only on the maximal wavelength in the absorption spectrum of λ_{gr} of the photoactive compound. Its magnitude is considerably less than unity for two reasons: (1) those photons with $\lambda > \lambda_{gr}$ are not absorbed by the substance, and (2) the energy of the photons with $\lambda > \lambda_{gr}$ is partly dissipated upon transitions from the higher excited states of the photoactive form. Compounds with the absorption limited to the UV-region have quite low values for the useful fraction of the incident energy η_A.

Taking account of (6.23), the efficiency of photochemical energy storage systems may be represented as follows

$$\eta_w = Q\eta_A\eta_{abs} \tag{6.24}$$

The coefficient η_{abs} depends on the absorption spectrum of the compound and such characteristics of the system as the concentration of the working solution, thickness of the layer in the collector etc. This value is very near to unity for most compounds considered in Table 6.III. As is evident from the data in this Table, the compounds **XVII** possess a number of advantages over other systems shown, such as absorption in the visible region of the spectrum, which is absent in norbornadienes, and which constitutes their major disadvantage, and a much greater kinetic stability of the photoisomeric form as compared to the derivatives of diacetylindigo. Besides, the Z-isomer of diacetylindigo is, unlike **XVIIb**, still capable of a considerable absorption in the visible region of the spectrum, which leads to initiation by sunlight of both the forward and reverse reactions (6.9).

Inclusion in the scheme of the reaction (6.18) of the additional stage of thermal acyl transfer has, however, the adverse consequence of substantially reducing the thermal effect of the reverse reaction (Fig. 6.3), which even without this is, in systems based on the $Z-E$ isomerization mechanism, much lower than in the systems associated with the formation of strained cage compounds. Further modification of the compounds **XVII** with a view to improving their capacity for serving as solar energy storage systems should be directed toward an enlargement of the energy gap between the isomers **XVIIa** and **XVIIb**.

4.4. *Photoinduced Aryl Rearrangements*

The rearrangement reaction of the labeled diazonium salt [55] may serve as an example of the reversible photoinduced 1,2-migration of an aryl group

$$
\text{XXIa} \quad \xrightarrow[\;]{h\nu} \quad \left[Ph^+ \;\; N_2\right] \rightleftharpoons {}^{15}N{\equiv}N{-} \quad \text{XXIb} \tag{6.25}
$$

XXIa   XXIc   XXIb

The rearrangement (6.25) takes place also in the ground state. The results of careful kinetic studies and ^{15}N experiments as well as secondary isotope effects are consistent with the initial formation of an ion-molecule pair **XXIc** from which the solvated phenyl cation and dinitrogen are formed [56]. Other photoinduced 1,2-rearrangements of aryl groups [26] include the isomerization of phenyl iso-cyanide to nitriles and of nitroarenes to aryl nitrites.

The photochemical Smiles rearrangement first reported for the 2,4-dinitrophe-nyl ether **XXIIa** and a series of *s*-triazinyl ethers [57] may be considered as a 1,4-aryl migration proceeding through a typical associative nucleophilic substitu-tion at the *ipso*-carbon atom of the migrating aryl group.

$$
\text{XXIIa} \quad \xrightarrow{h\nu} \quad \text{XXIIc} \quad \longrightarrow \quad \text{XXIIb} \tag{6.26}
$$

XXIIa   XXIIc   XXIIb

A different mechanism is operative in the case of the photo Smiles rearrange-ments of *p*-(nitrophenoxy)-*ω*-anilinoalkanes. This reaction was found [58, 59] to involve the formation of a radical-ion pair **XXIII** and a spiro-type Meisenheimer complex **XXIIId**, both ascertained by laser flash photolysis experiments. This mechanism closely corresponds to the Scheme (6.8) considered above.

The presence of a phenyl substituent at the nitrogen atom in **XXIIIa** is quite important as it ensures stability of the intermediate radical-ion pair. Thus, the photolysis of *β*-(*p*-nitrophenoxy)ethylamine, which is an analog of the compound **XXIIIa**, $n = 2$ where the *N*-phenyl group is substituted by a hydrogen atom, causes no Smiles rearrangement [60].

The products **XXIIIb** can also be obtained by the catalytic action of a strong base such as sodium hydride on dimethylformamide. Thus, both the photoinduced and thermal Smiles rearrangements of **XXIIIa** afford the same product.

Photoinduced rearrangements associated with the *O,O′*-migrations of aryl groups were detected and thoroughly studied in the series of derivatives of

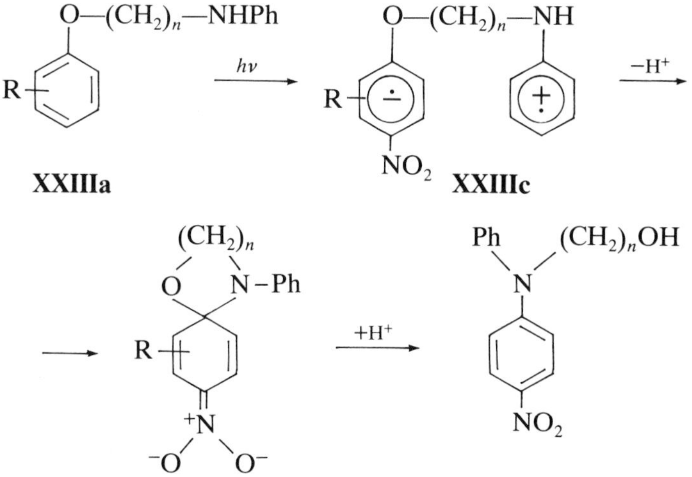

XXIIIa **XXIIIc** **XXIIId** **XXIIIb** (6.27) $n = 2–5$

1-aryloxy-9,10-anthraquinones **XXIV** and 6-aryloxy-5,12-naphthacenequinones **XXV** [61, 62].

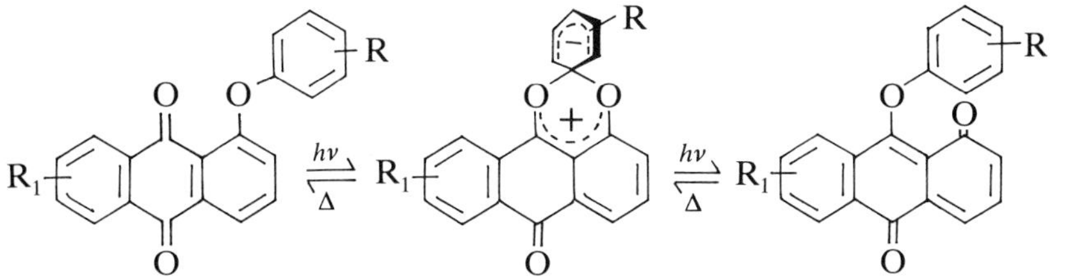

XXIVa **XXIVc** **XXIVb** (6.28)

XXVa **XXVc** **XXVb** (6.29)

These reactions belong to the photochromic type of transformations (6.9). The reverse reaction is initiated by heating or by visible light irradiation in the long wave absorption band of the photoisomers **XXIVb**, **XXVb**, and it is to be noted that the compounds **XXVb** exhibit a great kinetic stability ($k_{25} \simeq 10^{-8}$ s). Thus, the photoisomer 6-phenoxy-5,12-naphthacenequinone (**XXVb**, R_1, R = H) can be preparatively isolated from a benzene solution of the initial phenoxy derivative **XXVa** irradiated by UV light.

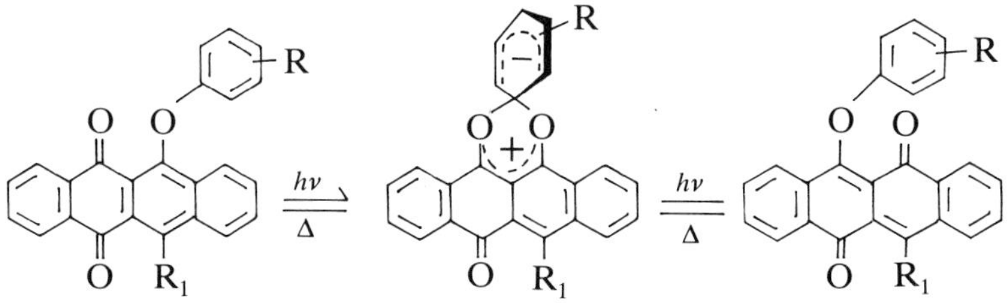

The mechanism of the photoreaction is thought [62, 63] to be based on associative intramolecular nucleophilic photosubstitution. However, the attempts to confirm by the ESR spectroscopy method the formation of radicals upon the irradiation of the initial *para*-quinones have been unsuccessful.

Isomerizations (6.28) and (6.29) can also be accomplished in a purely thermal manner in the presence of protonic and Lewis acids [64].

The migration of an aryl group under the action of UV-irradiation occurs much more readily than that of an acetyl group. So in isomerizations of 1-aryloxy-4-acetoxy-9,10-anthraquinone induced by UV light, the migration of the aryl group takes place in the excited state of **XXVIa** and **XXVIb**, whereas the responding shift of the acetyl group proceeds thermally [63].

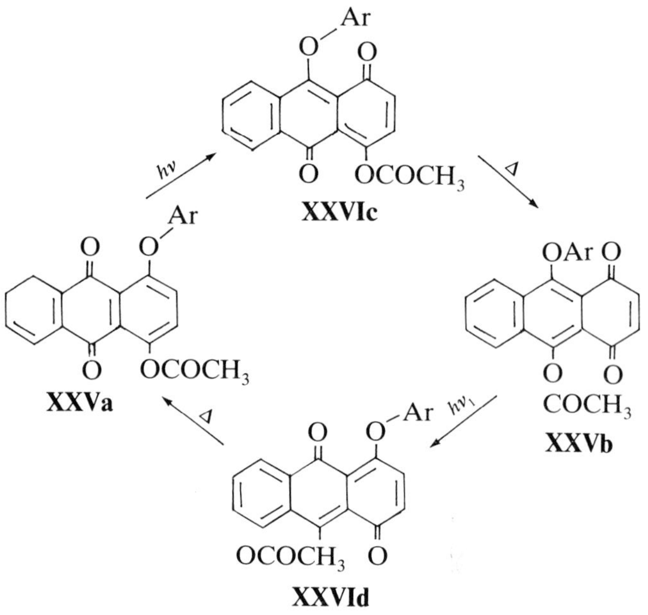

(6.30)

The photochromic systems (6.28) and (6.29) are of interest for practical application in view of the high kinetic stability of the photoisomers **XXIVb**, **XXVb** and a high photosensitivity of the initial *para*-quinones (not less than 10 cm^2/J at λ = 366 nm) which is retained also in solid polymeric matrices. Yet another valuable characteristic of the photochromic compounds **XXIV**, **XXV** is the big difference between the values of λ_{max} for the initial and the photocolored isomer (Table 6.IV).

Another example of the photoinduced migrations of aryl groups proceeding without the intermediate formation of radical or ion pairs is the rearrangements of 2-phenoxy-4,5-benzotropone [65].

Table 6.IV. Spectral Characterestics of Photochromic Aryloxyquinones **XXIV**, **XXV** according to data from [62, 63]

Structural Type	R	R_1	λ_{max}, nm before irradiation	λ_{max}, nm after irrad.
	H	H	364	480
	tert-Bu-p	2-OH	369	532
XXIV	H	2-NH$_2$	415	616
	tert-Bu-p	3-NH$_2$	415	434
	H	4-NH$_2$	478	581
	H	8-OPh	374	494
	H	H	400	480
	H	OCH$_3$	400	469
XXY	NHPh	H	496	558
	H	NHPh	396	476
	NO$_2$	H	394	479

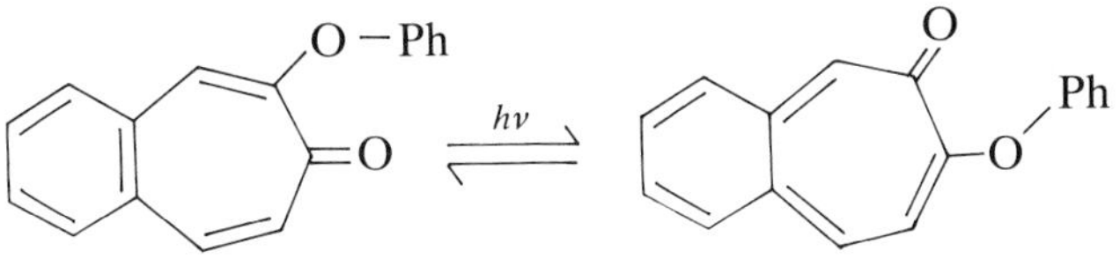

$$\tag{6.31}$$

XXVIIa   **XXVIIb**

4.5. *Photo-induced Rearrangements of the C_{sp^3}-centered Groups*

Instances of such rearrangements are sparse as is also the case with the thermal reactions. Owing to strong coordinative saturation of the sp^3-carbon atom, high values of ionization potentials and low values of electron affinity of these groups, the favored mechanism is in this case the formation of a rearranging radical pair through the α-cleavage of the migrating bond.

In addition to migrations of the substituted alkyl groups in the allyl system (6.6) examined earlier, we may point to other examples of the 1,3-shift of benzyl groups between two heteroatomic centers, *viz.*, the photorearrangement of polyalkylbenzyl thiocyanates **XXVIIIa** into isothiocyanates [66].

$$Ar-CH_2-S-C \equiv N \xrightarrow{h\nu} S = C = N-CH_2-Ar \tag{6.32}$$

XXVIIIa   **XXVIIIb**

and the reversible migration of the benzyl and naphthylmethyl groups between the ether and carbonyl oxygen [26, 27].

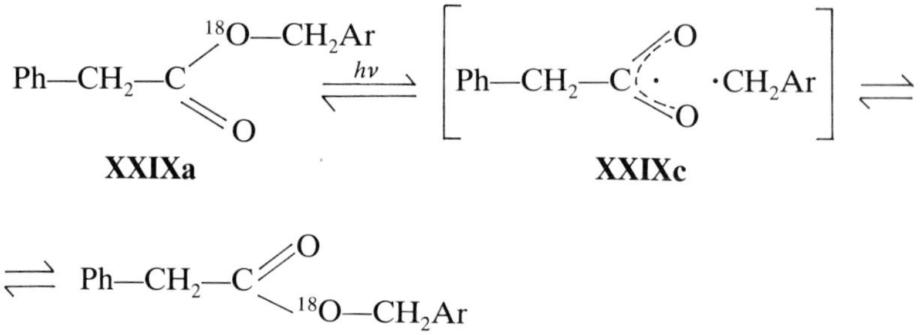

$$\text{(6.33)}$$

Ar = Ph, $C_{10}H_7$-1, $C_{10}H_7$-2

An intramolecular radical pair mechanism of type (6.5) has been proven in the case of recently discovered photochemical alkyl migrations of dialkylboryl acetylacetonate complexes [68].

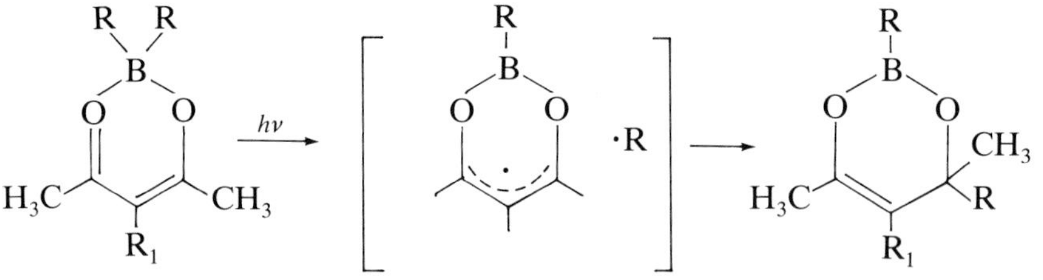

$$\text{(6.34)}$$

$R_1 = C_4H_9$, cyclo-C_6H_{11}; R = H, CH_3

The reaction is not quenched by oxygen or piperylene suggesting that the singlet is the reactive excited state. Crossover experiments have shown that the alkyl migration is essentially (95%) intramolecular. The singlet excited state of **XXXa** gives the radical pair **XXXc** in the solvent cage which only partly dissociates into free radicals. The latter species may recombine intermolecularly to yield crossed products or give typical free radical products such as cyclohexane and cyclohexene in minor quantities.

The data considered show that the photochemical initiation of the migratory tautomeric processes substantially broadens the mechanistic possibilities of these reactions and opens up new prospects in this field, thus following the developments already realized in the broader domain of preparative organic photochemical synthesis.

REFERENCES

1.   Goering H. L., Briody R. G., and Levy J. F., *J. Amer. Chem. Soc.*, **85**, 3059 (1963).

2.   Goering H. L. and Levy J. F., *J. Amer. Chem. Soc.*, **86**, 120 (1964).
3.   Beletskaya I. P., *Usp. Khim.*, **4**, 2205 (1975).
4.   Mikhailov I. E., Dushenko G. A., Minkin V. I., and Olekhnovich L. P., *Zh. Org. Khim.*, **20**, 2306 (1984).
5.   Kompan O. E., Antipin M. Yu., Struchkov Yu. T., Mikhailov I. E., Dushenko G. A., Olekhnovich L. P., and Minkin V. I., *Zh. Org. Khim.*, **21**, 2032 (1985).
6.   Kwart H. and George T. J., *J. Amer. Chem. Soc.*, **101**, 1277 (1979).
7.   *Ions and Ion Pairs in Organic Reactions.* Vol. 1. (Ed. X. Schwarc). J. Wiley. New York. 1972.
8.   Gordon J. E., *The Organic Chemistry of Electrolyte Solutions.* J. Wiley. New York. 1975.
9.   Buchwald S. L., Friedman J. M., and Knowles J. R., *J. Amer. Chem. Soc.*, **106**, 4911 (1984).
10.   Berson J. A., in: *Rearrangements in Ground and Excited States.* Vol. 2, p. 311. (Ed. P. De Mayo). Academic Press. New York. 1980.
11.   Baldwin J. E. and Brown J. E., *J. Amer. Chem. Soc.*, **91**, 3647 (1969).
12.   Shine H. J., *Aromatic Rearrangements.* Elsevier. Amsterdam. 1967.
13.   Rautenrauch V., Buchi G., and Wuest H., *J. Amer. Chem. Soc.*, **96**, 2576 (1974).
14.   Garst J. F. and Smith C. D., *J. Amer. Chem. Soc.*, **98**, 1526 (1976).
15.   Bunnett J. F., *Acc. Chem. Res.*, **11**, 413 (1978).
16.   Bigot B., Roux D., and Salem L., *J. Amer. Chem. Soc.*, **103**, 5271 (1981).
17.   Chanon M. and Tobe H. L., *Angew. Chem.*, **94**, 27 (1982).
18.   Minkin V. I., Simkin B. Ya., and Minyaev R. M., *Quantum Chemistry of Organic Compounds. Mechanisms of Reactions.* Khimiya. Moscow. 1986.
19.   Dyachenko A. I. and Ioffe A. I., *Izvest. Akad Nauk USSR, ser. Khim.* 1160 (1976).
20.   *Photochromism.* (ed. Brown G. H.). (Vol. 13, Techniques of Organic Chemistry). J. Wiley New York. 1971.
21.   Barachevskii V. A., Lashkov G. I., and Tscechomskii X. X., *Photochromism and its Applications.* Khimiya. Moscow. 1977.
22.   Nagaoka S., Hirota N., Sumitani M., Yoshihara K., Lipczynska-Kochany E., and Iwamura H., *J. Amer. Chem-Soc.*, **106**, 6913 (1984).
23.   Palui G. D., Lyubarskaya A. E., Simkin B. Ya., Bren V. A., Zhdanov Yu. A., Knyazhanskii M. I., Minkin V. I., and Olekhnovich L. P., *Zh. Org. Khim.*, **15**, 1348 (1979).
24.   Bellus D., *Adv. Photochem.*, **8**, 109 (1971).
25.   Nakagaki R., Hiramatsu M., Watanabe T., Tanimoto Y., and Nagakura S., *J. Phys. Chem.*, **89**, 3222 (1985).
26.   Kaupp G., *Angew. Chem. Intern. Ed. Engl.*, **19**, 243 (1980).
27.   Crump D. R., Franck R. W., Gruska R., Ozonio A. A., Paguotta M., Siuta G. J., and White J. G., *J. Org. Chem.*, **42**, 105 (1977).
28.   Hoffmann R. W. and Eicken K. R., *Chem. Ber.*, **102**, 2987 (1969).
29.   Veierov D., Bercovici T., Fisher E., Mazur Y., and Yogev Y., *Helv. Chim. Acta*, **58**, 1240 (1975).
30.   Houk K. N., *Chem. Rev.*, **76**, 1 (1976).
31.   Markov P., *Chem. Soc. Rev.*, 69 (1984).
32.   Russkich S. A., Klimenko L. S., Gritsan N. P., and Fokin E. P., *Zh. Org. Khim.*, **32**, 2224 (1982).
33.   Klimenko L. S., Gritsan N. P., Konstantinova A. V., and Fokin E. P., *Izvest. Sib. Otd. Akad Nauk USSR,* **17**, No. 6, 84 (1984).
34.   Gritsan N. P., Russkich S. A., Klimenko L. S., Plyusnin V. F., *Teor. Eksperiment. Khim.*, **19**, 455 (1983).
35.   Curtin D. Y. and Miller L. L., *J. Amer. Chem. Soc.*, **89**, 637 (1967).
36.   Hegarty A. F. and McCormack M. T., *J. C. S. Chem. Commun.*, 168 (1975).
37.   Puckett S. A., Greensley M. K., Paul I. C., and Curtin D. Y., *J. C. S. Perkin II*, 847 (1977).
38.   McCarthy D. G. and Hegarty A. F., *J. C. S. Perkin II*, 1085 (1977).
39.   Hartke K., Matusch R., and Krampitz D., *Liebigs Ann. Chem.*, 1237 (1975).
40.   Hartke K. and Wachsen E., *Lieb. Ann. Chem.*, 730 (1976).
41.   Hartke K., Krampitz D., and Uhde W., *Chem. Ber.*, **108**, 128 (1975).

42.   Wachsen E. and Hartke K., *Chem. Ber.*, **108**, 138, 683 (1975); **109**, 1353 (1976).
43.   Lyubarskaya A. E., Paluy G. D., Bren V. A., Zhdanov Yu. A., Knyazhanskii M. I., Minkin V. I., and Olekhnovich L. P., *Zh. Org. Khim.*, **12**, 918 (1976).
44.   Bren V. A., Knyazhanskii M. I., Lyubarskaya A. E., Orechovskii V. S., Paluy G. D., Rybalkin V. P., Izvest. Severo-Kavkas. *Nauchn. Centr. ser. estestv. nauk, No. 2*, 63 (1984).
45.   Aldoshin S. M., Atovmyan L. O., Minkin V. I., Bren V. A., and Paluy G. D., *Zeit. Krystall.*, **159**, 143 (1982).
46.   Aldoshin S. M., Dyachenko O. A., Atovmyan L. O., Minkin V. I., Bren V. A., and Paluy G. D., *Zh. Struct. Khim.*, **23**, 107 (1982); **25**, 106, 131 (1984).
47.   Aldoshin S. M., *Structural Aspects of Photochromic Transformations of Organic Compounds*. Doctoral Dissertation. Thesis. Chernogolovka. 1985.
48.   Paluy G. D., Bren V. A., Lyubarskaya A. E., and Minkin V. I., in: *Organic Photochromes*. (Ed. Eltsov A. V.) Publ. House Khimiya. 1982, pp. 233, 259.
49.   Aldoshin S. M., Atovmyan L. O., Sitkina L. M., Dubonosov A. D., Bren V. A., and Minkin V. I., *Khim. Heterocycl. Soed.*, in press (1986).
50.   Bren V. A., Andreychikova G. E., Krikov V. A., Minkin V. I., Aldoshin S. M., and Atovmyan L. O., *Zh. Org. Khim.*, **21**, 862 (1985).
51.   Scharf H.-D., Fleischhauer J., Leismann H., Ressler I., Schleker W., and Weitz R., *Angew. Chem. Intern. Ed. Engl.*, **18**, 652 (1979).
52.   Laird T., *Chem. and Ind.*, 186 (1978).
53.   Jones II G., Reinhardt T. E., Bergmark W. R., *Solar Energy*, **20**, 241 (1978).
54.   Jones II G., Chiang S.-H., and Xuan P. T., *J. Photochem.*, **10**, 1 (1979).
55.   Lewis E. S., Holiday R. E., and Hartung L. D., *J. Amer. Chem. Soc.*, **91**, 430 (1969).
56.   Zollinger H., *Pure Appl. Chem.*, **55**, 401 (1983).
57.   Matsui K., Maeno N., and Suzuki S., *Tetrahedron Lett.*, 1467 (1970).
58.   Mutai K. and Kobayashi K., *Bull. Chem. Soc. Japan*, **54**, 462 (1981).
59.   Mutai K., Yokoyama K., Kauno S., and Kobayashi K., *Bull. Chem. Soc. Japan*, **55**, 1112 (1982).
60.   Wubbels G. G., Halverson A. M., Oxman J. D., and Van De Bruyn H., *J. Org. Chem.*, **50**, 4499 (1985).
61.   Gerasimenko Yu. E., Poteleschzenko N. T., *Zh. Vsesoyuzn. D. I. Mendeleev Obschz.*, **16**, 105 (1971); *Zh. Org. Khim.*, **7**, 2413 (1971).
62.   Gerasimenko Yu. E., in: *Organic Photochromes*. (Ed. Eltsov A. V.) Publ. House Khimiya. Leningrad. 1982, p. 224.
63.   Fokin E. P., Russkich S. A., Klimenko L. S., and Gritsan N. P., *Izvest. Sib. Otd. Akad Nauk USSR, ser. khim.*, 117 (1979); 116 (1981).
64.   Gerasimenko Yu. E., Poteleschzenko N. T., and Romanov V. V., *Zh. Org. Khim.*, **16**, 1938, 2022 (1980).
65.   Griffin G. W. and O'Connell E. S., *J. Amer. Chem. Soc.*, **84**, 4142 (1962).
66.   Suzuki H., Uzuki M., and Hanafuga T., *Bull. Chem. Soc. Japan*, **52**, 836 (1979).
67.   Givens R. S. and Matuszewski B., *J. Amer. Chem. Soc.*, **97**, 5617 (1975).
68.   Okada K., Hosoya Y., and Oda M., *J. Amer. Chem. Soc.*, **108**, 321 (1986).

Subject Index